A Geography of the Third World

SECOND EDITION

John Dickenson
Bill Gould
Colin Clarke
Sandra Mather
Mansell Prothero
David Siddle
Clifford Smith
Elizabeth Thomas-Hope

London and New York

First published 1996
by Routledge
11 New Fetter Lane, London EC4P 4EE

Simultaneously published in the USA and Canada
by Routledge
29 West 35th Street, New York, NY 10001

Routledge is an International Thomson Publishing company

Typeset in Sabon by Solidus (Bristol) Limited
Printed and bound in Great Britain by
Butler & Tanner Ltd, London and Frome

British Library Cataloguing in Publication Data
A catalogue record for this book is available from the British Library

Library of Congress Cataloguing in Publication Data
Geography of the Third World
p. cm.
1. Developing countries–Economic conditions.
HC59.G3658 1996
330.9172'4–dc20 95-37939

ISBN 0–415–10672–9 (hbk)
ISBN 0–415–10673–7 (pbk)

A Geography of the Third World

SECOND EDITION

The Third World and its problems arouse much interest and concern. This book outlines the key themes and issues in the geography of the Third World at a level appropriate to the needs of the advanced students in secondary schools and at universities and colleges. It also provides a general and informative introduction to those interested in the field of development studies.

The major themes around which the book is organized include population growth and structure, agriculture and rural societies, industrialization, urbanization, and internal and external linkages. It brings together ideas, concepts and facts from a wide variety of sources, and draws upon the extensive field research of the authors.

This second edition incorporates new perspectives in geography, such as gender issues and renewed concern for the environmental impact of development. It also encompasses major developments in the Third World itself, such as the widening gap in economic performance between countries within the region, and the assertion of national cultures in the face of global processes. The authoritative text is complemented by a wealth of photographic and line illustrations, boxed case studies, chapter summaries and guides to further reading.

John Dickenson is Reader in Geography and Latin American Studies, University of Liverpool; **Bill Gould** is Professor of Geography, University of Liverpool; **Colin Clarke** is Lecturer in the School of Geography, Oxford University; **Sandra Mather** is Head of the Graphics Unit, Department of Geography, University of Liverpool; **Mansell Prothero** is Emeritus Professor of Geography, University of Liverpool; **David Siddle** is Senior Lecturer in the Department of Geography, University of Liverpool; **Clifford Smith** is Emeritus Professor of Latin American Studies, University of Liverpool; **Elizabeth Thomas-Hope** is Professor of Environmental Management, University of the West Indies.

Contents

Preface to the Second Edition

The first edition of this book was written to meet a perceived need for a basic introductory text on the geography of the Third World. The commercial success of that volume and its translation into four other languages suggests that it did fill a niche, and has provided some encouragement for a second edition. However, a new edition more than a decade later requires more than an updating of statistics and some cosmetic work on outmoded passages, for in the period since 1983 the Third World has changed. There have been direct changes in the circumstances in Third World countries, and also in the context of the Third World *vis à vis* the richer countries of North America, Europe, Japan and Australasia. New theoretical perspectives on development have also emerged, including changed views as to what might be the objectives of the development process. Any revision needs to reflect such changes.

Between 1982, when the first edition of this book went to press, and 1994, when this edition was being prepared, the population of the 'low-income countries', as defined by the World Bank, has increased by 800 million people, per capita gross national product (GNP) for those countries has risen from US $280 to $390, and life expectancy at birth has risen from 59 to 62 years. Yet for the very poorest country (Chad in 1982, Mozambique in 1992) per capita GNP was $80 in 1982 and had fallen to $60 in 1992; for the richest country (Switzerland) it has risen from $17,010 to $36,080. Although the average figure for the low-income countries has risen, it has risen much less in absolute terms than it has in the richest countries, though at roughly the same rate in proportional terms. The gulf between rich and poor countries has widened.

Between 1980 and 1990, per capita GDP for the Third World as a whole expanded at 3.2 per cent per year, much the same rate as the high-income countries and for the world overall. However, within the Third World there seems to be a widening range of experience in the rate of growth. During the 1980s the annual rate of growth in GDP was 7.8 per cent for the countries of East Asia but only 2.1 per cent for Sub-Saharan Africa. Some major regions have experienced sharply rising per capita income values and others seem to have stagnated or even experienced falling income levels. If there has been progress, it has been uneven; some regions still suffer from acute poverty.

The increasing diversity of conditions among countries of the Third World – indeed often within them – is significant for geographers, because of their interest in the spatial variations of places and countries, and this is a theme that will recur throughout this second edition. The contrast between the rapid economic progress and major social changes in some East Asian countries, which are now described as 'Newly Industrializing Countries' (NICs), and the persistent and generally intensifying poverty of many countries of Sub-Saharan Africa is particularly striking.

However, that differentiation in no way weakens our view of the essential validity of our conceptualization of the Third World as encompassing the 'poor' countries of the world. Our acceptance of the term 'the Third World' as a valid and valuable generalization, given justification by the historical and contemporary relationships of poor countries with the rich world, remains unchanged (see Figure 1.1). Though there is a baffling heterogeneity of environmental, economic and social circumstances in these countries, we identify them as belonging to the Third World, brought together, as they were in 1982, by the strength of the unequal relationships of underdevelopment that persist in the world-system.

The use of the term 'Third World' only makes sense when considered in the broader context of international economic, social and political relationships. But these relationships have themselves changed since 1982, when the Third World could be identified against a First World of developed capitalism and a Second World consisting of socialist states. The taking down of the Berlin Wall in 1989 has had both symbolic and material consequences. The break-up of the Second World emphasizes once again the uncertain terminology we use to divide up the globe in politico-economic terms. The end of the Cold War held out the promise of a 'peace dividend', in which resources deployed for military expenditure on defence or aggression could be redirected to improving the lot of the poor in Third World countries. So far, however, the New World Order, dominated by the one superpower, the USA, and with the increasing economic strength of the European Community and Japan, has shown limited willingness to turn defence expenditure into aid or other productive use to assist the poor countries. Political unrest within the former USSR and its satellites, such as in Yugoslavia, has sustained demands for military precautions, and for UN and NATO peace-keeping. The need to sustain refugees from these disturbances and to rejuvenate the fragile economies of Eastern Europe and the ex-Soviet Union has diverted aid and private investment from capitalist countries which might have gone to the Third World, while the unification of West and East Germany has placed strains on one of the First World's richer economies.

Elsewhere the spread of militant Islamic fundamentalism beyond the traditionally defined Middle East, from North Africa into western Asia and the former Soviet republics in central Asia, has created a new dimension within that region which requires distinct attention. This is compounded by warfare and friction within the Middle East, including the Iran/Iraq War during much of the 1980s and the Gulf War between Iraq and the USA and its allies in 1991, as well as the long-standing problems between Israel and its neighbours.

The end of the Cold War brought an increased global awareness of the dangers of regional 'hot' wars (e.g. in Iraq, Yugoslavia, Somalia) and the need to direct the attention of the international community to them, mainly through the United Nations. It also heightened interest in non-defence aspects of global relationships. The increased concern for international trade, and the costs and benefits of new patterns of trade, have come into a sharpened focus with the growing economic rivalry between the USA, Japan and Europe, both in bilateral trading relationships, and also in the broader trading relationships of the developed countries and the Third World in the General Agreement on Tariffs and Trade (GATT). The protracted discussions of the so-called Uruguay Round of GATT talks, initiated in 1986 and only completed in late 1993, has kept international trade and its impact on poor and rich countries alike well to the fore in global concerns.

The emergence of the environment as an issue of global concern has also been of considerable significance, culminating in the UN Conference on Environment and Development in Rio de Janeiro in 1992. Here the interests of the Third World were clearly at odds with those of the First World and of the former Second World. The environment has emerged as a Third World issue that cannot be treated in isolation from the more traditional political and economic aspects of global relationships that were more prominent in the first edition of this book. Increasing concern for environmental issues, as exemplified in global warming, ozone holes, acid rain and deforestation, and the fact that these are problems associated with economic activity, clearly has implications for the form and nature of further Third World development. Discussion of economic development in the Third World has now to take account of environmental consequences, and of the views of poor countries about the nature of and solution to their environmental problems.

Awareness of the depths of Third World

poverty has been greatly heightened in recent years by the media, particularly television. Recurring images of men, women and children starving and dying as the hapless victims of civil war and environmental catastrophe have provoked considerable public concern, generating greater support for the activities of non-governmental agencies (NGOs) and for specific events such as Live Aid. The now all too familiar images of victims of environmental catastrophes such as droughts, cyclones and earthquakes, have fostered some resurgence of beliefs about the role of environment as a determinant in creating and maintaining patterns of poverty.

Theoretical perspectives on the study of the Third World have also changed. The crumbling of the Second World, and the need to restructure the economies of new states in Eastern Europe, diminish the support given by the former USSR to its Third World satellites, but perhaps call into question the very nature of the Marxist or neo-Marxist perspective on which much of the 'dependency' approach prominent in the 1970s and early 1980s was based. A major new element in the analysis of the global economy and its component parts comes from the development during the 1980s of the world-systems approach (see Chapter 1). This approach builds on the previous global perspectives of dependency and underdevelopment by explicitly recognizing a basic geographical point: that countries are different from each other and may respond to the global systems of unequal exchange in different ways. Heterogeneity within the Third World is recognized as inevitable and necessary to an appreciation of the whole system. In particular, a global systems approach differentiates between the periphery (the poorest countries) and the semi-periphery (including resource-rich states and NICs) (see Figure 1.18). For geographers this has been an important development, as it allows more effective consideration of internal aspects of the geography of each country with the larger external questions. It requires more explicit consideration of a number of issues, including: the environment (at the scale of local environmental management in traditional and evolving commercial systems); the nature and quality of governance (how the economy and society are managed); the economic and social processes which affect the nature of rural and urban production systems; and the changing role of gender relationships in the evolution of these systems.

Within the discipline of Geography itself there have been important changes since the early 1980s that have fed into the approach that was adopted in the first edition and continues into the second. At that earlier period much geographical work was still driven by a spatial science, positivist approach that tended to seek regularities rather than differences within and between countries. The attraction of dependency theory, which suggested that Third World countries were essentially passive participants in a global order that kept them poor, emphasized the similarities at the expense of recognizing the heterogeneity. There was also a strong sense of 'modernization', through which the structures and values of developed societies were diffusing into the Third World, such that everywhere was becoming similar, though at a variable rate (and in some countries very slowly), to the economic and social pattern of Western countries. In the 1990s, however, geographers have generally rejected crude diffusionism and are rediscovering a 'sense of place', with the result that they are less constrained about considering the increasing heterogeneity within the Third World. After the attempts at homogenization and a search for common solutions, there is a renewed acknowledgement of variety – that El Salvador is different from Somalia, and both are different from Sri Lanka, and the causes of these differences are a proper concern of Geography.

The dominant perspective of the book is 'developmentalist' – its systematic approach explores demographic and economic themes which are relevant to the Third World experience. It follows a broad agenda relating to issues of population growth, rural and urban development, agricultural and manufacturing activity, and internal and external linkages which, implicitly or explicitly, might ameliorate or improve the condition of the Third World and its peoples. The demise of a Second World, and the rising hegemony of the World Bank on national and international development policies implies, in the 1990s, a dominating role for market forces in the 'development process'. The desirability of 'development' and the influence wielded by the World Bank implies common patterns and goals for the countries of the Third World. It might be assumed that market-driven development would give rise to increasingly uniform patterns of capitalist development and its affluent landscapes.

Yet Geography's rediscovery of a 'sense of place' suggests that places are not 'the same' – neither in their natural and cultural landscapes, nor in their levels of development and its human consequences. Spencer and Thomas (1978) suggest that conditions on Earth in the twentieth century are among the most complex that have ever been present on the planet. They assert that the variety of human living systems is greater now than ever before: they talk of 'the mosaic of a heavily populated world' (1978, 337). It is therefore essential for the reader to realize that, although this book is concerned, at the broad scale, with goals of improvement in the condition of the Third World and its peoples, the 'Third World' is itself highly variable. Its historical experience (Chapter 2), demography (Chapter 3), settlement and economy (Chapters 4–7), and its internal and external linkages (Chapters 8–10) are highly varied. Such variety should be evident in much of the text; it should also be apparent in the case studies provided by the boxes, in the patterns of the maps, and in the illustrations and their captions.

Another important development since the late 1970s has been the emergence of feminist geography and the geography of gender. Studies in the Third World have made a significant contribution to this field. Following Esther Boserup's pioneering *Women's Role in Economic Development* (1970), there have been a number of major studies by geographers exploring gender issues in the Third World (Momsen and Townsend, 1987; Brydon and Chant, 1989; Momsen and Kinnaird, 1993). It has been suggested that women have been 'invisible to geography' for a long time, and Brydon and Chant claim that the 'modernization' and 'dependency' approaches to development gave little or no place to women; women were either ignored altogether or assumed to be attached to men, such that their situation would automatically improve in line with economic progress. Yet, 'Women constitute half the world's population, perform nearly two-thirds of its work hours, and receive one-tenth of the world's income' (Seager and Olson, 1986, 101). In most countries of the Third World females outnumber males, yet by most criteria, such as nutrition, education, income or employment, they are disadvantaged relative to men.

In preparing this second edition we are more aware – objectively and through an extended range of personal experiences – from China, Korea and Indonesia to Zimbabwe and Guyana – of the heterogeneity of experience and performance of the Third World, and as a result are better able to consider it within the broader framework set by the book. However, since we reaffirm the continuing bases for defining the Third World, we have decided not to restructure the overall logic of the text. This deals sequentially with broad historical and conceptual relationships (Chapters 1 and 2), population and production (Chapters 3–8), and with the basic spatial structures (Chapters 8–11), but it integrates the changing issues and perspectives that have emerged within that structure. Thus readers will not find the chapters on 'environment' or 'gender' that may have been expected in a 'politically correct' revision. We have preferred to integrate these critical and cross-cutting issues more firmly into the structure, for they are symptomatic of the overall thrust of the argument.

The basic structure of the book is indicated in Figure P.1. As with the first edition the maps and photographs are an essential and integral part of our argument, and the amount of illustrative material has been increased. As before, we have included examples and longer case studies, mainly drawn from our own field experience, in our narrative. We have also added a series of 'boxes', to provide other examples, to illustrate specific points, or to draw attention to themes which are significant, but which we do not have space to explore fully. Key words or phrases which are defined in the Glossary are emboldened at their first occurrence. There is a list of further reading on the principal themes which gives a range of other potential examples, and

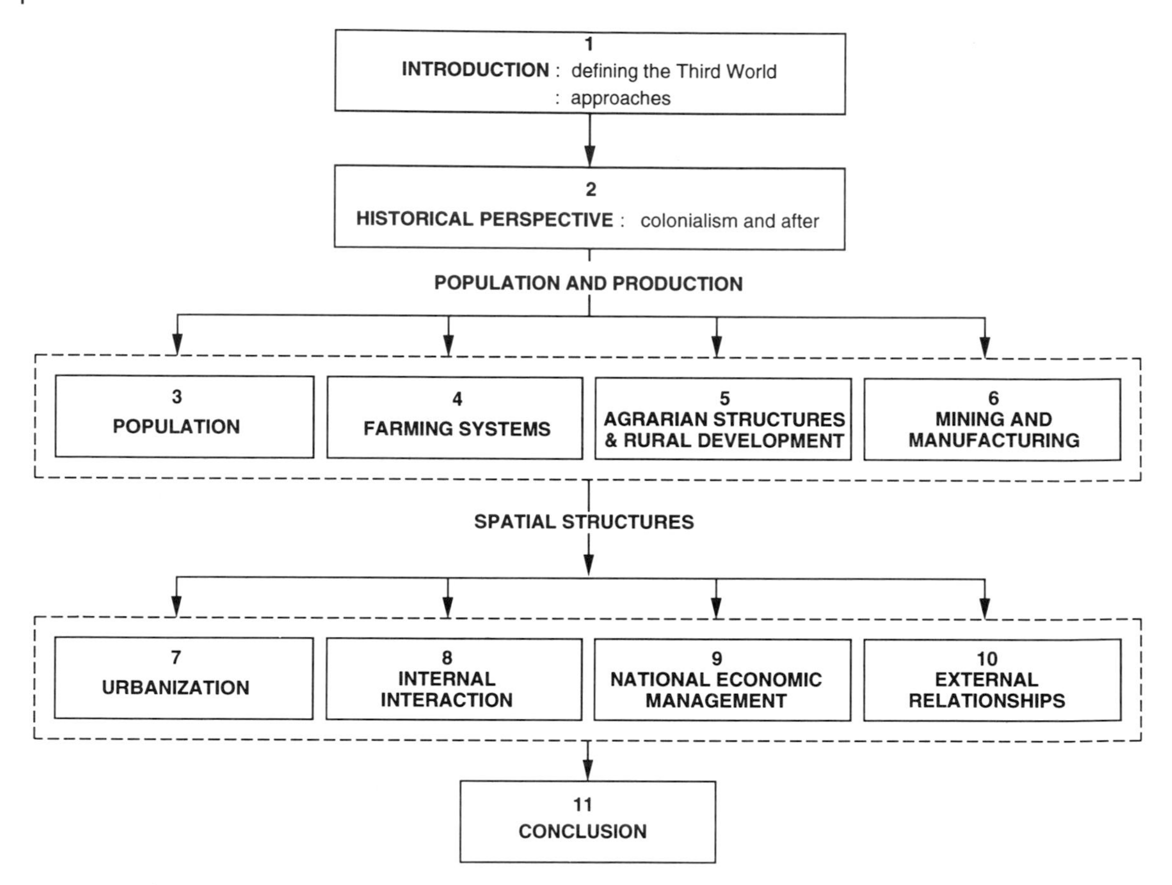

Figure P.1 The structure of this book

also seeks to guide the reader to material on these more specialized topics. Our method of preparing the text has followed that of the first edition.

In completing this edition we must reiterate our many debts of gratitude: to the people of the Third World with whom we have lived and worked; to past and present colleagues and students in Liverpool, Oxford and elsewhere in the First World and the Third upon whom our ideas have been honed; to our families who have tolerated further disappearances to the Third World, to our studies and to editorial meetings. Finally, we wish to acknowledge our indebtedness to Robert Steel, a pioneer in the study of the geography of the Third World and mentor of our interests in the region.

Acknowledgements

PHOTO CREDITS

Colin Clarke: 3.8, 7.4, 7.5, 7.6, 7.8, 7.9, 10.1
John Dickenson: Chapter 2, 2.1, 2.2, 2.4, 3.1, 4.7, 4.9, 5.3, 5.4, 5.6, 6.3, 6.4, 6.5, 6.8, 6.9, 7.10, 7.11, 7.12, 7.13, 10.3, 10.4
Bill Gould: 2.3, 2.5, 3.2, 3.5, 3.7, 4.2, 4.3, 4.4, 4.5, 4.8, 5.5, 5.7, 7.1, 8.1, 8.2, 8.4, 8.6, 10.2, 10.6, Chapter 11
Mansell Prothero: 7.2, 7.3
Ian Qualtrough: 8.3
David Siddle: 5.8
Clifford Smith: 4.1, 4.6

Ian Qualtrough was responsible for the production of the above illustrations.

Brazilian Gazette/CVRD: 6.1
Brazilian Gazette/Petrobras: 6.6
Oxfam Publishing: 10.8
COPRI, Peru: 9.1
Financial Times: 10.5
Howard J. Davies/Panos Pictures: 3.6
Ron Giling/Panos Pictures: Chapter 7, 7.7
Crispin Hughes/Panos Pictures: Chapter 10
Rhodri Jones/Panos Pictures: 8.5
Heldur Netocny/Panos Pictures: Chapter 3, Chapter 5
Sam Ouma/Panos Pictures: 5.1
Julia Parry: Chapter 1, 3.3, 3.4, 5.2, Chapter 6, 6.2, 6.7
Paul Smith/Panos Pictures: 1.1
Sean Sprague/Panos Pictures: 1.2, Chapter 4, 10.7
Chris Stowers/Panos Pictures: 10.9
Penny Tweedie/Panos Pictures: Chapter 8

CARTOONS

Gordon Stowell: 4.1, 8.1
Worldaware (formerly Centre for World Development Education): 1.2, 7.1, 10.1
P.J. Polyp/Leeds Postcards: 5.1
Observer: 4.2
Punch: 3.1, 9.1, 10.2
The Times: 1.1

Introduction

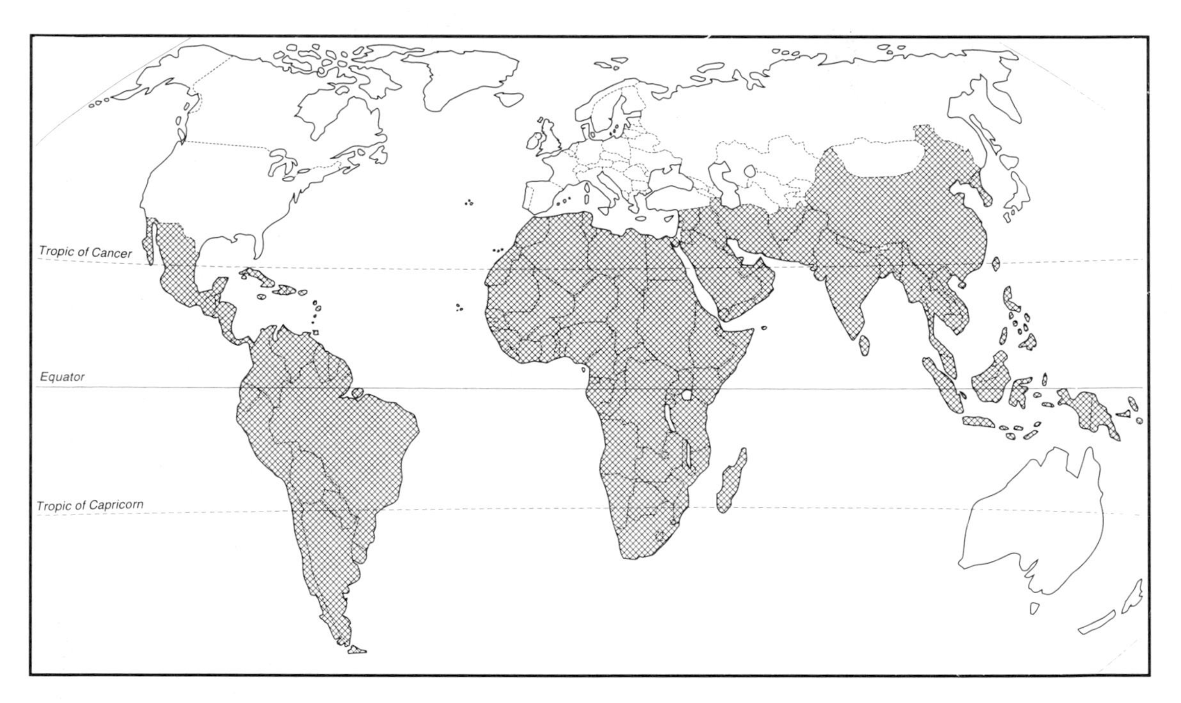

Figure 1.1 The Third World

DEFINING THE THIRD WORLD

It has never been easy to assess the differences between one area of the world and another in terms of relative well being. Judgements are inevitably value-laden. Is wealth to be assessed by the standards of a Texan or Saudi Arabian multi-millionaire or by the averages familiar to an ordinary citizen in the USA or Western Europe? Is poverty to be defined by the standards of a beggar on the streets of Calcutta or a woman farmer struggling to make ends meet in rural Zimbabwe or Honduras? Should wealth be measured purely in terms of money and the material possessions it will buy, or should spiritual and cultural values be taken into account (Plate 1.1)?

In fact the most commonly accepted method of assessment during the last half century has been the relative value of each national economy measured by its **Gross National Product** often expressed in relation to its population (Figure 1.2). Using this approach, the less affluent countries have been called in sequence

◀ Children, Zimbabwe

UNITED COLOR
OF BENETTON

'backward', then 'underdeveloped' or 'less developed' or 'developing'. More recent attempts to describe and measure national differences on a world scale have tried to recognize the problem of cultural prejudice and to address it by using new terminology. Two recent terms have been the 'Third World' and 'the South'. The term 'Third World' has, in fact, a long pedigree, having been first used in France ('*Tiers Monde*') in the 1950s to define the structures of disadvantage which echoed those of the peasantry (the 'Third Estate') within the social order of pre-revolutionary France. But increasing attention given to the problems of poor countries has changed the view of its meaning. By the 1960s the term was used more specifically to identify a three-fold division of the world on principally political and economic grounds: there was the First World of industrialized, market economy countries, broadly the capitalist or Western world; the Second World of centrally planned economies, variously called the Communist Bloc or the socialist camp; and the Third World of poorer countries, many of them recently politically independent from colonial rule. By the end of the 1970s, the scale of the problems of an unbalanced world economy consisting of industrialized nations in the northern hemisphere and a disadvantaged and dependent southern tropical sphere came clearly into focus with the publication of *North–South: A Programme for Survival. Report of the Independent Commission on International Development Issues*, more generally known as the Brandt Report (1980), in which rich and poor zones were starkly and clearly defined as 'North' and 'South' and the interdependence of the world economy clearly specified (Figure 1.3). Though the Brandt Commission did not use the term 'Third World', its generalization of 'South' largely corresponds in its geographical range and in its criteria for this generalization to our use of 'the Third World'.

The World Bank uses monetary criteria, expressed in Gross National Product (GNP) per capita per annum, to categorize the countries of the world into four groups (Figure 1.4). The less affluent countries are divided into two groups – the 'low-income' countries whose GNP per capita in 1991 was below US $650, and 'middle-income' countries whose 1991 GNP per capita ranged from $650 to $8000. The middle-income group was further subdivided into 'lower-middle-income' ($650–$2520) and 'upper-middle-income' ($2530–$7620). The fourth group comprises 'high-income' economies, mostly Western countries and members of the Organization for Economic Cooperation and Development (OECD). In the basic World Bank listing of countries with populations of more than 1 million people, 40 countries are in the low-income group, 43 in the lower middle-income group, and 22 countries are in each of the upper middle-income and in high-income groups.

The World Bank lists a further 73 states with populations of less than 1 million people, together with some states where economic data cannot be usefully estimated (such as Somalia and Iraq where civil disruption precludes meaningful estimates, and socialist countries, such as Cuba, where data on economic activity are not comparable with those of other states). This listing includes countries as diverse as Guinea-Bissau and Comoros with GNP per capita below $300, Iceland and Luxembourg with GNP of over $15,000 and oil-rich states such as Qatar and the United Arab Emirates where GNP per capita is above $20,000. These have been excluded from the data discussed below in Figure 1.11.

If the oil-rich states of the Middle East are excluded on the grounds of their peculiar circumstances, the USA, Sweden and Switzerland are generally accepted as rich, developed countries against which others might be measured. However, income is only one measure used to indicate levels of development. Generally, the 'poor' countries have other social and economic characteristics in common, some of which are indicated in Figures 1.5–1.9. When compared to the 'rich' countries, the people of the poor

Plate 1.1 Street scene, Maputo, Mozambique A billboard advertising the stylish clothes of the Italian Benetton company in the capital of the world's poorest country

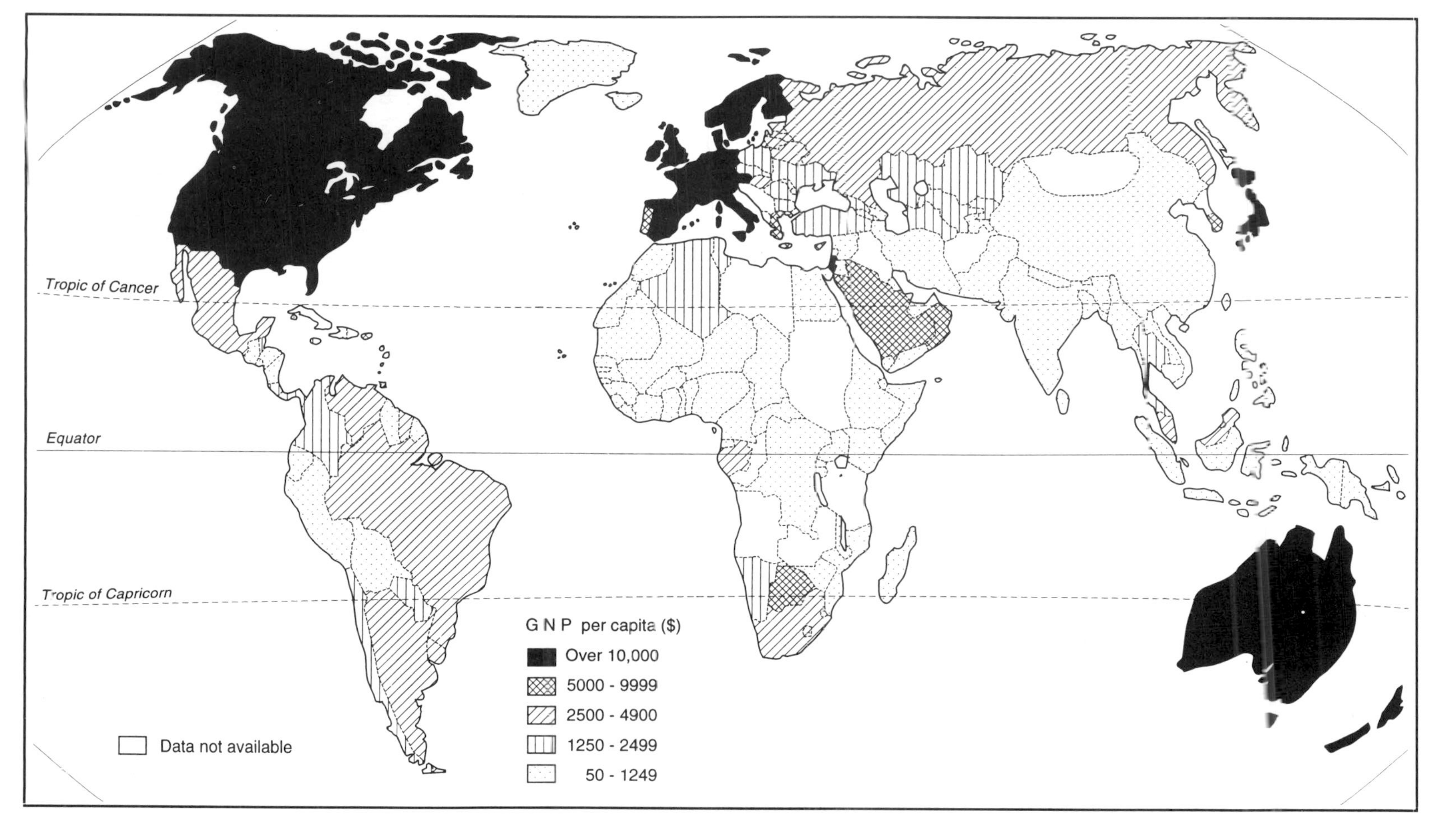

Figure 1.2 Gross National Product per capita, 1991
Source of data: World Bank, 1993

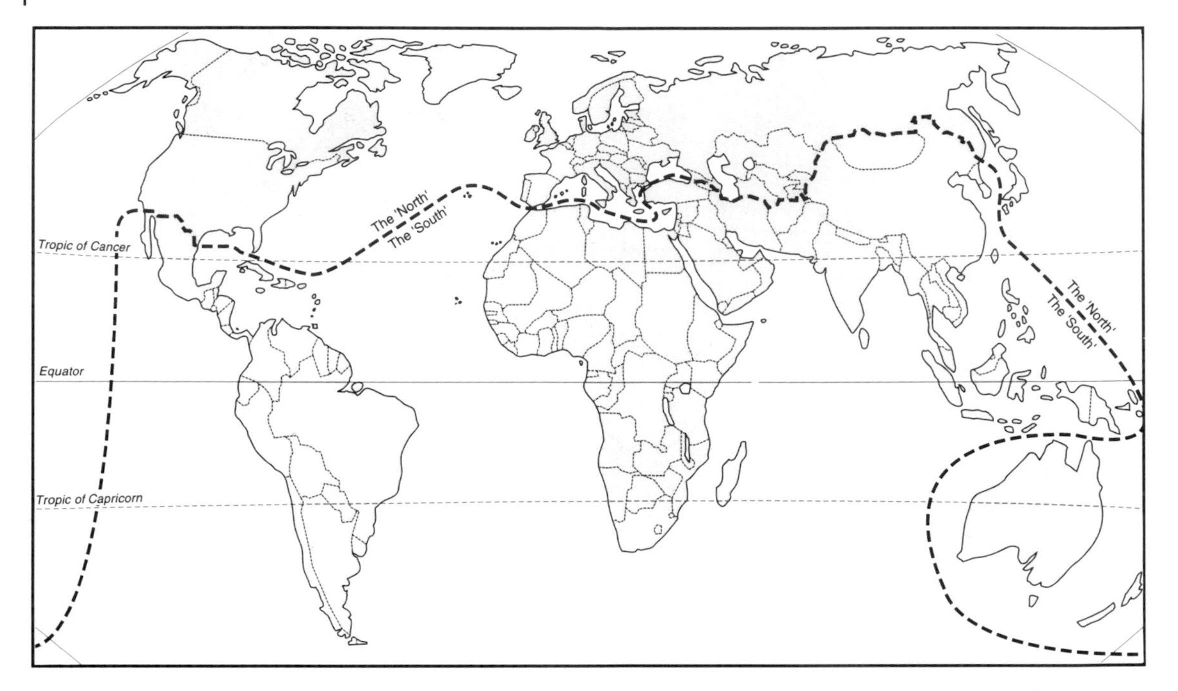

Figure 1.3 'North' and 'South'. The term 'The South' came into common use with the publication of the Brandt Report (1980), *North–South: A Programme for Survival. Report of the Independent Commission on International Development Issues,* which distinguished between the 'rich', 'developed' 'North' and the 'poor', 'developing' 'South'. It was often mapped on the then recently developed Peters Projection, so that the 'South' became a distinctive image

world have diets which are deficient in quality and quantity. There tend to be more people working in agriculture than in manufacturing. The average duration of life tends to be lower. A higher percentage of the population is illiterate. The export economies depend on primary products, from agriculture to mining, and often only one or two of these generate the majority of the export earnings. These and other measures can be used to try to define what has been called a 'commonwealth of poverty'.

A brief reflection on these maps suggests that there is a broad group of countries that share characteristics of low income and disadvantage in provision of basic needs, and which might be identified as the 'Third World'. A closer look, however, reveals that these measures do not always include all of the same countries, so that the precise boundaries for the Third World are difficult to demarcate. Furthermore, the number of countries included or excluded can be altered by minor manipulations of the criteria used.

It should also be apparent from the maps that there are some striking variations *within* this Third World, for example in life expectancy or levels of literacy. It is possible to identify differences which now separate the poorest states within the world (mostly in Africa) from those with less disadvantage in the indices used. Of the 40 countries with over 1 million people with GNP figures below $650 per capita, identified as the 'poorest' countries, 28 are in Africa, 9 in Asia and only 3, Haiti, Honduras and Nicaragua, are in Latin America. On such evidence one might claim that Latin America is less poor or less underdeveloped than Africa and Asia.

These variations draw attention to the very important theme of scale in any discussion of the Third World. At the global scale we have identified contrasts in the level of development between the First and Second Worlds on the one hand and the Third World on the other. At a different scale, within the Third World, we can see other contrasts: some parts of it are poorer than others; some Third World countries are richer than others (Figure 1.10). There is a very heavy concentration of countries at the poorest end of the spectrum, with 48 of the 127

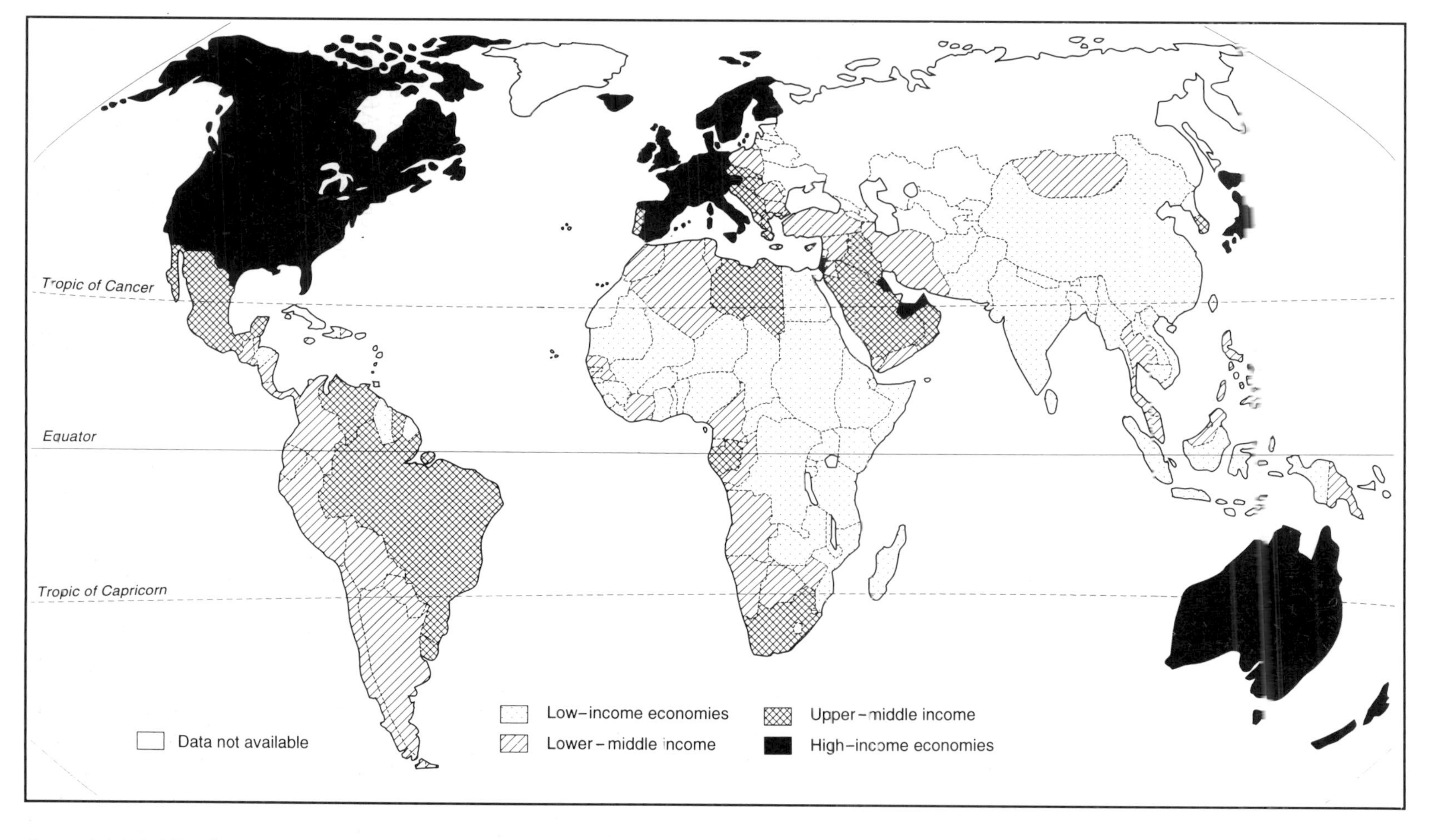

Figure 1.4 World Bank groupings, 1992
Source of data: World Bank, 1992

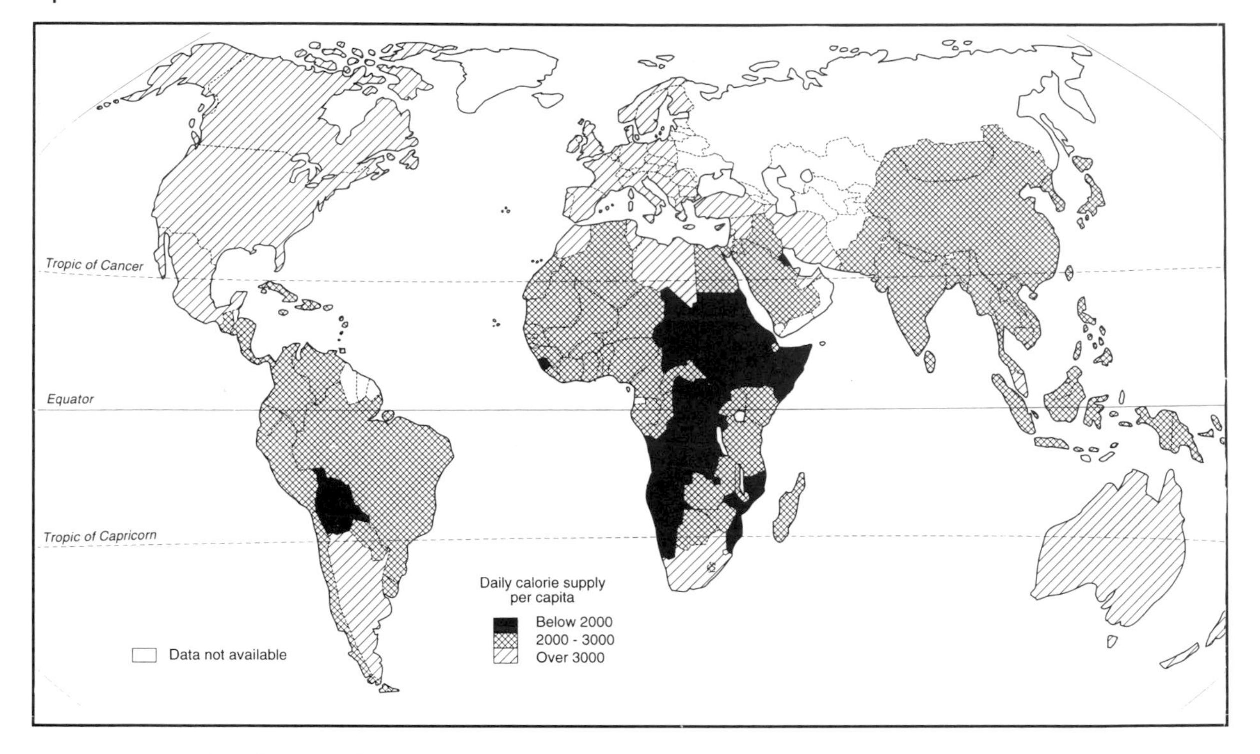

Figure 1.5 Daily calorie supply per capita, 1989
Source of data: World Bank, 1992

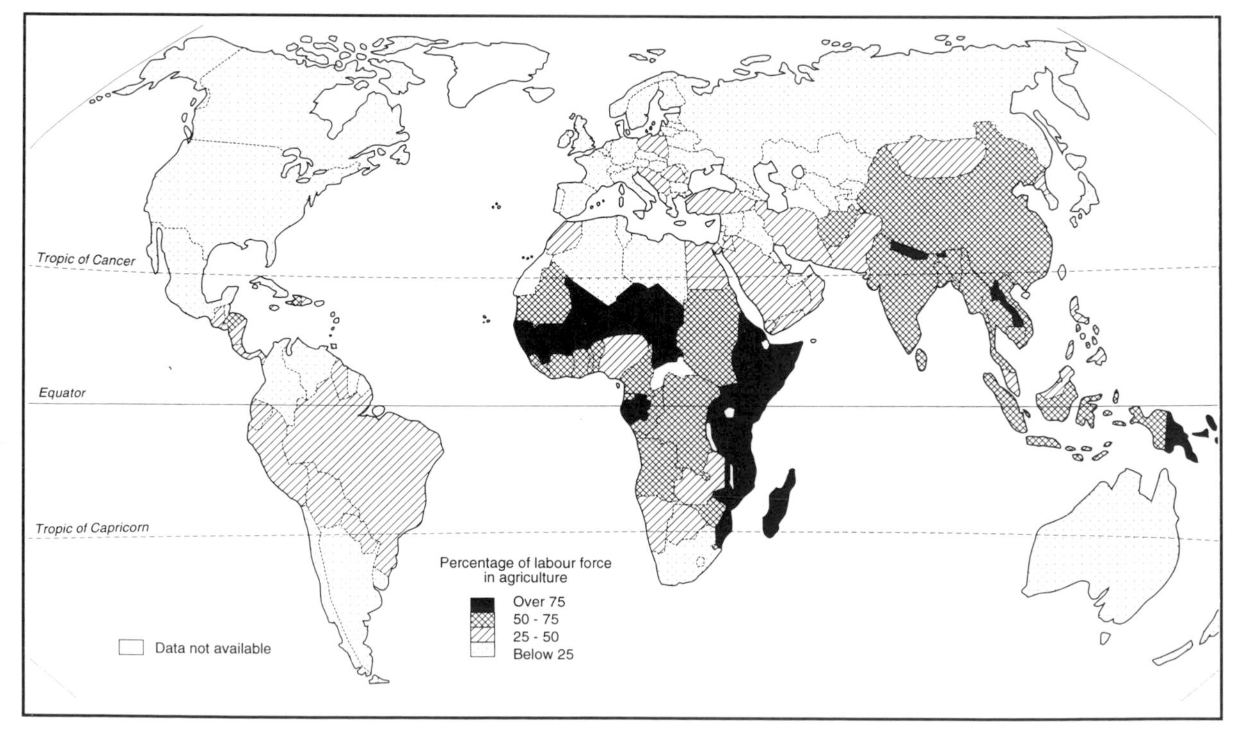

Figure 1.6 Percentage of labour force in agriculture, 1986–89
Source of data: UNDP, 1992

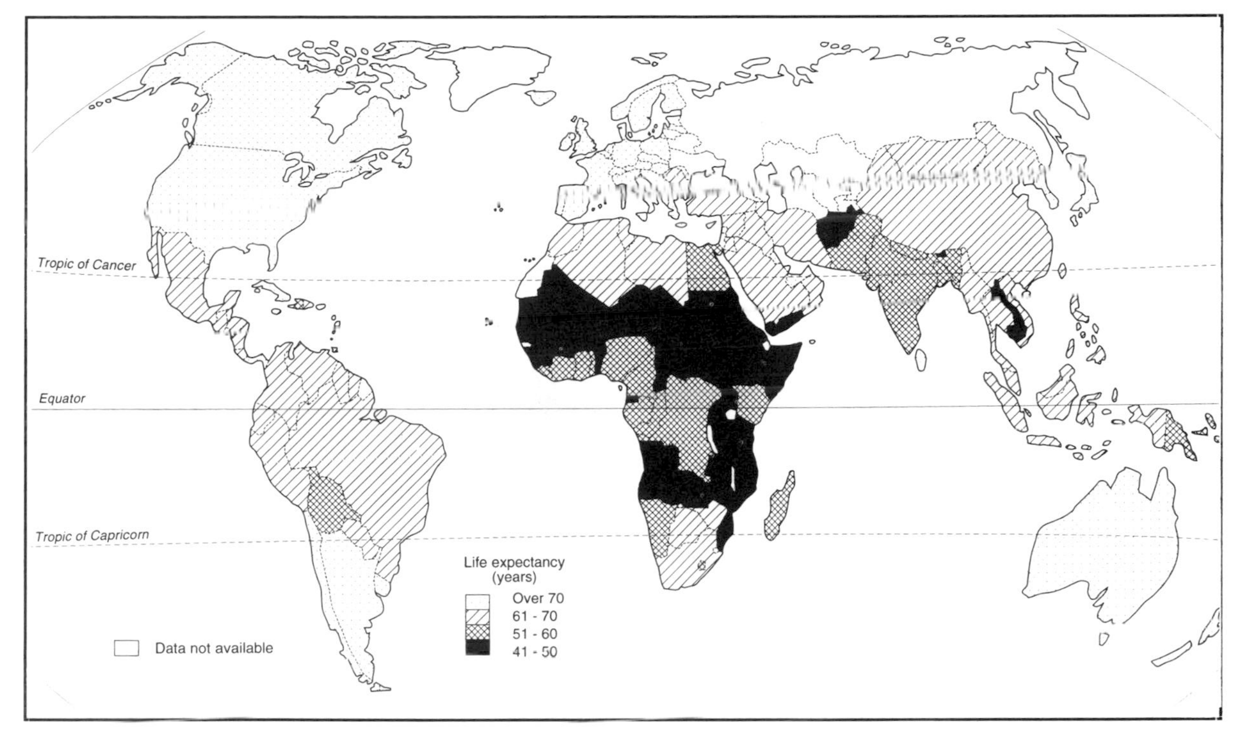

Figure 1.7 Life expectancy at birth, 1990
Source of data: World Bank, 1992

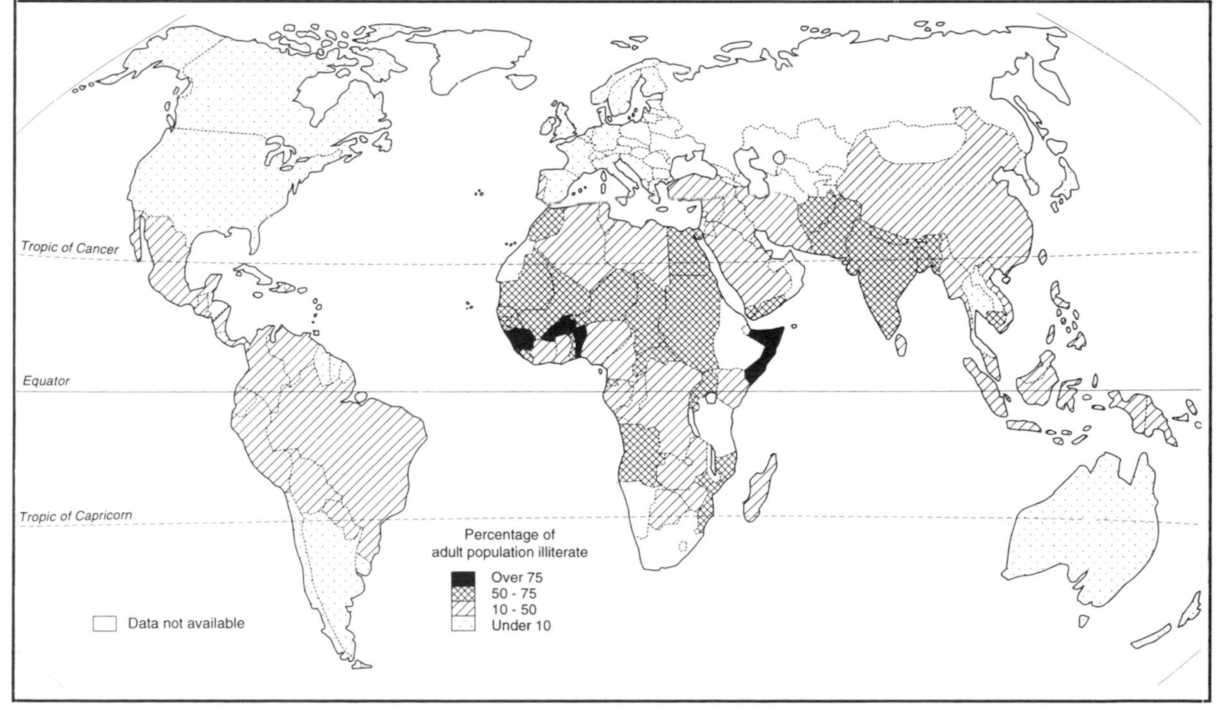

Figure 1.8 Adult illiteracy, 1990
Source of data: World Bank, 1992

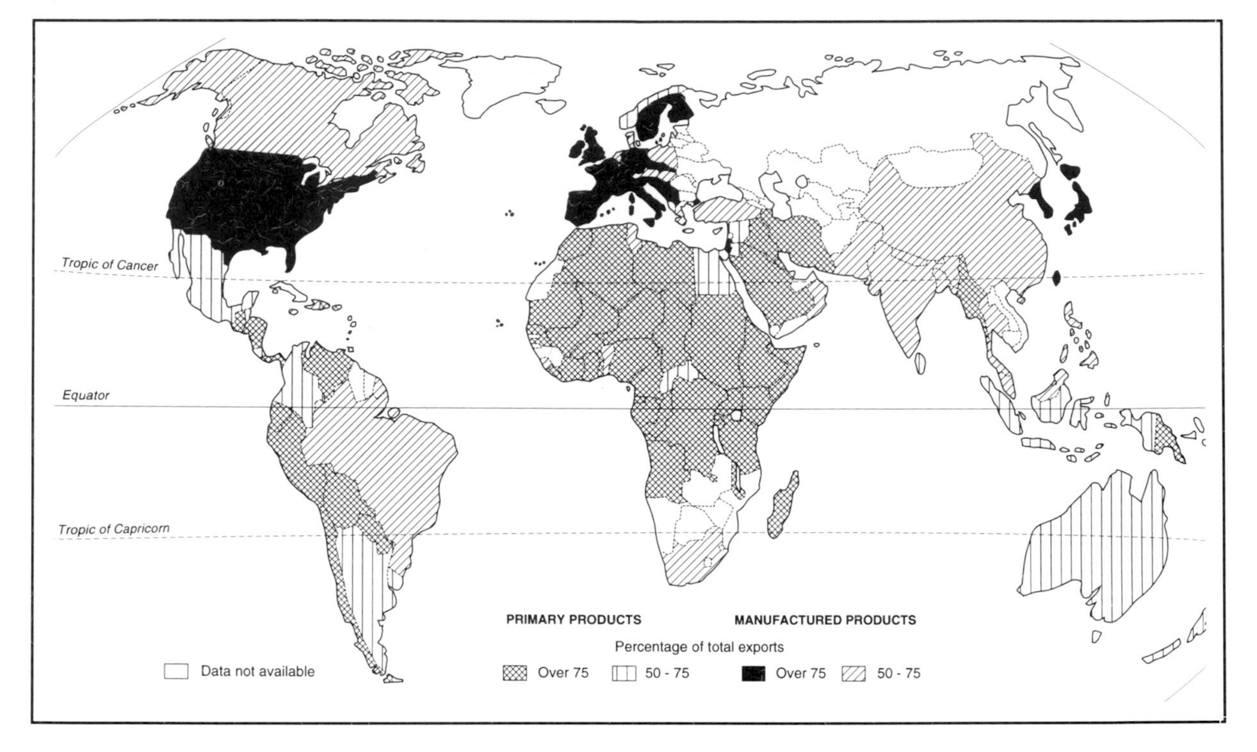

Figure 1.9 Structure of export trade, 1990
Source of data: World Bank, 1992

countries having a GNP per capita of less than $1000 in 1992 and the others spread out over the range up to $22,000, with 22 countries having figures above $10,000. The 1992 patterns show substantial variation within the Third World and there have been important changes since 1981 (Box 1.1).

The United Nations identifies 44 countries as the 'least developed'. This definition is based on 'one or more of the following constraints: a GNP per capita of around $300 or less, land-locked, remote insularity, desertification and exposure to natural disasters' (UNDP, 1992, 208). Only one of these criteria is economic: the remainder might be termed environmental. What a map of this group of countries reveals is the concentration of the 'least developed' in Africa (Figure 1.12). Why should this be? Is it there that the economic, geographic and environmental criteria identified by the UN operate most in concert?

The use of such criteria do, however, raise more general points about defining the Third World. The economic criterion is unambiguously a measure of poverty, but a map based on the other criteria would yield curious results, since several of the world's high-income countries are islands or land-locked; a number of oil-rich states have desert environments, and natural hazards such as earthquakes and hurricanes are not phenomena confined to the Third World.

These contrasts also operate at other spatial scales. Using indices of economic and social development we can demonstrate that some parts of individual countries are richer and more developed than others. It is also possible to demonstrate that, in general, the population of Third World cities and towns is richer and better provided for than are the people living in rural areas (Figure 1.13). The simple measure of per capita income, at a global, national or regional scale, also presumes that measurable wealth is equally distributed amongst the whole population. It can easily be shown that this is not so (Figure 1.14). Income distribution is 'skewed'; shares are unequal; some people are richer than others, whether in New York,

A cartoon interpretation of potential conflict between 'North' and 'South'

Moscow, Asunción or Ouagadougou. There is also frequently inequity on the basis of gender. Although women in the Third World (as in the First) have greater life expectancy than men, they tend to have lower levels of education and literacy than males (Figure 1.15), and to be less engaged in wage labour.

At these varying scales there is inequality. There is relative poverty in one country or region compared to another, between town and country, between one segment of society and another, and between males and females. In some places there is absolute poverty – the poverty of people whose existence is 'that of, at best, half men, living poorly and living briefly, living in the twilight world of the illiterate, living in the brutalizing certainty that half their children would perish of hunger or preventable disease before adolescence' (Buchanan, 1964, 108) (Plate 1.2).

EXPLAINING GLOBAL PATTERNS OF INEQUALITY

A crucial question is 'Why should this be?' Why are some countries richer and more developed than others? The fact that much of the Third World lies between the tropics led to suggestions that the 'harsh' environment could explain the lack of development. Ideas of 'environmental determinism' have a long history, and European perceptions of harsh climates and associated diseases were advanced to explain tropical backwardness. In recent years media images of the ravages of drought in Africa and floods in Asia have given some stimulus to such interpretations, though usually couched in more fashionable 'environmentalist' terms.

However, several of the countries identified as 'low-income' countries – Bangladesh, Egypt, Pakistan or China for example – lie partly or entirely outside the tropics. On the other hand there are oil-rich states, and trading and manufacturing countries in tropical South-East Asia, whose economies have prospered to a high degree, while living standards in parts of Eastern Europe, distinctly non-tropical, approximate closely to those in many areas of the Third World.

It has been claimed that the Third World is deficient in the natural resources necessary to sustain development. Yet many Third World countries, dependent in their export trade upon a few primary products, are major sources of raw materials to the economies of developed countries, such as oil, iron ore, bauxite and other 'essential' minerals.

A recurring image of the Third World is of the 'teeming millions' of people. Could overpopulation explain their backwardness? While it is true that countries such as China, India, Brazil and Nigeria rank among the world's more populous countries, many Third World countries have small populations. Most of the countries defined by the World Bank in the low- or middle-income categories had populations between 1 million and 10 million, while of the 72 countries with below 1 million inhabitants, at least 50 could be regarded as Third World countries. Nor is population density a critical common factor. Some Third World countries have very high population densities – above 200 per square kilometre in El Salvador, Burundi and Vietnam, but many others, including some with large total populations, such as Brazil, Indonesia and Nigeria, have low overall densities. Furthermore, some of the economically successful NICs, such as Singapore and Hong Kong, have very high densities.

Because most of the richer countries have 'white' populations while many Third World countries are inhabited by 'coloured' people, racist arguments were used to suggest that the 'natives' are 'inferior' or 'indolent'. This was a common theme during the colonial period, but such an interpretation ignores the evolution of early human civilization and economic advance in Egypt, Mesopotamia, India and China, and the important technical advances made in China, India and the Islamic countries during the European 'Dark Ages'.

Most Third World countries have been colonies of Europe at some time during the past 500 years; could the explanation be colonialist exploitation? Yet, if being a European colony is a prerequisite of poverty, how do we explain the present ranking of the USA, Canada or Australia, or even Hong Kong, a British colony until 1997? Third World countries experienced the rule of different colonial powers, Britain, France, Belgium, Spain and Portugal and others: the nature of colonial rule varied. So too did its duration: some countries have been independent of their colonial masters for over 150 years; many have achieved independence only since 1945; a dwindling few retain some form of colonial status, mostly, unlike Hong Kong, economically insignificant though often with strategic significance, such as mid-ocean islands like Ascension Island or St Helena in the South Atlantic. Furthermore, a few Third World countries – Thailand, Liberia, China – have never been colonies in any formal sense.

None of these explanations are adequate or capable of common application across the Third World, and it is unlikely that there is any one single factor which can be identified as the cause of Third World poverty. There are many countries, many contrasts, many causes.

In recent decades there have been increasing

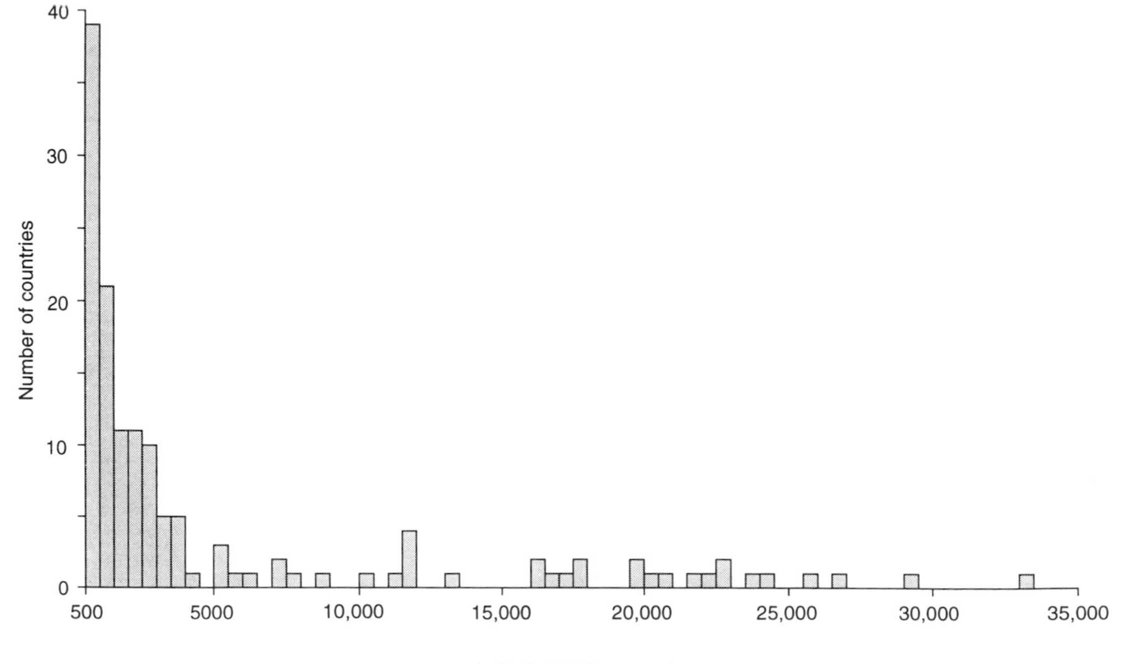

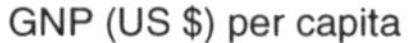

Figure 1.10 The global distribution of income, 1992
Source of data: World Bank, 1994a

BOX 1.1 A WIDENING INCOME GAP WITHIN THE THIRD WORLD

The distribution of GNPs by country in 1991 clearly shows a distinct bias towards the lower end of the income scale (Figure 1.10). The majority of countries have per capita income levels at below $1000, but the rest are well spread between $1000 and $20,000. This is a pattern that has persisted, even though the absolute levels of these measured incomes have risen over time. A comparison of the 1981 and 1992 distributions (a direct comparison of groups of countries is not possible, due to different definition of the groups at the two dates) illustrates the patterns of change by major income groups as defined by the World Bank.

The mean per capita income of the low-income group has risen least, by less than 50 per cent, whereas the mean income of the next group, slightly larger in both 1981 and 1992, has risen by 76 per cent. However, the mean income of the high-income countries almost doubled in this period. The global distribution is even more elongated in 1992, but the large number in the poorest category remains. Clearly, the impact of global income improvements has concentrated disproportionately in the already rich countries, with the poorest countries being left behind.

The regional breakdown of these data is indicated in Figure 1.11. Sub-Saharan Africa has the most skewed distribution, with the modal group at $200–400 in both years. Asia is the other continent to contribute to the poorest group in both years, though its 1992 distribution has improved over that for 1981, with a few countries at over $1000 per capita. For Latin America and the Caribbean (the Table includes data for only those countries with over 1 million people and so excludes most Caribbean countries) there is a clear improvement at the middle-income levels. The presence of a few oil-exporting high-income countries accounts for the distinctly bimodal distribution for the Middle East and North Africa.

The overall pattern that emerges from this comparison confirms the widening income gap within the Third World between the poorest and the middle-income countries, and particularly between Sub-Saharan Africa and South Asia on the one hand, and Latin America, the Middle East

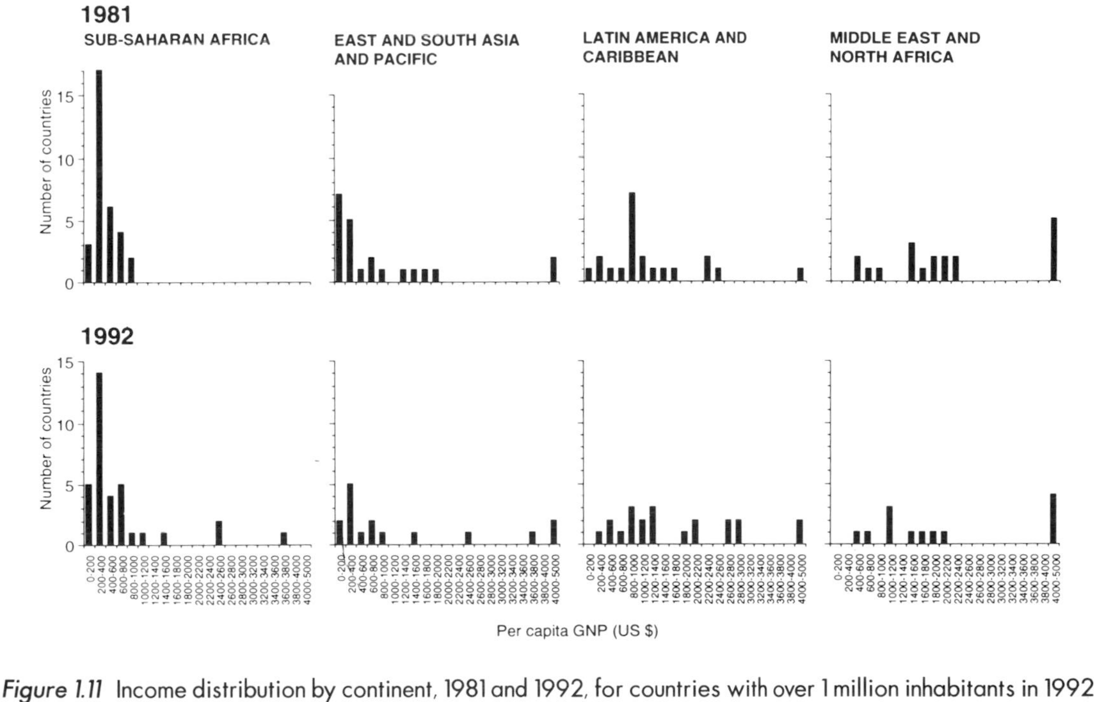

Figure 1.11 Income distribution by continent, 1981 and 1992, for countries with over 1 million inhabitants in 1992
Source of data: World Bank, 1994a

BOX 1.1 *continued*

and East and South-East Asia on the other. There has been development as measured by rising per capita incomes but it has been differential, and has not occurred where it is needed most. This widening income gap is paralleled by similar differential patterns of growth in food production, industrial output and most other familiar indices of development. There is increasing differentiation of economic performance and social improvement within the Third World.

Table 1.1 Mean income by World Bank region, 1981 and 1992

Country group	*1981*		*1992*		*% change 1981–92*
	Mean income $ pc	*n*	*Mean income $ pc*	*n*	
Low-income	270	34	390	42	+44%
Lower-middle income	850	39	1500*	45	+76%
Upper-middle income	2490	21	4020	21	+61%
High-income	11,120	19	22,160	22	+98%

Note: * = estimate
Source of data: World Bank, 1983, 1994a

efforts, by both the more developed and the less developed countries to foster development within the Third World and to ease the poverty of their peoples. In the last ten years it has become increasingly clear that since there is no single factor which is the cause of world poverty, there can be no single solution. Yet more and more people and organizations are now grappling with the problems of the Third World. These include international agencies such as the International Monetary Fund (IMF) and its sister institution the World Bank, the

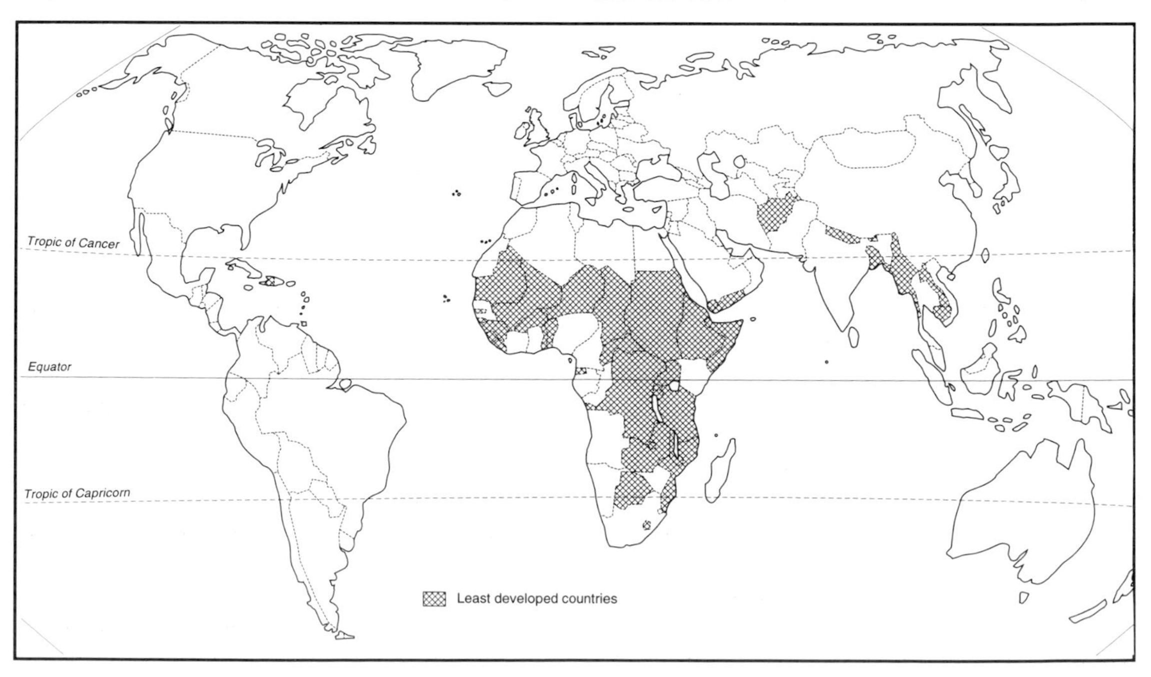

Figure 1.12 The poorest of the poor
Source of data: UNDP, 1992

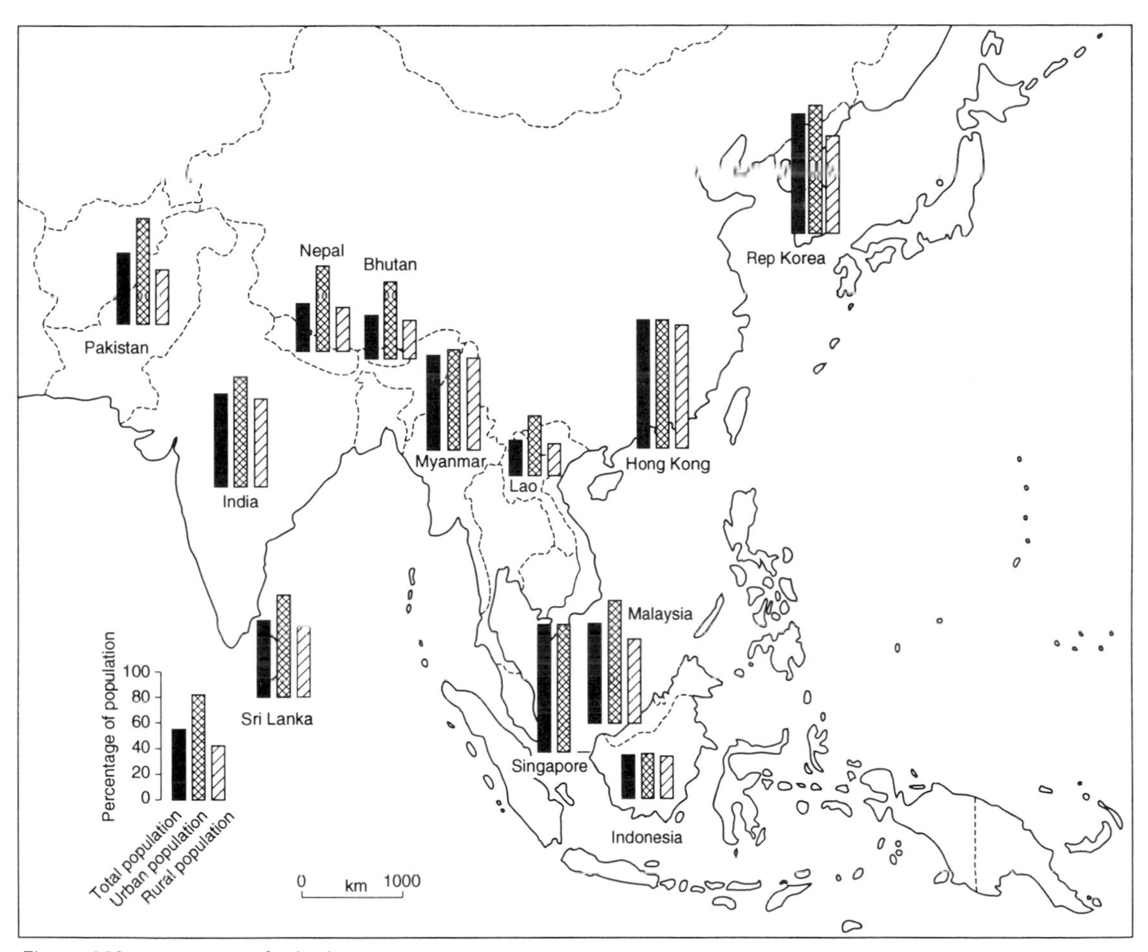

Figure 1.13 Access to safe drinking water in Asia, 1990
Source of data: World Bank, 1992

largest development organization, and other aid agencies, particularly associated with the United Nations. Trade groupings like the European Community and OPEC, and governments of independent states in the North and South also have aid programmes, both individually and through the groupings formed in the United Nations, while the role of Non-Governmental Organizations (NGOs) has increased dramatically in proportion to the improvements in flows of information in the West concerning the plight of the Third World poor. Perhaps even more significant have been efforts made by the poor themselves, male and female, of the Third World who organize themselves locally to confront the problems of increasing population, the impact of disease or the challenges of rural development.

APPROACHES TO THIRD WORLD DEVELOPMENT

Given the changing conditions of the Third World and the changing international context which we have outlined above, it is not surprising to find that approaches to development have changed considerably over the last 50 years. It is possible to argue that there have been three main phases in the evolution of attitudes, though subsequent developments have never entirely removed either the attitudes or the effects of the earlier phases.

In the first phase, Western theories and strategies for the development of the Third World emerged as a product of two major events in recent world history: the great recession in trade in the 1930s and the end of the Second World

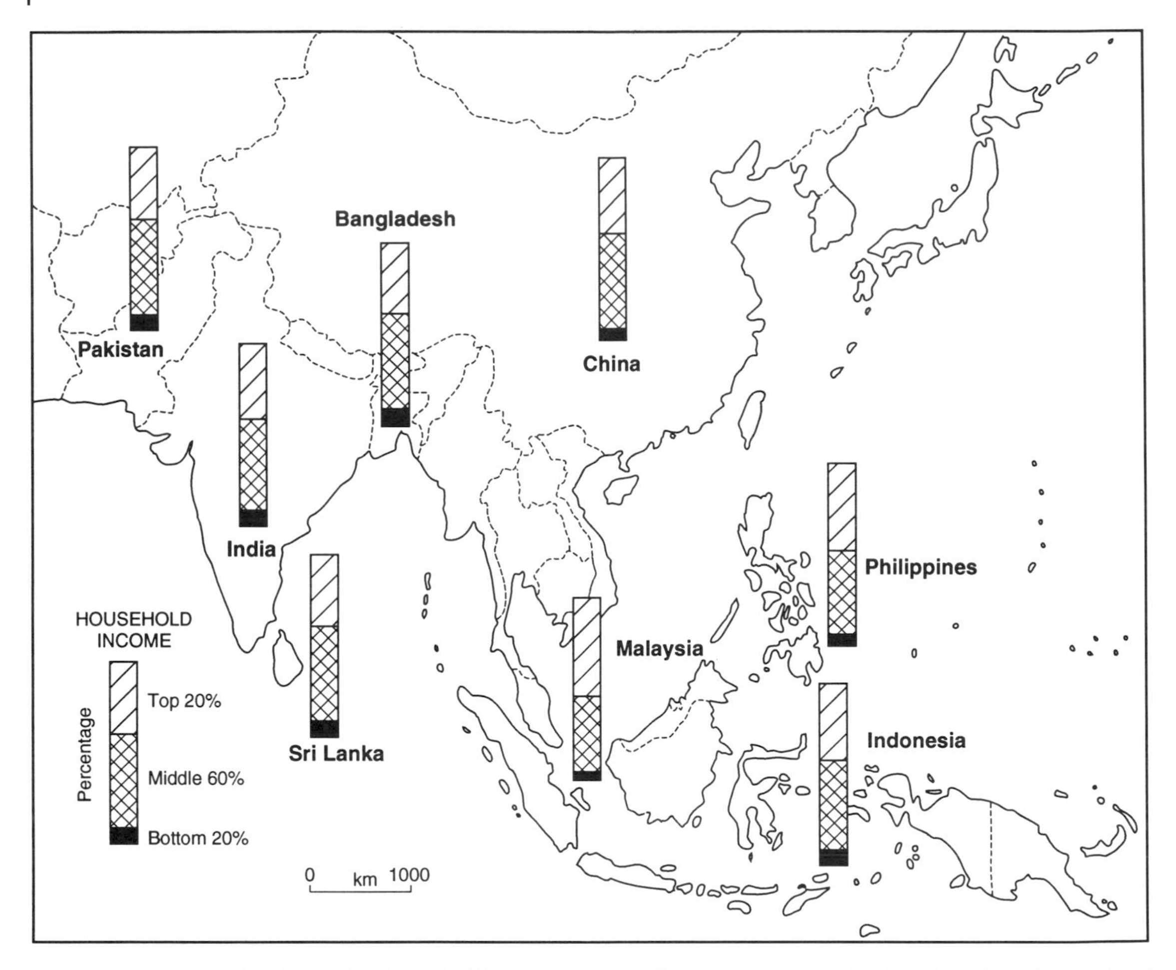

Figure 1.14 Income distribution by household in Asia, *c.* 1990. If income were equally distributed, each quintile of households would have one-fifth of income. As the map indicates, the poorest 20% of households have well below one-fifth of income and the top 20% more than one-fifth
Source of data: World Bank, 1994a

War in 1945. The first allowed free rein to the ideas of John Maynard Keynes and his followers, the second created an atmosphere of freedom, idealism and optimism and provided the seedbed of the colonial independence movement. Keynes (1936) believed that the free play of the money markets would lead to booms and slumps of increasing severity, so that in the modern world no sophisticated social, political and economic system could survive without careful regulation of the economy. This did not imply any lack of enthusiasm for capitalism itself, but a belief that economic growth was more likely to be continuous and to benefit more people if it were stage-managed.

The evidence of world-wide economic depression in the 1930s brought this argument into sharp focus, and the post-war reconstruction of Europe was the first major attempt by European governments to use Keynesian economic strategies to regulate domestic economies and to plan 'aid' intervention on an international scale. The initial success of these measures through the Marshall Plan (the American Aid-to-Europe programme which allowed massive capital investment in new industries and successfully stimulated the resurgence of European industry after the devastation of the Second World War) encouraged development theorists to follow a similar reasoning process when they turned their attention to the poor nations of Latin America, Africa and Asia.

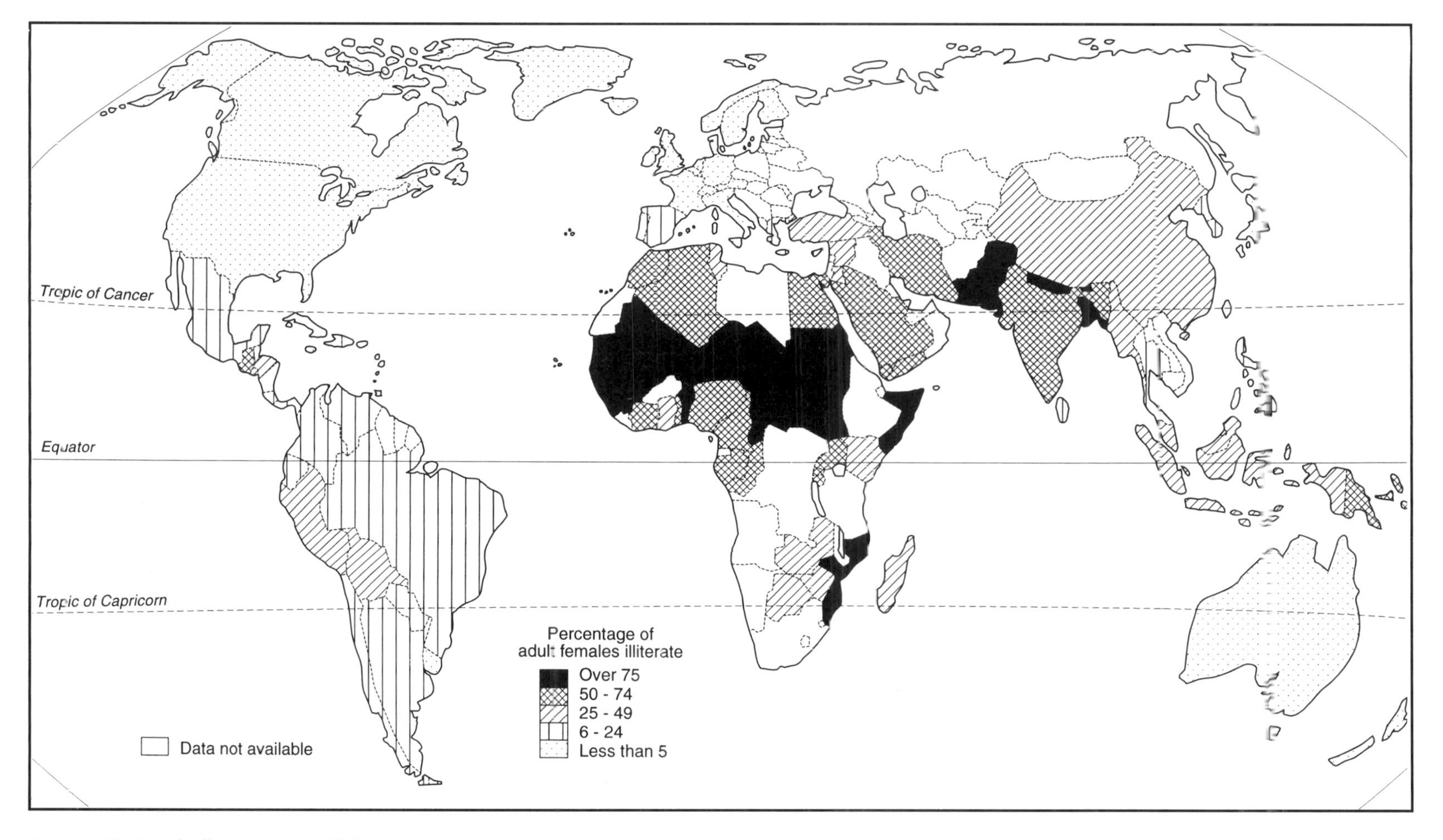

Figure 1.15 Female illiteracy rates, 1990
Source of data: World Bank, 1994a

The first theories of Third World development were, then, the product of three major influences: the Euro-American experience which had generated an enthusiastic belief in managed capitalist economic and social development; a new spirit of optimism about the future; and growing determination within these poor countries, still under colonial rule in many cases, to secure economic advance.

In the 1950s political 'winds of change' were sweeping the continents of Africa and Asia, and many new and exciting ideas about social and economic progress were in the air. Yet, at the same time, most Western observers would have had no difficulty in identifying the Third World as a group of economically backward countries, generally in the tropics, which had been left behind by the socially, economically and politically 'more sophisticated' nations of the northern hemisphere and southern temperate regions. This highly value-laden view would have raised few eyebrows in the 'white' Western areas of the world. What is more important, few who had dealings with Third World countries would have baulked at the inclusion of such words as 'ignorant', 'superstitious', 'corrupt' and 'primitive' in such a definition. There was a highly charged sense of cultural and (at base) racial superiority, moderated only by a genuine 'missionary' zeal to improve the lot of the under-privileged. Moreover, a firm belief in the virtues of cultural patronage characterized attitudes and values both on the part of the rulers and also on the part of the ruled. Even the radical Third World intellectuals who served a political apprenticeship to independence in colonial jails readily embraced the lineaments of this philosophy when it came to adopting development strategies for their new nations. Perhaps there was no other choice. As indicated above, there were strong historical currents feeding early development theory. It is therefore not surprising that the first theories of Third World development were characterized both by a strong belief in the power of interventionist economics and the superiority of the then current Western model of economic (and by implication cultural) development, as a linear process directed towards the same kind of economic, social and political structures which characterized Western Europe and North America.

Walter Rostow (1960) gave an impetus to such thinking by likening the process of economic growth to that of an aeroplane setting off down a runway. After a long 'pre-industrial' taxiing process the economic engines are revved for a rapid surge towards 'industrial' take-off into the clear blue sky of sustained economic growth. Although this was quickly recognized as a simplistic model, based on too limited a view of the European Industrial Revolution, the underlying assumptions of comparability and convergence were embodied in most early theories of Third World development. The argument was quite straightforward. Progress had been achieved in the West (and in the European East) by eliminating the rural character of economic and social relations, while the Third World was characterized as having peasant social and economic attitudes and institutions of precisely the kind condemned by Western theorists. It was these 'primitive, backward, conservative, superstitious' elements which had all been removed in the progress towards an urbanized and 'developed' society.

Having identified a clear continuum of progress from rural poverty to urban affluence in the West, it was inevitable that the first development strategies should approach the problem of inducing change in the Third World by concentrating on the 'successful' end of this spectrum. Since Western development had taken place through the generation of an urban-industrial system increasingly geared to a sophisticated and capital-intensive technology, this was the direction in which Third World economies should go. By this means the interaction of negative effects or 'the vicious circles of poverty', which perpetuated the adverse condition of poor countries and poor people, could be broken. If a man was poor, he could afford little food; if he was under-nourished, he was more prone to disease; if he was ill and under-fed his capacity to work was limited; if his work ability was limited, his income was low; and in consequence he could afford little food. There was then a process of circular causation of poverty. The Swedish economist Gunnar Myrdal (1963) and others argued that if this process could be broken and reversed by industrialization then economic progress might be achieved. There

Plate 1.2 Child sleeping in a doorway, São Luis, northeast Brazil Orphaned or abandoned children are often the most impoverished and deprived people in Third World cities

might then be a converse 'circular causation of cumulative development', a process which Myrdal linked to Rostow's 'take-off'. If some impetus could be given to a poor family, community or nation, there might be change for the better.

First World views of Third World population

For example, the location of a factory at a particular place might spur development there, providing jobs, higher incomes and new services of transport, education and health. The population, place and country would benefit, and there would be a diffusion of the benefits of growth.

This process of diffusion, identified by Myrdal as **'spread effects'** and by Hirschman (1958) as 'trickle-down' effects, would provide the stimulus to economic progress and an improvement in the lot of the poor countries. Precisely what were the best initiators of this process were unclear, though there was a strong preference for the urban-based industrial developments which appeared to have been vital in the process of European economic revolution. The French economist François Perroux (1971) suggested the notion of *industries motrices* (propellent industries) which would provide the necessary stimulus; once these had been identified and established, they would form the nuclei of 'growth poles'. Perroux's idea was enthusiastically taken up, not least because it appeared to have spatial implications: growth would diffuse from the growth pole to other places. However, Perroux himself was concerned primarily with an economy in the abstract, suggesting only that growth might be induced, not *where* it might take place.

There had emerged, then, a development philosophy based primarily on the experience of capitalist Europe, in which diffusion of industrial and urban growth was the key to progress; and a belief that a major stimulus was needed to break the vicious circles of poverty. There were some uncertainties as to quite what the best stimulus was. There was also, from an early date, a recognition that economic progress could not be achieved everywhere at the same rate. Both Myrdal and Hirschman had recognized countervailing forces, 'backwash' or 'polarization' effects, that would inhibit equality of development throughout a country. Hirschman noted that, in a geographical sense, growth would necessarily be unbalanced. He believed that, over time, the trickle-down effects would outweigh those of polarization; Myrdal felt that this was unlikely unless there was intervention in the economy by the state.

Such ideas provided the basis for ideas of a '**core–periphery**' model, most clearly expounded by John Friedmann, one of the most influential writers in the development field since the 1960s. He stated that

> economic growth tends to occur in the matrix of urban regions. It is through this matrix that the evolving space economy is organized. The location decisions of most firms, including those in agriculture, are made with reference to cities or urban regions.
>
> (Friedmann, 1966, 28–9)

Cities were thus the core of economic advance; around them was the area of more efficient agriculture, and beyond were the backward subsistence activities (Box 1.2).

So it was that the city and the city region were identified as the catalyst of a process designed to engage the whole national space. Major growth-poles comprising related industries would be identified and encouraged by a network of communications. Aid would be channelled down a hierarchy of minor centres. The focal growth impulses of industrial development would produce a flow of investment to trickle down this hierarchy of economic and spatial feeders to the smaller urban centres. Eventually even the smallest villages would receive the benefits. Like Rostow's runway analogy of economic growth, the idea of development trickling down a spatial infrastructure blurred a number of issues that are taken up later in this book. But it was enthusiastically adopted as a working spatial model for development planning and is still an implicit feature of many national plans. Indeed the process was given specific structure by planning theorists. Major industries could be implanted in major centres. They would provide jobs for the urban poor, create wealth which, when invested in subsidiary industrial enterprises, would generate more jobs. These in turn would generate further wealth. From this wealth taxes could be raised to invest in schemes of public health and education. It all seemed quite straightforward. All that was needed was to put what economists call the 'multiplier' to work.

Why is it, then, that things had gone so badly awry by the 1980s? Thirty years of investment, technical expertise and national planning had not been successful in achieving development objectives in most countries. The gap between rich and poor, nationally and internationally, continued to grow. The main cause of the failure had been an inability to appreciate the effects of such a policy 'on the ground'. Myrdal was one of the first to notice that 'backwash' was a more significant consequence of development intervention than trickle-down or spread effects, and that the tide of human migration towards major urban centres was both a symptom and a cause of rural poverty and urban imbalances, with many Third World cities developing a parasitic role in relation to the surrounding countryside. The diffusionist assumptions of the Rostovian development model were everywhere being questioned as a result of the apparent polarization of development in favoured countries and regions.

Moreover, important shifts of emphasis had also taken place in the world. By the mid-1970s the United Nations had become dominated numerically by newly independent and fiercely nationalistic countries of the Third World, who had a voice on the international stage and were prepared to use it. Oil had been recognized as a scarce resource with important reserves in the Third World. The exploitation of this situation, in the Middle East 'oil crisis' of 1973–74, with a five-fold increase in the oil price from $2.50 to $12.50, had a significant impact on First World economies and gave political and economic weight to the OPEC countries. The growth and consolidation of China as a world power also had an impact on the thinking of those who live in poor countries. Socialist revolutions replaced some Western-oriented post-colonial regimes by those seeking development by an alternative route, and particularly through prioritizing rural development.

Matching these changes was the re-evaluation of the colonial (and post-colonial) development period in theories, not of development, but of *under*development. Many observers had noted that 'modernization' benefited mainly the elites of Third World countries. The

BOX 1.2 DEVELOPMENT REGIONS (THE FRIEDMANN MODEL)

John Friedmann identified five types of region relevant to national planning policies. Figure 1.16 expresses them spatially. The core region is the focus of economic growth, with an urban nucleus. Around it upward-transitional areas are in the process of development. Their resource endowments and proximity to the core make them likely beneficiaries of further growth. Resource frontier areas represent the frontier of settlement or even leapfrog into virgin territory to exploit new resources of land or minerals, often with lines of penetration provided by new highways and mineral railways. Downward-transitional areas represent areas of settlement that are stagnant or declining because of decline in their resource base or industrial structure. They are areas which lose population and capital to more dynamic areas. Friedmann also identified regions with special problems, such as those at national boundaries, or where there are conflicting resource uses. At a less complex level, the core and upward-transitional regions could be identified as the 'core' and the remaining regions of the country the 'periphery'.

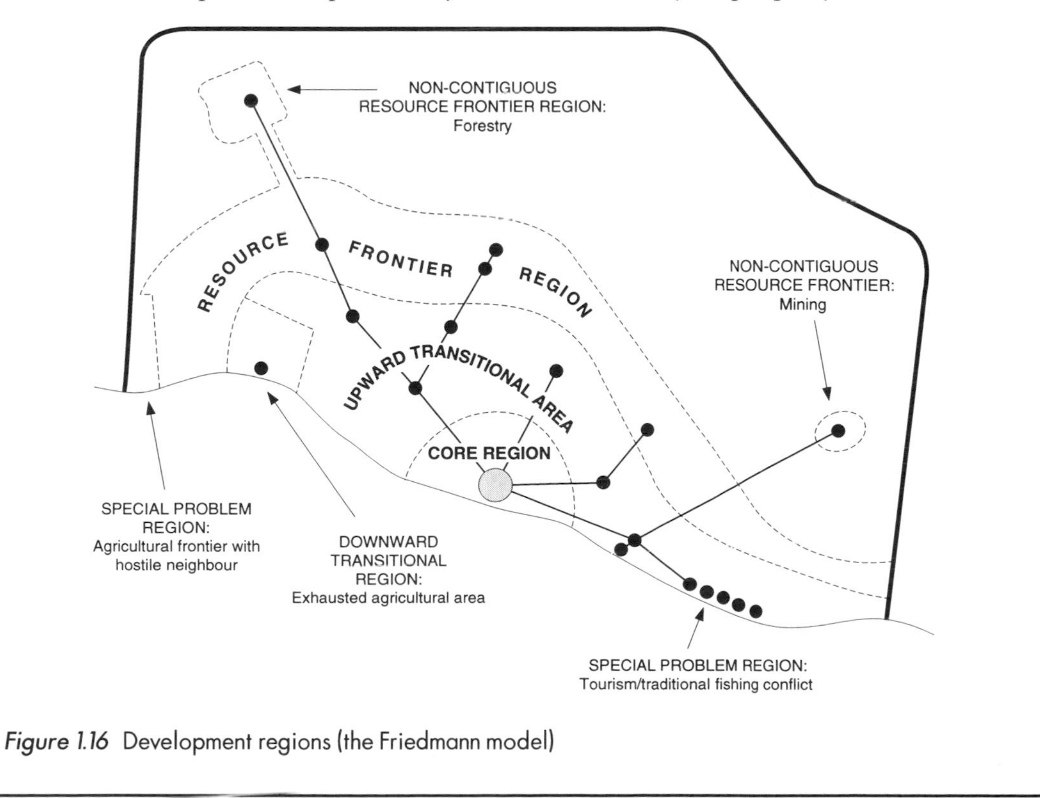

Figure 1.16 Development regions (the Friedmann model)

new state apparatus tended to create an elaborate administrative hierarchy (a civil service). This in itself created a wage and salary-earning class and a market for Western-style consumer goods. Imports of such goods rose, and in order to save foreign exchange Third World countries 'substituted' such imports by manufacturing locally. The local industrial economy was soon dominated by factories producing beer, soft drinks and spirits, cigarettes and tobacco, confectionery, transistors and refrigerators. Obviously only those with 'urban' incomes could buy such goods in any quantity. Most of the industrial plants were constructed with imported equipment and used foreign expertise. Spare parts were part of the deal and in this way suppliers could ensure that projects were only serviced by external technologies. The manipulation of Third World economies by individual large-scale (multinational) companies with the connivance and encouragement of local elites was part of a wider malaise which some identified as the domination of the 'periphery' of the global economy by those who live in the 'metropolitan-core areas' of Europe and America. At a national level, the core area of the **primate city** dominated the rural periphery.

The second phase in changing attitudes to development began towards the end of the 1960s. Disillusion with the ineffectively slow impact of interventionist and laissez-faire approaches led to the growth of alternative theories emphasizing the dependence of Third World economies on the structure of international capitalism. The dynamism of the Chinese socialist state had encouraged a number of new countries, especially in Africa and South-East Asia, to adopt a socialist agenda for change. New states like Tanzania and Cuba seemed, at least at first, to respond better to socialist experiments. Moreover, Marxist social scientists in the West, usually with Soviet and East European experience in mind, began to suggest that development planning and aid intervention could be seen as a process of exploitation by capitalist centres of their client Third World peripheries. This was envisaged as taking place at several scales. Lipton (1977) argued that this was largely a matter of urban exploitation of a rural poor in terms of trade and concentration of relative advantages, an 'urban bias'. A.G. Frank (1969) argued that these connections between satellite and metropolis were channels through which the centre appropriates part of the satellites' economic surplus, which thus gravitates up towards the economic core of the capitalist world (Figure 1.17). He envisaged the process operating at a variety of scales: at a world scale with the metropolis of eastern USA exploiting other parts of the world, and at a national scale with a national metropolis exploiting its satellites. The pattern is shown in Figure 1.17 as a diagram of a metropolis–satellite hierarchy, and as a national and international map of the relations between the USA and Brazil. Dependency theorists also claimed that some regions stifle local entrepreneurial initiative by an uncritical adoption of external assistance.

Rodney (1972) and Amin (1973) took the view that in Africa this process of **dependency** had a much longer root. It began with the slave trade and had only gained momentum rather than changed its character since that time. Wallerstein (1974) took this argument a stage further by claiming that this process of exploitation was defined by the earliest appearance of capitalism as a mode of economic organization, and that thereafter all local and regional systems were subverted within its ever-growing hegemony. His 'world-systems' theory takes a global view of the ways in which all parts of the world had been integrated, politically as well as economically, into the world-system dominated by the 'core', North America and Western Europe, and operated through the creation of a 'semi-periphery', those richer Third World states with mineral exports and/or limited industrialization for export, and the 'periphery', the poorest countries that had been in the past and were continuing to be exploited by their involvement in the global systems of exchange (Figure 1.18). In this period, up to the early 1980s, dependency and underdevelopment theorists were dominant in creating the prevailing climate of opinion that emphasized international structural constraints on the development of individual Third World countries.

The third phase of changing attitudes, during the 1980s and into the 1990s, has seen a period

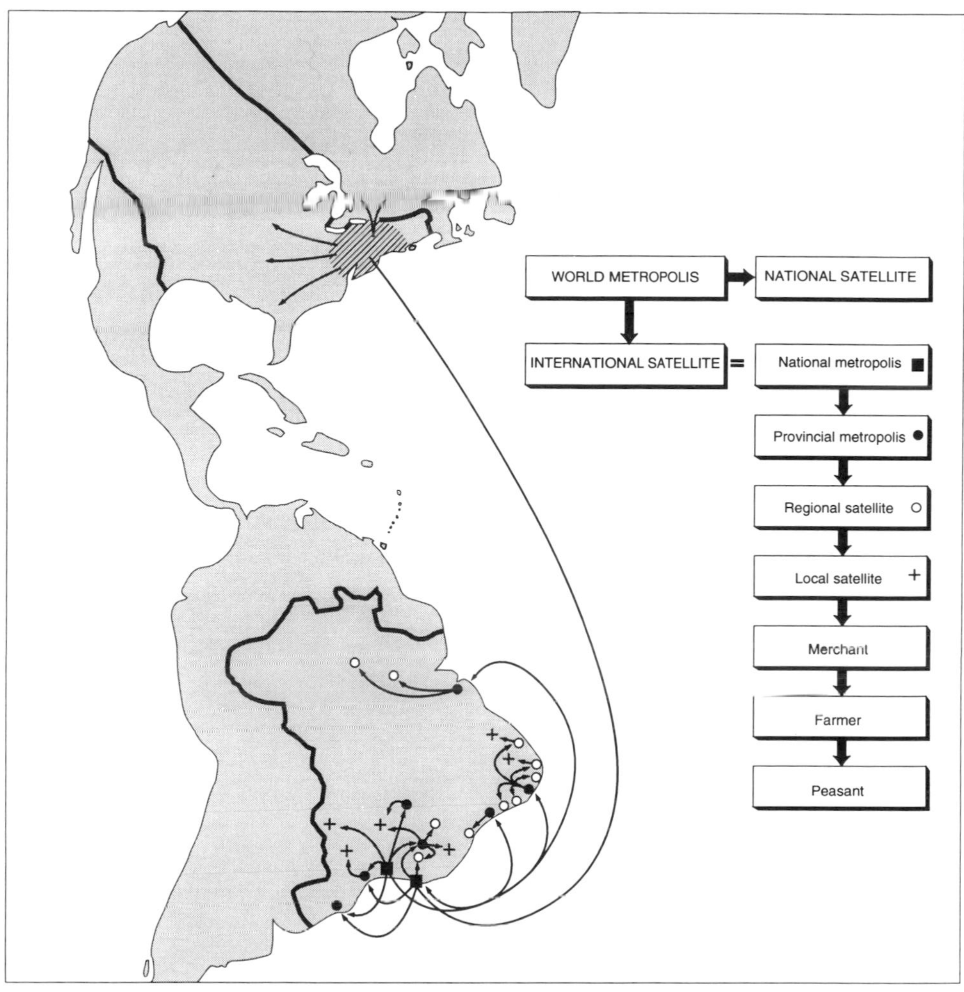

Figure 1.17 The Frank model

of great uncertainty. There has been a recognition of the limitations of grand intervention planning, both socialist and capitalist. We have also seen the impact of a new view of global interdependence through the workings of international trade and the IMF, with the assumption that it is through greater involvement in the world economic system, not less, that countries develop. After more than a decade of mismanaged statist intervention, there have been new policies for targeting areas of high potential and encouraging market forces under the terms associated with **structural adjustment** programmes, first developed by the World Bank and IMF. These have become an article of faith for these agencies, an approach which has emphasized the power of the market and the ways in which even the humblest farmer is part not just of a nebulous local system but of the network of international trade. Loans are given to develop and encourage **backward** and **forward linkages** in the economy. This may work well when economies are growing and there is an active 'market', but they are not strategies for economic growth in the poorest countries in a period of world recession when prices for **primary sector** products are low, or in periods of drought when there is little cash crop surplus. In addition, many people continue to maintain their livelihoods outside the 'market' within community and kinship structures.

On the other hand, the spread of the new laissez-faire economics of monetarism has been encouraged by the loan conditions, so-called

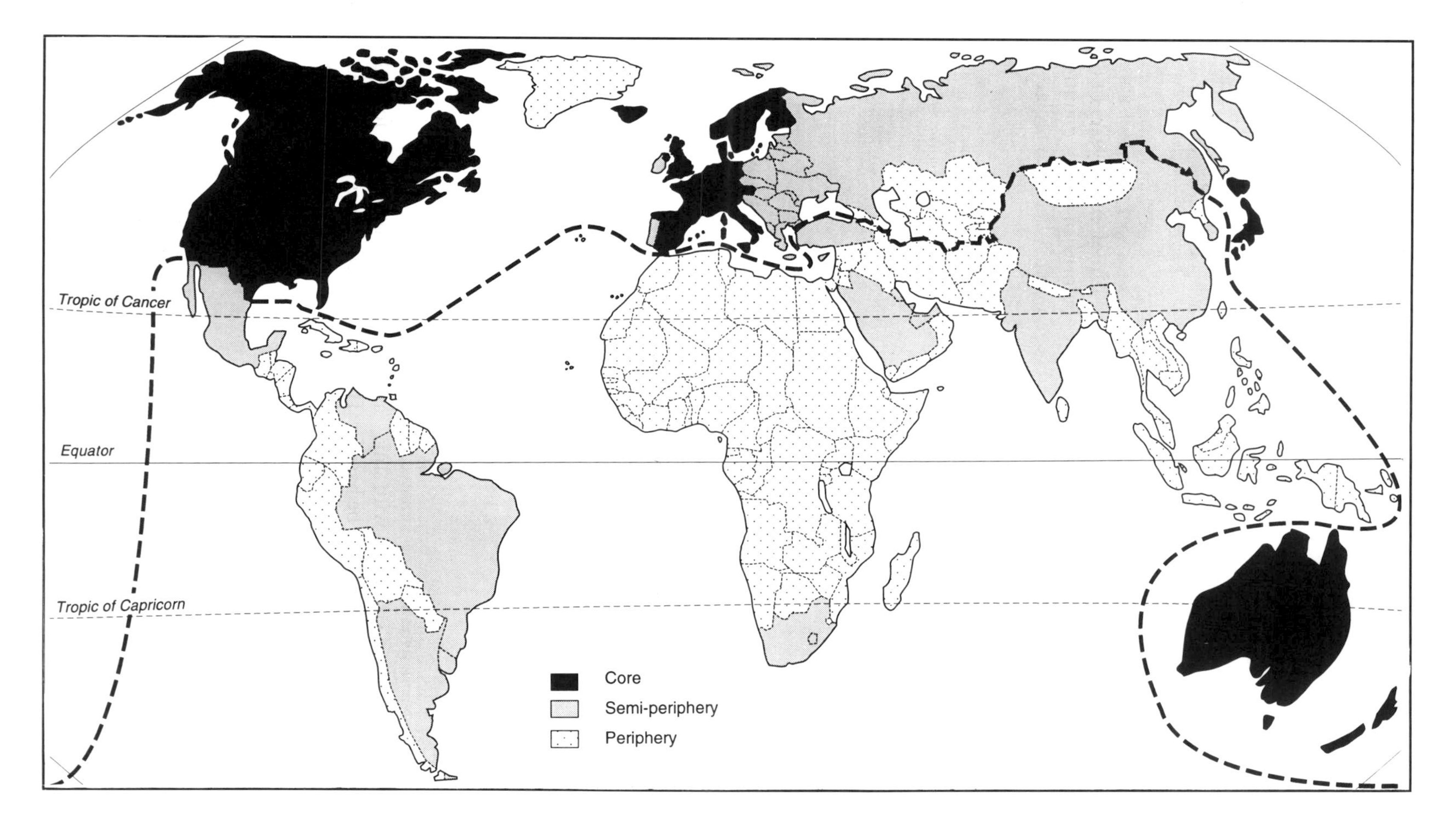

Figure 1.18 The world-system: core, periphery and semi-periphery (the Wallerstein model)
Source: After Knox and Agnew, 1994, 21

'market **conditionality**', laid down by the international agencies. As a consequence, in some areas of the Third World the interventionist role of the state is now much lighter than it was in the past. Its advocates claim that it allows local enterprise and capital to be released towards the solution of local and national development problems. Critics argue that weakening of central control might result in social breakdown, drug trafficking, factionalism or even civil war.

In the last resort the debate between those who favour 'radical socialist transformation', those who favour a mixed economy with some state control, and those who believe in (more or less) unfettered capitalism, is not just a Third World affair, and it will be no surprise to find differences of attitude reflected (as they should be) in the pages of this book. Indeed the experience of development in all countries provides us with ammunition for debate between those who occupy different ideological, theoretical and political positions.

The arguments concerning the merits and demerits of unfettered free enterprise or collective control of the means of production are no longer presented in terms of empty theories. Socialist attempts to reduce poverty have now been tried and tested in the same furnace of experience. These sometimes bitter experiments have taught radical regimes that political dogma cannot replace economic reality, that circumstances alter cases. Not all countries are 'ripe for socialism'. Not all regions within a country are likely to respond in the same way to uniform national policies. Sometimes the political machinations of superior forces (or superpowers) subvert honest attempts to achieve a local initiative. Progressive strategies with any hope of long-term success must always be based on a recognition of the art of the possible. The arguments between those who favour free enterprise and those who prefer collective controls are often presented in the abstract or at a national or international scale of reference. But it is not only a question of either dictatorship or peoples' democracy, totalitarian police state or free and open elections, centralized 'top-down' economic management or 'bottom-up' and grassroots initiatives. Against this large-scale ideological backcloth, real decisions are made at a local scale, where compromises are more important than ideological confrontation. The resultant systems will surely develop some of the advantages of being able to choose the best from alternative forms' social and economic policies, but equally, because of human fallibility, may incorporate some of the disadvantages of each ideology.

There are at least some hopeful signs that the poor of the Third World are to take an increasingly important part in the decisions about their own future. It took many years of development initiative before, in some countries at least, 'barefoot' doctors began to replace large prestigious hospitals and rural schools began to teach rural science; before the shanty town was recognized as a potentially viable social entity rather than an administrative inconvenience. It has taken many years of socialist planning for some communist countries to begin to recognize that personal initiative and private incentive are a vital part of life. At the same time a new self-assertiveness has been seen in the cultural realm, in indigenous music, arts and crafts and national cultural consciousness. It has also been evident in the religious sphere, and particularly in the growth of Islamic self-consciousness – a religious radicalism that has transformed the self-image of those who were previously culturally subservient, with explicit rejection of both Western and socialist economic and cultural models.

One key aspect of the shift is clearly from emphasizing the international constraints on development, which characterized the dependency era, to emphasizing the internal opportunities for development using models, assumptions and objectives formulated within Third World countries themselves. While the Brandt Commission (1980) had only one of its chapters devoted to internal aspects, 'The task of the South', the South Commission Report of 1990 was almost entirely devoted to the opportunities that the countries of the Third World, individually and together, could use to promote development. A recognition of the possibilities for indigenous development models and strategies is not an approach that is favoured by the IMF and World Bank, which cling to a belief in the superiority of a global but Western-based model and in the virtues of international exchange; but even these institutions have become more aware than they had been in the past of the local context of development, and the virtues of harnessing local potential and economic and cultural relationships.

Hence there has been a resurgence of concern for the role of indigenous social institutions as well as economic relations in Third World countries. Focus on these institutions has required renewed consideration of the nature of the family and intra-household structures, including generational and gender relationships, and how these affect and are affected by development. Of particular importance in this respect, and a theme that recurs frequently in the subsequent chapters of this book, is the renewed concern for gender divisions and the changing relationships between men and women. Development needs to be about people and livelihoods as well as, perhaps even more than, about economic relationships.

In consequence, the separate economic and social roles of men and women and the relationships between these roles is clearly central. Specifically, how are gender relationships defined in poor, predominantly rural households and communities, and how have these changed with 'development'? Have women been the victims or the beneficiaries of 'development'?

Of similar significance has been the resurgence of concern for the relationships between environment and development in the Third World. In the earlier periods, which saw development as economic growth and modernization (up to the 1970s), and during the dominance of the dependency school (until the mid-1980s), the environment as a factor in development was given little attention. Although geographers had maintained their traditional interest in the environment, they had ceased to see it as a major constraint on development and saw it more in terms of offering a range of possibilities within which development could occur. Their approach to development issues was shaped primarily by the broader theoretical debates about the nature of the international system and the economic and social conditions for development, and particularly for distributional aspects of that development. However, they remained aware of the role of environmental variation in affecting patterns and types of development within countries. They therefore found it difficult to accept many of the assumptions which seemed to be at the heart of models of economists and others, in which the environment was taken as a given, unvarying from country to country or even within countries, or that development strategies could afford to ignore the environmental context in which they were set. Such a view is now no longer acceptable, if it ever was.

The main impetus for linking the environment and development emerged in the 1970s, associated with the first UN Conference on the Environment and Development in Stockholm in 1972. It was then largely a First World concern generated by growing problems of pollution and resource depletion caused by unprecedented industrial growth in Europe, North America and Japan, part of a neo-Malthusian concern for overpopulation and overconsumption. This was associated at the time with pessimistic views of a future collapse of the world economy due to over-exploitation of the world's resources, and approaching 'Limits to Growth', which was the title of an influential book at the time (Meadows *et al.*, 1972).

This perceived threat to the global resource base seemed to retreat in the late 1970s, with extensive new oil finds being exploited to reduce the prices charged by the OPEC cartel, thus effectively destroying it as a critical world force. However, by the mid-1980s awareness of new global environmental problems, notably depletion of the ozone layer and associated global warming brought the issue of the environment surging back to global prominence. The Brundtland Report of 1987, *Our Common Future*, shifted concern over the environment/development relationship from a primarily First World and global scale focus to one which recognized its effects on local livelihoods in all societies, especially among poor and marginal peoples. It therefore has particular relevance for the Third World. '**Sustainable development**' became a familiar phrase and objective in all countries, but it had by then become clear that many of the most serious problems of non-sustainability and environmental deterioration had been evident in Third World countries. The environment had become a Third World issue at the global scale. However, it had never ceased to be an issue for a great many people at the local scale.

The primary focus for the new concern was the 1992 UN Conference on the Environment and Development (UNCED) held in Rio de Janeiro, around which two clear views polarized. The dominant view from the Third World was that the global problems of climate change, sea-level rise and global warming were caused primarily by carbon dioxide, sulphur dioxide and CFC emissions in developed countries, and could only be resolved by better management of resource consumption in rich countries – and possibly even a reduction of such levels of consumption. It was also claimed that this overconsumption was having such serious effects on Third World areas that substantial compensation was required to remedy losses of land and biological productivity.

By contrast, the dominant view from the First World was that overpopulation and excessive deforestation in the Third World were equally important as causes of environmental deterioration, and that the countries of the Third World needed to adopt stronger policies to curb population growth and the over-exploitation of resources. In practice, there was a recognition of the interconnectedness of the First and Third Worlds in the development of measures to alleviate the problems and to better manage the relationship. It was agreed by most participants at the Rio conference that there is no fundamental incompatibility between environment and development, but that the style of development and its impact on resource use needs to be better managed in all countries. Future development cannot proceed without environmental management; but adequate environmental management requires capital and skills that can be generated as a result of development. For Third World countries the need is clearly for additional wealth and skills to permit better environmental management, both at the national scale in the development of pollution regulation policies, and at the local scale to provide farmers with appropriate inputs of water and fertilizer that will permit long-term sustainability with rising levels of consumption by poor people.

That questions such as those of new models of development, of gender relationships, and of environment and sustainable development are now important to a geography of the Third World has come about as a result of the changing approaches to its study. The Third World itself is constantly changing, and its study must reflect the changing circumstances. Changes in the way in which geographers and others have approached the study of the Third World even extend to renewed discussion of the value of the term itself.

What is most clear when all these factors and controversies over the last 50 years are taken into account is that the Third World has gone through what is by any standards a long cycle of dependency, not only on economic and political forces beyond their borders, but on theories and ideologies which were not rooted in their own experiences. In addition, the conscience of the northern rich has been much affected by media reporting of the general poverty and the regional crises of the Third World. The generation of emergency aid to meet the immediate demands created by tragedies like the Ethiopian famine in the mid-1980s, or the Rwanda catastrophe of 1994, and the response in spontaneous charity initiatives involving young people, then creates a flow of funds for investment which is in the hands of donor organizations rather than international agencies or governments. Increasingly this is then directed into programmes which represent both the idealism of the donors and the needs of the recipient in something closer to an equal partnership. The sense of shared environmental, economic and social problems is growing. Many of them are explored in one way or another in the pages of this book.

IS 'THE THIRD WORLD' STILL A VALID GENERALIZATION?

A wide variety of terms are used to describe those countries with which this book is concerned. The general use of the term, 'The Third

BOX 1.3 THE UNITED NATIONS HUMAN DEVELOPMENT INDEX

In 1990 the United Nations introduced the notion of a 'human development index', defining human development as the process of enabling people to have wider choices. Much of the thrust of debate about development has been couched in terms of improving economic well-being, and measuring progress towards that goal in terms of increasing the GNP of Third World countries and raising the GNP per capita levels of Third World citizens. However, the UN argued that though income was an important element in people's lives, it was not the sole one; it suggested that education, health, a satisfactory physical environment, and freedom of action and expression are also important opportunities and assets. The UN suggested that generating and accumulating wealth should not be the only goal, but that the objectives of development should be that people can enjoy long, healthy and creative lives.

The Index was formulated to indicate these broader considerations, incorporating measures of life expectancy, adult literacy and income, as a more comprehensive measure of human development and by which to compare countries, rather than GNP alone, and these criteria have been extended to take note of environmental issues and political freedoms. It has become incorporated in the development strategies of Third World countries, is used in discussion of aid allocations, and forms an important component in the UN's International Development Strategy for the 1990s. The Index does not measure absolute levels of development; instead the composite measures used provide a ranking of countries from the highest to the lowest levels of achievement in the various criteria. It is unsurprising that First World countries such as Canada, Norway, Switzerland, Sweden and Japan rank highly, and that countries such as Burkina Faso, Afghanistan and Sierra Leone appear at the bottom of the ranking.

Though more sophisticated than a simple GNP per capita measure, it should be recognized that the HDI is still a *generalized* measure; as well as questions as to the reliability of the data upon which it depends, it does not fully take account of variations in levels of human development depending on gender, rural–urban disparities, regional differences, etc. Even at a generalized

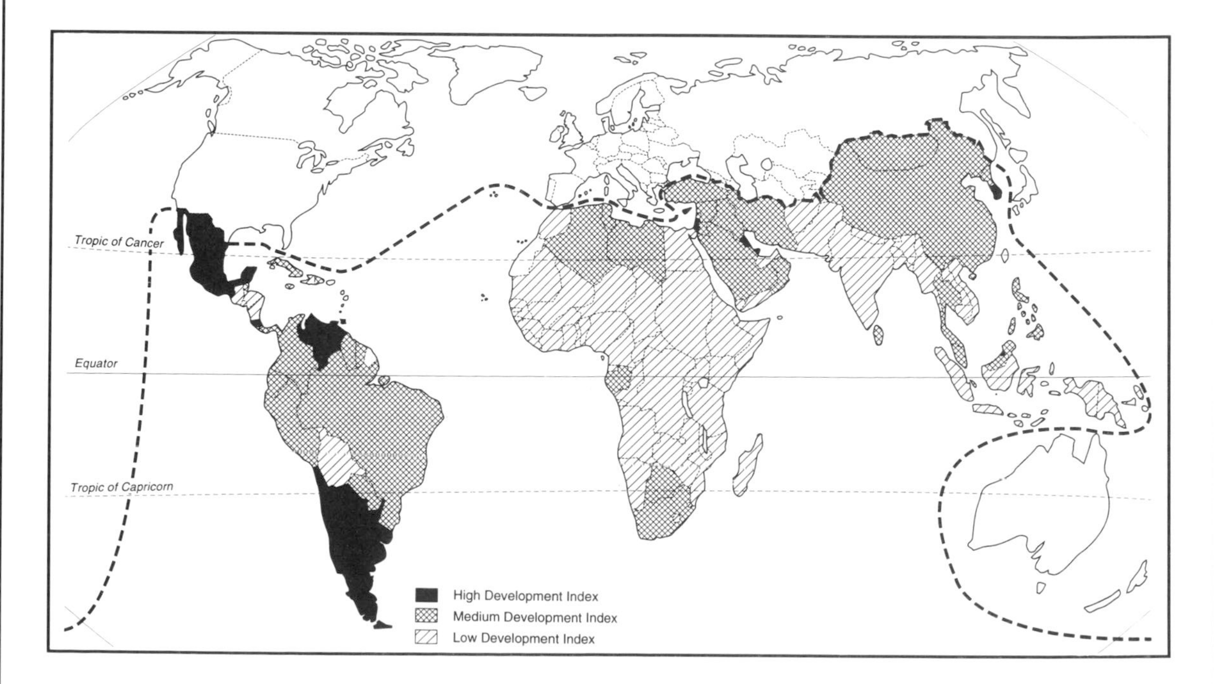

Figure 1.19 The 'South' as classified by the UN Human Development Index
Source of data: UNDP, 1992

BOX 1.3 ***continued***

level, however, it does reveal some of the variety in human development levels within the Third World (Figure 1.19). A few countries, mainly in Latin America, some oil-rich states, and NICs such as Hong Kong, Singapore and Korea, rank with the High Index countries of the 'North'. Most of the remainder of Latin America, together with North and South Africa, the Middle East and China, form the middle ranked countries, with the core of the Low Index countries lying in Sub-Saharan Africa and the Indian sub-continent.

World', was justified near the beginning of this chapter as a meaningful generalization, with advantages over the several alternatives that have been and continue to be used. The term (always used with capital 'T' and 'W' to identify a geographical entity like country – e.g. Chile – or a region – e.g. Africa) acquired common currency particularly in the 1960s and 1970s, at a time when the Cold War was at its height, and the superpower rivalry clearly identified a First World of capitalist economics and a Second World of socialist economies. The Third World, initially of Non-Aligned States, stood apart from them, but most were or became client states of one camp or the other, for complete political and economic independence in an essentially interdependent world of alliances and trade is unrealistic. However, with the end of the Cold War and the effective collapse of the Warsaw Pact and the Russian socialist state (but not the Chinese socialist state), the idea of a Second World, a cohesive international community of socialist countries, no longer has any validity. Does this, of necessity, imply that the Third World has also lost its essential rationale and cohesive power as a worthwhile generalization?

There are strong and authoritative voices that argue that the term no longer has the meaning it once had, and for two distinct reasons. First, international conditions have altered very considerably. Not only is the world of international political alliances very different in the 1990s from the way it was in the 1960s, but the world economy has also changed, with the much greater importance of multinational companies in regulating patterns and types of international trade, and the increased significance of trading blocs, such as the European Community, and of international trading relationships through various bilateral and multilateral agreements. These conditions have meant different things and have had different effects in different countries, such that it is difficult to generalize about the nature and the effects of trading relationships. The place of an NIC in the international economy, and its trading and political relationships with the rich countries, is clearly very different to the place of one of the poorest low-income countries in international relations. Since the differentiation within the Third World has been growing (see Box 1.1), the value of generalizing at all about those countries has become more questionable. Furthermore, the changes in approaches to Third World development have emphasized the importance of national and local priorities rather than a global development model. No two countries are the same, and the rejection of a linear path to development that was attractive to the modernizers and diffusionists of an earlier period means that countries will not strive to be the same. Grassroots and decentralized approaches in national development strategies will further accentuate the differences within as well as between countries, rather than maintain the similarities.

The second reason is more associated with theory and ideology. John Friedmann, mentioned earlier in the text as a leading exponent of urban-based development strategies in the 1960s (see Box 1.2), now maintains that while the idea of the 'Third World' had a sound validity in an earlier period, by the 1990s he, like several other leading development theorists, had rejected these earlier ideas in favour of a more radical approach to development that sought directly to target development to those he calls the 'disempowered' – those marginalized and impoverished as a result of development that creates wealth in favoured countries and regions (Friedmann, 1992). He prefers a focus on these disempowered people rather than on geographical entities like regions or countries, for only then can the targets be accurately identified. For him the Third World is too soft a euphemism, identifying places not people.

The arguments in this book reject both of these reasons, and the use of the term 'The Third World' is sustained. There remains a sound justification for making worthwhile and meaningful generalizations about such a large proportion of the world's states and population. Drawing sharp boundaries around the Third World is impossible, and it should be apparent, from the evidence above, that there is considerable internal variety within this region. Some continents, countries, areas, social groups are less deprived, or more affluent, than others. There is clearly enormous contrast within the 100 countries which the World Bank defines as being of low- and middle-income, and where per capita GNP ranges from $60 in Mozambique to $7500 in Saudi Arabia. The boundaries of the Third World are imprecise and its constituent countries demonstrate great social, economic and cultural heterogeneity (Box 1.3). Within these countries there are some highly developed areas and some rich people, living at First World levels of affluence. Their existence does not, however, deny the essential deprivation of the bulk of the population or large tracts of territory. This gives some cohesion to the continuing notion of the Third World, and there is a set of meanings attached to the term that are broader than 'developing' or 'poor', and more geographically sustainable than 'South'.

The focus on places rather than people comes as no surprise in a *geography* of the Third World, but the emphasis in the book is clearly on the people who live in these places. People present equally difficult problems of generalization as do places, and a focus on the disempowered people, as Friedmann suggests, has its own problems of identification, measurement and interpretation. To continue to identify a *Third* World has no connotation of cultural or social superiority, but identifies an economic dimension, relative to the First World, that is at the heart of the concept. While it is certainly true that the Second World is no longer a meaningful generalization, this need not undermine the strength of the First/Third dichotomy. Indeed, it may allow a stronger focus on the whole world-system and the nature and strength of the reciprocal relationships between the two groups. While there are problems of identifying criteria for generalization and of symbolic meanings implied by the continuing use of the term, the material presented in this book supports the argument that 'The Third World' is the most appropriate and most widely understood term to define and discuss the majority of countries and the majority of the world's population as a meaningful generalization.

SUMMARY

This introductory chapter has defined the Third World as consisting of those countries that are given a common sense of identity through their deprivation relative to the rich countries of the First World. The mass of their populations are mostly, but not exclusively, poor. They are mostly, but not exclusively, poorly provided with education, health and other basic needs. Third World countries are mostly, but not exclusively, tropical. They are mostly, but not exclusively, severely indebted to international commercial banks and to individual governments and multilateral lending agencies. They are given a cohesion and common identity through their relationships with a global economy dominated by the rich countries of Europe, North America and Japan. Ideas about the causes of the existence of the Third World as an idea and as a major world region are many, and have changed in relative prominence over the last 50 years. So too have the approaches adopted by Third World countries and the international community for their development. Increasingly many of the objectives of development are being generated internally rather than being imposed from outside, but Third World countries remain constrained by structures of unequal exchange within the international economy.

2

Historical Perspective

A common theme for most of the Third World is that the majority of the territories within it have been subject to control by European countries for some part of the past 500 years. Only Iran, Afghanistan, parts of Arabia, Liberia, China and Thailand had never been under formal European government. Nevertheless, European expansion affected them all, and when European empires came to an end, the Third World still bore the legacies of the preceding five centuries. It inherited political, economic and social structures that, even after the dissolution of formal empires, were still influenced by the forces of modern imperialism.

European countries were not the first to create empires, nor were they the only imperial powers between the sixteenth and twentieth centuries. Hindus from India colonized South-East Asia in earlier times and, with the Chinese, controlled much of its trade. Muslim influence, based in Arabia, extended from southern Spain and North Africa to India by the thirteenth century. The Turkish empire, stretching from the Mediterranean to the Indian Ocean, survived until the First World War.

Not until the nineteenth century did European empires come to affect the entire globe but, unlike their predecessors, European expansion created a world order from which virtually no country could exclude itself. A new global system evolved with Europe as the centre of military power, wealth and technological invention, and, with few exceptions, the rest of the world constituting a relatively weaker and less advanced periphery.

The exceptions included the former European colonies in North America and Australasia, where large-scale European settlement produced a totally different form of colonization. They became self-governing European societies overseas and were effectively extensions of Europe. Subsequently they evolved with their own identities, though still part of the global centre. A further group of colonized territories that do not form part of the Third World are the Asian territories of Tsarist Russia, which were incorporated into the USSR. Japan, which was also never colonized, became an expansionist power in the late nineteenth century, and has achieved a technological pre-eminence which currently rivals the industrialized countries in the West.

The historical processes associated with the changes that led to the emergence of the gap between the Third World and the First are the subject of diverse interpretations. The question that many have raised is: did European expansion and influence cause Third World economies and societies to retain traditional and outdated methods of production and social institutions? If so, then it would be pertinent to ask whether this was due to European exploitation of overseas territories, or whether traditional methods of Third World production were so deeply entrenched that they were unable to respond promptly and fully to the modernizing influence of Western development. An alternative view questions whether expansion of the world capitalist economy drew the countries which lay outside the central core of global developments

◀ Monument to the Discoveries, Portugal

into an international system of capitalism, within which they were not only exploited, but with which their internal systems were unable to cope.

The impact of European expansion was not uniform throughout the world. The pattern of expansion varied, so did the motivation for it, and the countries affected were themselves varied in character, historical background and exploitable resources. However, one thing is clear: the gap in wealth and technology which currently exists between Europe and the European settler colonies on the one hand, and the countries of the Third World on the other, is relatively new.

THE THIRD WORLD ON THE EVE OF EUROPEAN EXPANSION

At the time of the European 'Age of Discovery', some indigenous peoples in the Americas, Africa, Asia and Australasia had highly developed sophisticated cultures and technologies. In each continent there existed a range of cultural levels, from simple hunter-gatherer tribes to more advanced, urbanized civilizations.

In Latin America the Aztecs had developed a complex, highly stratified society, with a nobility which held military, secular and priestly authority. Skilled craftsmen worked gold and other precious metals. Intensive agriculture supported a dense rural population and a large, sophisticated urban system, culminating in the capital city Tenochtitlan, which so impressed the Spanish conquerors on their arrival in 1519. In western South America the Incas had also developed an extensive empire. Agricultural techniques included irrigation, terracing and use of the foot plough. Contact between the widely scattered rural communities and the imperial centre at Cuzco was maintained by a well-developed system of roads. Impressive monuments and fortresses further demonstrated the cultural level achieved by the Incas.

Advanced civilizations had also evolved in Asia. In China the Ming dynasty of the fifteenth and sixteenth centuries inherited a long tradition in philosophy, art and literature. Few technological achievements elsewhere could compare with China's abacus, fine silks, lacquers, porcelains and explosives. Throughout much of South-East Asia Hindu trading empires were contemporaneous with the European Dark Age. At a time when Europe was a collection of small feudal principalities, much of South-East Asia was integrated into a loosely structured, but culturally and commercially sophisticated unity.

Early development in Africa, as in America and Asia, varied greatly. There were, for example, marked contrasts between the societies of the Egyptian, Ethiopian and Sudanese empires on the one hand, and the hunter-gatherers of the Congo or the Kalahari Desert on the other. Some states had well-established trading and agricultural systems, as in the kingdoms of West Africa and Zimbabwe.

Muslim culture, centred on the Middle East, made important contributions to learning in the fields of mathematics and astronomy. In medicine, considerable progress was also made and hospitals were a feature of the principal Muslim cities as early as the eleventh century. Knowledge of chemistry and other natural sciences was advanced and Muslim cartography greatly influenced medieval European map-making in the Mediterranean.

Thus, by the time European expansion began, the continents which now comprise the Third World had nurtured highly complex civilizations with advanced levels of cultural, technical and social development. Some of these areas were not 'undeveloped' or 'underdeveloped'. Rather they were what Buchanan (1967, 21) has called 'pre-developed' (in the sense of having a non-European conceptualization of being 'developed'), with societies which had achieved a high degree of cultural, economic and political development before the arrival of Europeans, and in some cases in advance of progress in Europe itself. Their position in relation to the West today is due, at least in part, to the impact of European colonialism on their development, which created a particular pattern of relationships between colonizer and colonized, and profoundly modified the internal structures and patterns of the colonized territory.

EARLY EUROPEAN EXPANSION

The era of European overseas expansion lasted from the late fifteenth century to the mid-twentieth. This period falls into two distinct phases: the first between 1450 and 1800, and the second from the nineteenth century up to 1945. The first phase was largely located in the Americas and is associated with considerable settlement by European migrants. It laid the foundation for later expansion and, by the end of the eighteenth century, the relative power of Europe in military, scientific and financial spheres was such that the establishment of a core–periphery pattern at a global scale could begin. The second phase of expansion was more rapid than the first. It focused upon Africa, Asia and the Pacific, and within 120 years Europe had claimed most of the rest of the world. In these new colonies Europeans were important in various administrative and commercial capacities, but, with some exceptions such as Algeria, Angola, Mozambique, Rhodesia and Russian Central Asia and Siberia, these were not colonies of settlement.

Europe in the Americas

Conquest of Latin America was carried out by the Spanish, Portuguese, French, Dutch and British. Spain secured the largest, richest and most diverse territory, and did so with great rapidity. Within 70 years of Columbus's arrival in the Caribbean, Spain had overthrown the Aztec and Inca empires, and occupied most of Middle and western South America. Portuguese occupation of Brazil was less rapid, and initially remained close to the seaboard. It is significant for this process of European conquest that the lands of the New World, discovered or undiscovered, were divided between Spain and Portugal by the Pope in 1494. The other European powers came later to the region, securing smaller territories in the Caribbean and Guianas.

These invasions were motivated by economic considerations. The Spaniards in particular sought precious metals, exploiting gold and pearls in the Caribbean before 1510, and major sources of silver in northern Mexico and Upper Peru from the mid-sixteenth century. Mining was the key to the early colonial economy, providing wealth for export to Spain, the impetus to urban growth, and fostering extensive development of ancillary activities to serve the mines – producing food, pack animals, timber, and salt and mercury for use in silver processing. This economy was exploitative not only of land but also of labour. The mines required substantial labour forces, which were obtained by imposing forced labour on the indigenous population. Initially, the Indians were also required to provide services and tribute to Spanish settlers; later they were induced or required to work on the substantial landholdings created. The colonial territories were exploited in the interests of metropolitan Spain, which exercised close administrative control, urban development and trade, and restricted the development of manufacturing.

Portuguese occupation of Brazil was less rapid. After initial small-scale extraction of forest products, a plantation sugar economy was established in the north-east coastlands. The large landholdings were worked initially by Indians, but after these had fled or died, slaves were imported from Africa and provided a major element in the Brazilian economy until after 1850. Portuguese envy of the mineral wealth of Spanish America prompted exploration of the interior, and resulted in the discovery of gold in the late seventeenth century. This led to the first significant inland settlement in Brazil, and also contributed to the westward push of Portuguese territory beyond the 1494 demarcation.

Sugar was also a major element in the colonies of the other European powers, in the Caribbean islands and the Guianas, again based on plantations and African slave labour. The wealth generated made these possessions the most valuable parts of the emerging empires of eighteenth-century Britain and France. However, sugar monoculture had a stultifying effect on other sectors of the rural economy, limiting the production of food crops and fostering the need to import essential foodstuffs. The plantation system had a pervasive influence on both the economy and society in the Caribbean.

Plate 2.1 The colonial town of Ouro Preto Gold was discovered in interior Brazil at the end of the seventeenth century, and prompted a 'gold rush' which resulted in a number of mining camps growing into prosperous settlements. Ouro Preto's full name translates as the 'Rich Town of Black Gold', and its wealth sustained a rich civilization of architecture and the arts. When the gold was worked out, the town stagnated and its elegant Baroque churches, public buildings and town houses are now a World Heritage Site

Plantation society was dominated by Europeans, with African slaves forming the lowest echelon of society. Even after emancipation, their descendants found themselves at the base of a highly stratified and status-conscious society.

Throughout Latin America three centuries of European rule brought profound change. Vast areas of land had been brought into cultivation; major mineral resources were exploited; cities, routes and ports had been built; and trade, albeit closely controlled and narrowly based, had been developed (Plate 2.1). Yet most of the benefits accrued to Europe, or to European settlers in the region; consequences for the indigenous population were limited and often adverse. European lifestyles, language and religion were, with varying degrees of success, imposed on the aboriginal populations. In many areas the demographic consequences of conquest were considerable. As a result of warfare, enslavement and disease the Indian populations of Middle America, the Andes and Brazil were substantially reduced or, in the case of the Caribbean, eradicated. Substitution of the indigenous labour force by black slaves also profoundly modified the demographic pattern. Slavery was most important in Brazil, the West Indies and the Guianas, which received an estimated 7–8 million slaves during the period from 1550 to 1850. In consequence, in much of the Caribbean and in the canelands, goldfields and early coffee plantations of Brazil, blacks became numerically dominant in the population. In addition miscegenation between whites and blacks, and with the surviving Indian population elsewhere, gave rise to the substantial proportion of people of mixed race which exists in much of the continent.

European colonization in the Americas reached its peak by the late eighteenth century. Shortly afterwards Brazil and the mainland colonies of Spain became independent. In the Caribbean, by contrast, although French St Domingue secured its independence (as Haiti) in 1804, colonial rule has persisted into the 1990s.

Independence in Latin America did not bring immediate and rapid economic progress. In the former Spanish territories there was strife, disorder and dictatorship. It was not until the 1850s that economic transformation began to take place, and the continent's resource assets to be realized. Countries of Western Europe, particularly Britain, provided the capital, coal, manufactured goods and technical skills the continent lacked, in return for raw materials. In consequence, though there was political independence, a new economic dependence on non-Iberian Europe emerged. Economic advance, in the building of railways and telegraph lines, the introduction of steamships, and improvement of the ports and urban services, was primarily in the interests of Europe not of the Latin American countries, and some of the economic and spatial patterns that evolved were essentially to serve the emerging European core, rather than the hinterlands of the newly independent states (Plate 2.2).

EUROPEAN EXPANSION IN THE NINETEENTH CENTURY

As the European empires in America began to crumble, in Africa and Asia only the foundations of empires had been laid (Figure 2.1). The Spanish and Portuguese had trading and naval bases in the Far East, and Dutch, English and French trading companies had begun to operate. In Africa European penetration was still in its early stages and confined to coastal trading posts.

The period from 1800 to 1914 was the age of European imperialism. The countries of Western Europe, motivated by political rivalry, militarism, and a desire for markets and for sources of cheap raw materials, sought control of territory in the tropics. These incursions brought advantages in the establishment of law and order, the introduction of medical skills, and the creation of large-scale economic activities, such as mines, plantations and transport systems. Yet if imperialism was a stimulus to economic and

Plate 2.2 Steam locomotives, São João del Rei, Brazil Railway building was an important aspect of colonial and neo-colonial activity in the Third World. Most of Brazil's railways were built by British capital, but these locomotives were American-built. They have now been superseded by diesel locomotives and form part of a railway museum collection

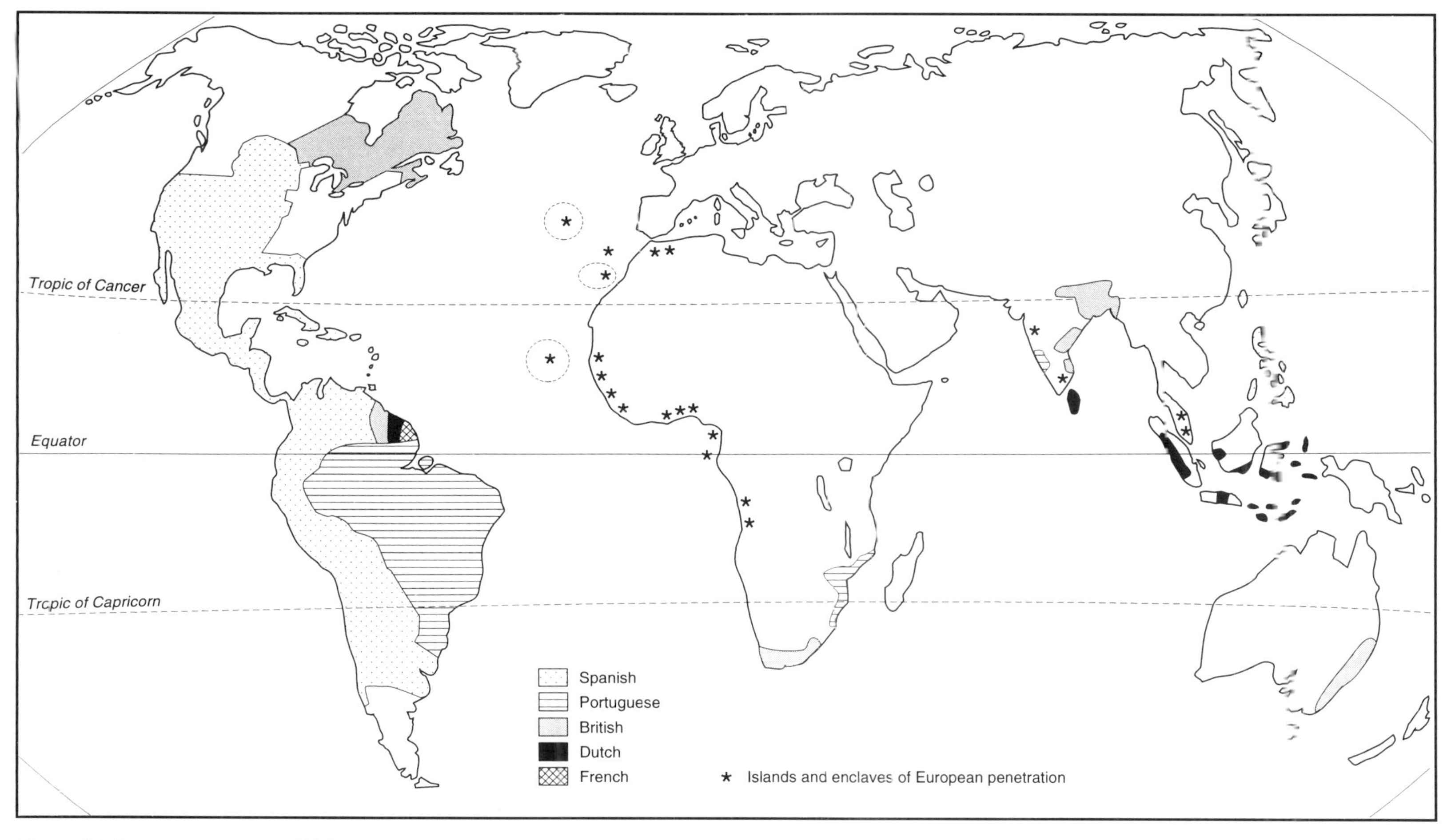

Figure 2.1 European empires *c.* 1800

social change, it also created a distorted pattern of development. The new colonies were not seen as areas to undergo balanced development; instead they were sources of raw materials for European cities and industries, and potential markets for the products of those cities and industries. The direction of development was, therefore, in the interests of the imperial power, not of the colonial territory and population. The industrialization of Europe made tropical colonies desirable as markets for manufactures, advantageous places for the investment of surplus capital, and assured sources of raw materials. Overseas expansion was also an expression of European nationalism and a growing sense of superiority over the rest of the world. Christian missions were established everywhere and explorers penetrated what they regarded as 'dark continents' and 'unknown lands' (Plate 2.3).

European empires went on to expand faster during the nineteenth century than at any previous time. In 1800 Europe and its possessions (including ex-colonies) covered about 55 per cent of the land surface of the world; in 1878, 67 per cent; in 1914, 84 per cent and yet more by the outbreak of the Second World War in 1939 (Figure 2.2). By the end of the nineteenth century the 'scramble for Africa' had resulted in virtually all of the continent being divided between Britain, France, Germany, Portugal, Belgium, Spain and Italy. Territories in the Pacific had been claimed by Britain, France and Germany, the USA and Spain. Russia occupied Central Asia as far as the Afghanistan border and the British increased their control over the Indian sub-continent as far as the north-west frontier. The latter two powers agreed to preserve the neutrality of Afghanistan and Persia (now Iran) as non-colonized buffer states. Similarly, Siam (now Thailand) lost its empire and was subjected to unequal economic treaties, but remained independent as a buffer between British and French possessions in Indo-China. With this one exception, South-East Asia was completely divided between the Netherlands, Britain, France and the USA by 1914. The Chinese empire survived expansion of the European empires as well as the expansionist designs

Plate 2.3 Anglican cathedral, Georgetown, Guyana Religion was an integral part of the colonial process, whether as a mission to native peoples or as part of the fabric of the colonial city. Georgetown Cathedral was built in the early twentieth century, using a local raw material – timber, but in a characteristic English style. It is claimed to be the largest timber-framed structure in South America

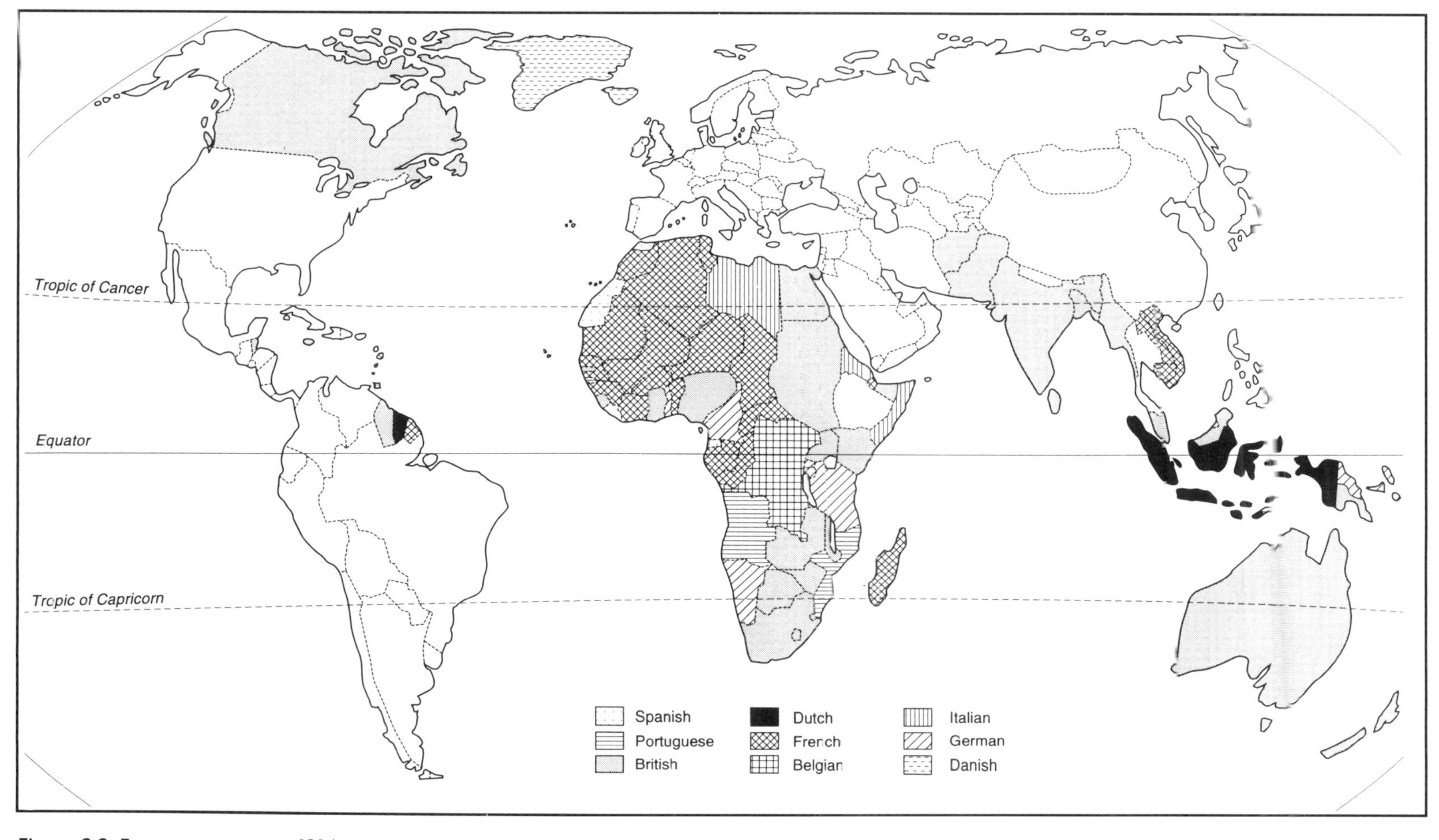

Figure 2.2 European empires in 1914

of Russia, Japan and the USA. While diplomatic deadlock preserved China's political independence, economic imperialism prevailed through informal control over the government. The colonial powers also secured extra-territorial rights in the Treaty Ports.

Africa

In Africa the precise type and impact of colonial expansion varied. After 1870 European countries sought to expand their economic and political influence. In some cases, such as Kenya and Algeria, settler colonies were established. In others, such as the Ivory Coast and German Cameroons, plantation economies controlled by metropolitan import–export companies were created. Elsewhere, mineral resources provided the basis for exploitation in Rhodesia and the Congo. In some cases essentially negative influences prevailed – in Uganda and Chad political control was taken to prevent the expansion of other colonial powers.

However, common to all was the establishment of a dual economy, with a modern, European export-oriented sector alongside (and often undermining) the traditional self-sufficient economy. Internal conditions and local peoples were no longer the determinants of events. The African territories became part of an international economic and political system controlled from Europe (Plate 2.4).

Plate 2.4 Shop front, Oporto, Portugal This shop in Portugal's second city captures many elements of the colonial relationship. The *Casa Oriental* (Oriental House) advertises the sale of tea (*chá*), coffee and chocolate, and other 'colonial foodstuffs'. Hanging in the window is *bacalhau* (dried salt cod), which is an important element in Portuguese cooking, and was a trade good between the North Atlantic fisheries, Portugal and its colonies. The faded painting also conveys the colonial relationship, in which a black servant waits on a white colonial

In the Islamic states of north-west Africa the French, and later the Italians and Spanish, acquired land primarily by conquest. French settlers engaged in the production of three traditional Mediterranean crops for export: olives, wheat and vines. These occupied the best land and used machinery, though the pattern of exploitation varied. In the earliest colony, Algeria, the French encouraged large-scale immigration of small landowners; in Tunisia and Morocco the creation of very large commercial enterprises was more common. Throughout these territories though, modern, export-oriented agriculture stood in contrast to the traditional cultivation and extensive pastoralism of the native peoples.

In contrast, European activities in Egypt were primarily commercial. Financial indebtedness brought Anglo-French control over governmental and financial affairs, and then British occupation in 1882. Exports of cotton drew Egypt into the world economy, with Europe providing the industrial and financial skills and Egypt the raw materials. The British encouraged the development of irrigation and an extension of the cultivated area. Cotton output increased substantially, but it continued to be an export crop, rather than the basis for a local textile industry, despite the fact that cotton textiles constituted one-third of imports.

In West Africa European impact was markedly different. It occurred initially through

BOX 2.1 CULTURAL IDENTITIES IN AFRICA

The 'scramble for Africa' of the 1880s brought all of Africa, except Liberia and Ethiopia, under colonial rule. The British and French, the dominant European powers, took the major share, and rather smaller sections were granted at the Congress of Berlin to Germany, Portugal, Spain and Belgium. The contemporary map (see Figures 2.2 and 2.4) almost entirely reflects the decisions made by the European powers over a century ago, for the winning of independence by the former colonies, mostly between 1957 and 1965, confirmed and strengthened the boundaries created by colonialism. At the period of independence there was some sense of Pan-Africanism among the political elite, such that common background and political identity might encourage political unity between states. However, this soon gave way to an intense nationalism focused on the new nation states, regardless of the complexity of traditional cultural patterns, of many tribes and cultural groups each with their own pre-colonial identities and languages. In Nigeria, for example, the new state tried to create a distinctively Nigerian identity, rather than a Yoruba or Hausa identity, and was able to strive towards that through the national education system, a national language (the colonial language) and a national media, often at the expense of the identities of smaller groups within the country's cultural mosaic.

National identity has developed not only out of the identities of the various groups that form the population of a country, but has also been associated with the colonial inheritance. This is most obvious in the use of the colonial language – English, French or Portuguese, with only one formally bilingual country, Cameroon, the result of a merger between former French and British territories. Other common colonial legacies include education, literature, music and sport. This has created in Africa a very clear distinction between Francophone countries, the former French and Belgian colonies, whose political and cultural ties remain very clearly with Paris; the Anglophone countries, linked to London; and, to a lesser extent, Angola and Mozambique, with their links to Portugal. Patterns of international airline routes and telecommunications continue to reflect the colonial links, and remain a potent force in the continuing 'balkanization' of Africa.

These influences have created some cohesion in the new nation states, shaped by colonialism. There have been internal rivalries within some states as a result of tribal and regional differences, and threats of succession, as in Nigeria and Sudan, fiercely resisted by the central state. Internal frictions in some states, however, have created considerable violence, as in the cases of Angola, Mozambique, Rwanda and South Africa.

missionaries and explorers whose activities aroused considerable public interest in their home countries. Until the 1860s European activity was associated with suppressing the slave trade, and there was little expansion inland. However, a flourishing commerce began to develop in vegetable oils and groundnuts. These products were grown and collected by Africans, with little intervention by Europeans, but the trade required extensive commercial contacts and led to rivalries between British, French and German trading companies, and prompted increased European control over the region. Gradually treaties made with local rulers became more binding until they led to the virtual loss of indigenous control. In both British and French West Africa economic policy subordinated African interests to those of the metropolis. Though there was less alienation of land than in other parts of Africa, and peasant farmers produced most of the groundnuts and palm oil, the colonial system forced them to concentrate on export crops, as well as on producing subsistence foods.

East Africa was drawn inevitably into the network of world trade after the opening of the Suez Canal in 1869. Rivalry between the European powers ensued, with Britain eventually emerging as the dominant influence. Early interest was in the suppression of the slave trade, but by the early twentieth century efforts were made to make the colonies profitable additions to the empire, mainly by agricultural development.

New systems of land use and land tenure were introduced, particularly in Kenya. Large-scale immigration was encouraged and substantial areas of land were alienated. The pattern was not uniform, however, for in Uganda export cotton was primarily a peasant crop. In those areas where land was sold, leased or given to Europeans by the authorities, traditional African economic activities were disrupted. The pastoral and nomadic tribes were severely affected and some put up considerable resistance, which the British quelled, frequently by force. Labour shortages were overcome by a system of taxation which forced the Africans to work for wages, while at the same time preventing the accumulation of capital within the African reserves.

In southern Africa the local economy and social systems were irreversibly altered with the penetration of Dutch settlers from the seventeenth century onwards. Traditional seasonal grazing lands were appropriated, and disease, particularly smallpox epidemics, decimated large numbers of the population. The Africans had no option but to work for the Europeans or to retreat into less hospitable desert areas, for their cattle were seized or killed and a system of taxation forcibly imposed. The Rural Reserve policy, generally applied in the second half of the nineteenth century (except in Rhodesia (now Zimbabwe) where it came much later), introduced a new element into the territorial relations between Europeans and Africans. The best land went to the Europeans, with the native reserves generally on the more marginal areas, and suffering from overcrowding.

Discoveries of gold and diamonds after 1860 brought even more fundamental changes in the political economy of southern Africa. A system of migratory labour evolved and men left the reserves for varying periods of time to work in the mines. This gave the mining magnates full control over the workers. It also greatly influenced the character of the urban areas which developed and the social and economic relations between Europeans and Africans, and between the European-dominated commercial farming and urban areas and the 'traditional' farming areas, styled the 'native reserves' in British colonies.

The Indian sub-continent

There were major contrasts between the colonial regimes that developed in Africa and in the Indian sub-continent, and also within the latter. These differences reflected the varying conditions within which the colonial systems operated, as well as the diverse purposes the colonies were intended to serve.

In India the British consolidated earlier trading interests and extended their influence until, in 1849, they controlled the whole country. The Indian Empire became an important source of agricultural exports, particularly of cotton, jute, hides, oilseed and tea, and large irrigation works were begun, for example in the Punjab. Development of a transport system was crucial to this agricultural expansion, and prompted the construction of a major rail network (Box 2.2). Conversely, easier access to the Indian market for manufactured cotton goods from Lancashire destroyed the existing artisan cotton textile industry.

English became the medium of official communication and of education, and India eventually adopted British political institutions. Acceptance of alien rule was fostered principally by making very few changes in the basic structure of Indian life. Moreover, members of the established social groups, the aristocracy, the merchants, bankers and educated professionals, were absorbed into the expanding colonial civil service, which gave them the illusion of power. The advantages of British rule probably lay most in its imposition of law and order over a vast, hitherto divided sub-continent; the provision of a system of centralized government; and the construction of a comprehensive system of roads, railways and canals which could facilitate subsequent large-scale industrial and agricultural growth.

Sri Lanka's colonial experience (as Ceylon) was dominated by the development of a plantation economy. While rubber plantations were able to attract peasant labour as the work was intermittent, local people refused to work on the coffee and tea estates. Tea production was largely carried out by Tamil immigrants imported from southern India. The system of compulsory labour for maintenance of existing

BOX 2.2 RAILWAYS IN INDIA

In developing an export economy in a country of some 4 million square kilometres, a good transport system was essential. In the pre-colonial period transport was limited to the bullock cart and the rivers. The British set out to develop a substantial railway network. The first line was opened in 1853, and by 1870 there were 6400 kilometres of railroad. Over the next decade this was doubled, and up to 1914 construction averaged some 1200 kilometres a year (Figure 2.3). By that date the system was virtually complete, later development being mainly infilling. The network was based on an early colonial proposal that the nodes of the system should be Bombay, Calcutta, Madras and Lahore, and it evolved to integrate the principal ports and cities, and the most productive areas, particularly the Ganges and Indus plains. The railways were built by a variety of government and private sources, but had come under government control before independence. However, much of the system was single track, and constructed at various gauges, inhibiting its efficiency. Nonetheless, the railways were a major British legacy to India, and remain a significant component in the contemporary transport network.

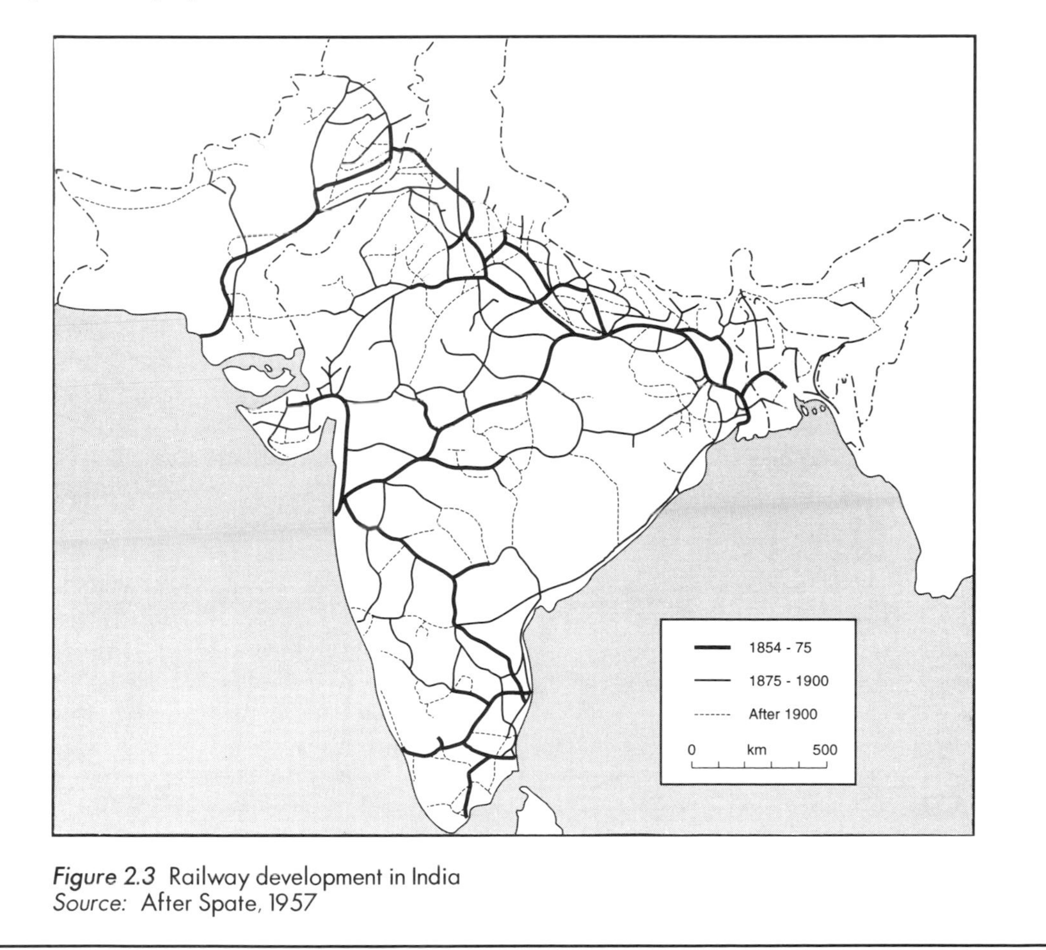

Figure 2.3 Railway development in India
Source: After Spate, 1957

irrigation systems was effectively destroyed by the British administration, so that self-sufficiency in food production was lost, and rice, the staple food, had to be imported, especially in the dry zone.

The expatriate plantation sector grew rapidly and created great wealth. Little of this wealth ever reached the Ceylonese, though there were some exceptions. An upper class of indigenous landowners had existed long before the arrival of the British, making their profits by renting out rather than cultivating land. The rise of the coffee industry gave them an impetus for engaging in production; and later many began to produce tea, coconuts and rubber. Because the owners of such estates were local, the profits remained in Ceylon.

South-East Asia

The countries of South-East Asia were once basically self-sufficient, but were changed into suppliers of primary commodities for Western markets by the impact of colonialism. They were thus brought into almost total dependence on the fluctuations of a world market over which they had no control. Where a labour supply was not forthcoming, immigrants were brought in or came spontaneously to occupy positions as middlemen in colonial commercial activities. Meanwhile, the majority of the region's indigenous inhabitants became increasingly marginalized within their own society, and dependent upon the new monetary economy into which they were forced.

In Indonesia Dutch control under the 'culture system' established a planned economy on a vast scale and, especially in Java, resulted in large-scale production of sugar, coffee, tea and other foodstuffs for the Dutch market. This system led to the diversification of crops, the introduction of scientific methods of production, including modern irrigation works and a great improvement and extension of the communication network. The culture system involved gross exploitation of the Javanese cultivators, and forced labour when voluntary workers could not be recruited. Less intensive colonial contact in the outer islands of the archipelago meant that people were not so involved in the production of export crops and remained relatively unaffected by the rising demands and expectations of a monetary economy.

The experience of the British colonies in South-East Asia differed only in detail from that of Dutch Indonesia. The wealth of Burma (now Myanmar), which had developed through the production of rice, tea, lead, rubber, copper and other export goods, was largely in alien hands. Laissez-faire colonial policies permitted free trade, British control of resources, and the immigration of Indians who acted as commercial middlemen and money-lenders. However, for the peasantry, whose self-contained subsistence economy had focused upon the social life of the village, circumstances deteriorated and by 1940 at least 50 per cent of them had become landless agricultural tenants and labourers. British rule had liberated the Burmese from some of the restrictions of traditional society, and there had been some improvements in health conditions. Balanced against such benefits was the fact that the Burmese had a disproportionately small share in the prosperity of their country and, for the majority, life had become marginalized as they strove to survive within the framework, though on the fringe, of the new international monetary economy.

Malaya's transformation into a leading producer of tin and rubber was associated with large-scale Chinese and Indian immigration, so that in its peninsular states a multi-racial society emerged where the immigrants outnumbered the indigenes. The port of Singapore became the centre of British financial and commercial interests in the region, and virtually a city of Chinese traders. While development on the peninsula was rapid in the regions of Chinese-dominated tin mining, the rest of the country was little changed and the social imbalance between the immigrants and indigenous Malays became very clear. The development of the west coast states was further increased by the British construction of roads and railways linking the mining areas, plantations and ports.

In contrast to the situation in most of South-East Asia, in Malaya there existed the preconditions for a balanced agrarian policy. There was an abundance of land relative to population size and a highly prosperous tin industry that financed public works and other developments without resort to a rigid taxation system. Furthermore, the high level of Chinese and Indian immigration obviated the need for coercing Malays into working on the plantations. Smallholders were not seen to be in competition with the estates, so that in contrast to other colonial situations the British authorities in Malaya also encouraged the development of small-scale farming.

In Sarawak, Brunei and Borneo, European enterprise was minimal. The countries were poor in resources and, in the case of Sarawak, the paternalistic administration deliberately discouraged modern commercial exploitation. The result was a far slower rate of economic growth or development of an infrastructure than was the case elsewhere in the region. In 1940 Sarawak was little changed from what it had been a century before, with much of the land still covered in virgin forest and the greater part of the population engaged in subsistence farming. Of its known resources only oil was of any real significance for the outside world. The advantage was that large numbers of people had not become marginal within their own society and economy.

Contrasts in colonial activity also existed among the French South-East Asian territories of Vietnam, Cambodia and Laos. However, the differences were more of degree than of kind. The pre-existing differences in the development of the northern and southern parts of Vietnam were further exaggerated by the colonial regime. The densely populated north was largely left to its traditional self-sufficient subsistence economy based on rice production. The south bore all the imprints of the classic colonial situation: large-scale commercial enterprises and the creation of a distinct dual economy. Development, which was concentrated in the more sparsely populated southern region, was associated with the appalling poverty of the peasants, their continual indebtedness to money-lenders and other middlemen and an ever growing population. Immigrants to Vietnam were mainly Chinese, who had been going there long before the beginning of colonial commercial activities, but who went in increased numbers during the colonial period. They played their typical role of middlemen operating between the Europeans and the Vietnamese peasants.

Cambodia and Laos, like the neglected eastern territories of Malaya, were perceived to be countries of low economic opportunity and were, therefore, omitted from the mainstream of colonial activities. They remained areas where a simple subsistence pattern of life prevailed. Cambodia contributed small amounts of rice, rubber, maize and pepper to the international market, and Laos produced some teak, tin and coffee. As elsewhere in the colonial world, even the small amount of commercial activity that developed was monopolized by outsiders – in this case, French, Vietnamese and Chinese. The simple, though highly refined, cultural life of the Khmer people of Cambodia and likewise the traditional world of the Laotians were, as a consequence, only minimally affected by the direct influence of colonialism. However, they were not to escape the many indirect implications of being poor, unsophisticated societies in an increasingly prosperous and technological world.

The Philippines shared many of the features of colonialism in the region. Typically, there was a dual economy: foreign-owned plantations produced a few primary goods for the world market, while the small-scale subsistence sector was characterized by poverty and indebtedness to a group of middlemen comprised mainly of Chinese immigrants. On the other hand, Philippine society differed from any other in South-East Asia in that at the end of the nineteenth century its history of colonization by Europeans – the Spanish – had already lasted for three centuries. By the time the island societies came under US colonial rule in 1898 they were in many respects Westernized in terms of culture and attitudes. This factor was of major importance to the pattern of US penetration in the twentieth century. The traditional Asian self-contained subsistence economy had already been influenced by one oriented towards the world markets. A basic infrastructure existed and the range of commercial agencies and credit facilities characteristic of a monetary economy were an accepted part of Philippine life. The major economic change under US rule was the spectacular rise of the sugar industry which turned the economy into a virtual monoculture.

The Pacific islands

Traders, whalers and missionaries accounted for most of the Europeans in the Pacific in the nineteenth century. These were mainly British and some French nationals. By the end of the century the US also had significant commercial interests in the region. The implications of colonization for the Pacific islands were broadly similar to those of Asian countries affected by the spread of the plantation economy. The earlier small-scale trade in sandalwood and sea-slugs for the Chinese market was replaced by a much larger export trade in coconut oil and guano. Both of these products were controlled by small settlements of Europeans in the islands. Demand for plantation labour led to many social and political disturbances, and, as a consequence, by the 1860s the political systems in Fiji, Samoa and other island groups were beginning to disintegrate.

THE IMPACT OF COLONIALISM

The process of colonization within any one country or group of countries was conditioned by the pre-existing historical circumstances and the attractiveness of its resources to the colonizers. In turn, these affected the degree of exploitation and the nature of the developments that took place and consequently the level of marginalization that occurred within the economy and society. The communities least attractive to European settlement, such as Laos, Borneo, Brunei and Sarawak in South-East Asia, remoter islands of the Pacific and the more isolated areas of central Africa, were among the least exploited. Yet, their greater retention of indigenous social traditions and economic practices did not enhance their possibilities for future development. On the contrary, they have remained among the poorest of Third World countries. In so far as they gained from their isolation, it has been in that Western material expectations have been only minimally aroused. On the other hand, Malaysia, well-endowed with resources for which there is international demand, or the Philippines with its cultural sophistication, are examples of countries which, though exploited by the colonial system, have, nevertheless, emerged with the facility to compete favourably by comparison with other Third World countries.

The net balance in the advantages versus the disadvantages of colonial penetration was not, in the last analysis, a measure of the total influence that European expansion had upon the countries of the periphery. It was being on the periphery of global events and, concomitantly, the internal marginalization of large sectors of the population with which this was associated, that made the vital difference to future developments.

It is not surprising that the preservation of political sovereignty in China, Afghanistan, Thailand and Iran was insufficient to isolate these countries from the effects of an evolving global power balance, which they were in no position to counter. They were all the victims of unequal trade treaties with European countries, and undoubtedly part of the colonial sphere of influence in the Middle and Far East. The Open Door Policy imposed upon China by Britain, France, the US and Japan virtually made China an economic satellite of the imperial powers until the communist revolution of 1949.

Other countries which maintained political autonomy also bore significant hallmarks of colonialism. Thailand's economy, for example, depended on the export of four primary products – rice, teak, rubber and tin – of which only rice was controlled by Thais. In return, manufactured goods were all imported. A typical colonial economy developed, with a foreign-owned export sector and a traditional agricultural sector, and even rice was transformed from being a subsistence crop to a commercial one. As in the formal colonies, peasants were forced into a marginal way of life in relation to the country's mainstream of production. While Thailand gained none of the apparent advantages of a formal colony, such as European medical and educational services or a communication system, its slower pace of change might well be viewed as a net advantage. Thailand escaped many of the socially disruptive effects of colonialism that were experienced by neighbouring Vietnam and Burma.

Colonial status was primarily a political

phenomenon but the economic aspects of European dominance in the world, which colonialism heightened, did not necessarily depend on formal relationships. Economically, therefore, all the world outside of the settler colonies of North America and Australasia came under the domination of Europe and, later, under the influence of these former settler colonies as well.

DECOLONIZATION

In 1939 the colonial empires in Africa, Asia and the Pacific were at their peak; by 1965 they had virtually ceased to exist. There is no simple explanation for this rapid decolonization: it was undoubtedly due to a combination of factors, and varied with different territories. The demand for independence came both from the colonies themselves and also from external forces which eventually militated against the continuation of empires.

The nationalism which crystallized the response in the colonies against alien rule was a European phenomenon until the Second World War, but through exposure of the colonial elite to Europeans and European education, nationalist ideas were ultimately transmitted to the colonies themselves. In addition, the period of European occupation of the colonies had brought about such fundamental change in the indigenous societies that the conditions which had made alien rule possible in the first place had been largely destroyed. Large-scale mining and agricultural operations produced both an urban and rural landless proletariat. Population growth led to a shortage of land, and improved communications broke down the isolation of village communities. Religious teaching and secular education created an awareness of the differences between Europeans and colonial subjects, and rising expectations developed through the increased awareness of Western forms of modernization and technology. Such factors combined to produce conditions for the rise of nationalism among the colonial population. Further, it is likely that within certain societies, in particular those of Islam, the preconditions for nationalism and opposition to Western, Christian domination were latent at the time of colonization and it was simply the opportunity for self-assertion that came later. However, it was not essential that these preconditions should exist within each society, for once the trend had been set by some countries the process spread to others. Just as it had been impossible for countries on the periphery to avoid the influence of European power during the expansion of overseas empires so, later, it was only a small number of territories that were not carried along by the waves of decolonization which swept the empires away.

External influences also contributed to the dissolution of empires. The Russian Revolution gave impetus to the notion of mass movements and class struggle. It generated anti-imperialist sentiments which had their impact on West European thinking, encouraging the view that colonial domination was intrinsically immoral. The USA also began to exert pressure on the European powers as it pursued its own expansionist goals in the mid-twentieth century. However, the Second World War had the most direct impact upon the relationship between colonies and the metropolitan powers. During the war France, Belgium and the Netherlands were cut off from their dependencies and it was difficult to return to full colonial rule after the war. Other colonies were less affected, and in the long-established Caribbean colonies, for example, the notion of Britain as the 'mother country' was greatly encouraged by the sense of allegiance that the war fostered.

Yet the war brought fundamental changes to the colonial world and during the 25 years after 1945, substantial decolonization took place in two main waves. Most of the states that achieved independence in the first phase of decolonization, before 1950, were in the Islamic Middle East or in the Far East. In the second phase, the majority were in Africa and the Caribbean (Figure 2.4).

All that remained of European empires by 1970 were either territories that had been fully incorporated into the colonizing states or those that were so small or poor that their independence was not thought feasible. France incorporated its Caribbean colonies as Overseas Departments, and its small Pacific dependencies

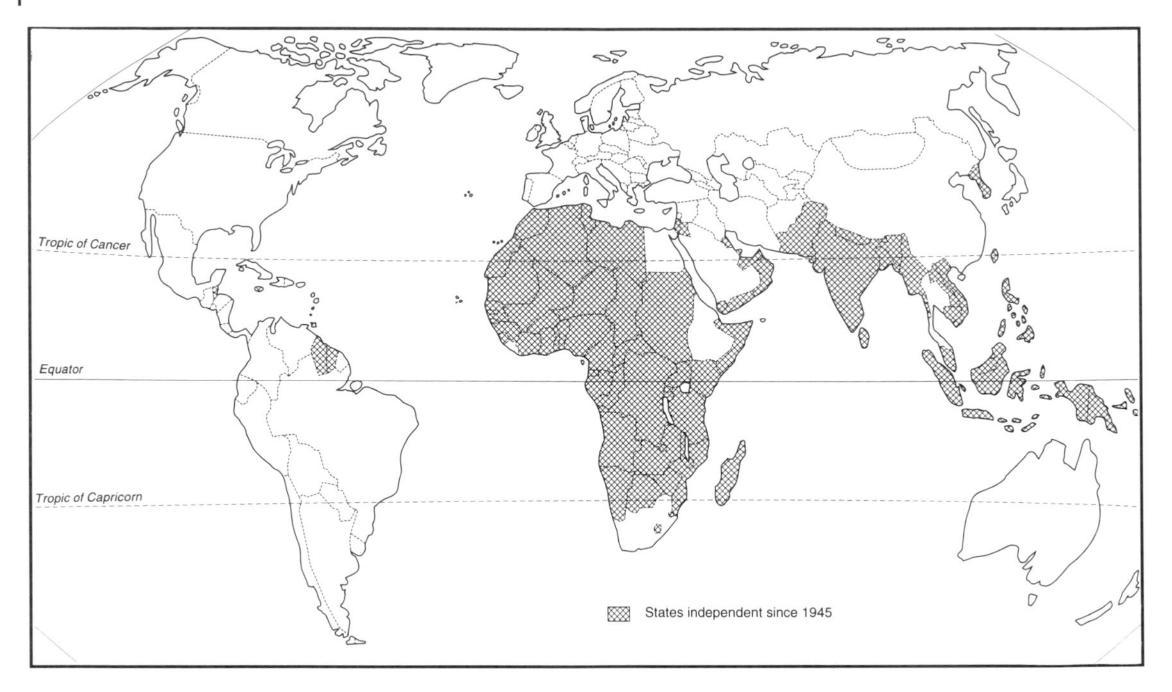

Figure 2.4 Countries obtaining political independence since 1945

as Overseas Territories. Portugal still held most of its empire and was only forced to relinquish Mozambique and Angola after the death of Salazar and the fall of the Caetano regime in 1974. The British and Dutch retained their island colonies, many of which became independent in the 1970s. Spain incorporated the Canary Islands and its small North African city enclaves, but its West African colonies either became independent (Rio Muni and the island of Fernando Po united as Equatorial Guinea) or were incorporated, more controversially, into Morocco (Spanish Sahara). Russia incorporated all her colonial territories as part of the USSR, and the USA incorporated Hawaii and retained its other possessions in the Caribbean and the Pacific as self-governing dependencies. The only countries outside these categories were Zimbabwe, South West Africa and the Republic of South Africa in which large European settlements had become established, but which differed from the settler colonies of North America and Australasia in that the European population never became a numerical majority. Similar situations existed in Algeria, Zambia and Kenya and their independence was resisted for some time. However, these became sovereign states before 1965. In the remaining anomalies, white minority rule prevailed until 1979 in Zimbabwe (though formal independence was in 1980), 1990 in Namibia, and 1994 in South Africa, though independent of British authority since 1910 (Box 2.3).

THE CONSEQUENCES OF COLONIALISM FOR THE PERIPHERY

The colonial experience had drawn the countries of the Third World into close linkages with the colonial powers of Europe, but the relationship that emerged was an unequal one; the benefits that accrued to the core were greater than those received by the periphery. Even after independence the unequal nature of the exchange tended to favour the continued economic advance of the North, to the detriment of the South.

The colonial experience should not, of course, be seen entirely as a negative process. We can only speculate what the experience of the colonized territories (and the colonizers) might have been without this major episode in world history. The period did witness a diffusion of ideas and technology. Colonial governments created an infrastructure of railways and ports, or provided the stability which encouraged private enterprise in these fields. If such projects were primarily in the interests of the metropolitan power, they did establish a basis, even if an imperfect one, for later independent development. Colonial governments instituted a system of law and order, provided a basic administrative structure and, together with the churches, improved medical and education systems.

The nature of the colonial experience has posed many problems for Third World countries. In the typical colonial administration non-Europeans were given little experience of the higher levels of control and authority. Balkanization of territory through the creation of colonial boundaries bearing little relationship to pre-existing ethnic and tribal groups brought together unrelated, and sometimes hostile, groups. With independence, this has brought friction, and sometimes open warfare, as tribal conflict is exacerbated by political aspiration and the domination of one group over the rest.

Patterns of land use were profoundly disrupted in many colonial areas. The creation of plantation economies to provide export cash crops left the producing countries vulnerable to the fluctuations of international markets, while the wealth generated went mainly to the metropolitan power. In several cases the plantation system also saw the large-scale movement of labour, as slaves from Africa to Brazil or the Caribbean or, after the abolition of slavery, as contracted labour from India to the Caribbean, East Africa and the Pacific islands.

Though medical services improved health conditions, particularly in combating tropical diseases, they also had implications for the demographic structure. The prolongation of the life-span, coupled with the continuing economic and social prestige of large families, resulted in rapid increases in population. Education created new opportunities, but since it tended to be on a European model, it encouraged imitation of Western ideas and values. The adoption of European languages by the local elite, and in some cases by the entire population, meant that numerous indigenous languages and tribal dialects became subordinate to those of Europe. Moreover, people came to be educated for roles which they could not, or would not, play locally, but could more readily assume in the metropolises of the centre. The migration of educated and skilled people from the Third World to the cities of the First, has created a depletion of trained people and a loss of investment in local human resources (Chapter 10).

In addition, empires were not built solely on economic relationships but involved cultural and psychological factors. Racism and a sense of superiority were an integral part of colonialism. In the late nineteenth century when colonialism, conquest and imperialism were at their height, Europeans no longer had any doubt about their superiority over the rest of the world, its peoples and their cultures. What was 'good' for the colonized was seen from the viewpoint of the colonial masters. Of equal, if not greater, significance was the fact that even the colonized peoples themselves came to believe it. Even after independence, and attempts to distance themselves from the colonial experience, the intrusion of Western values and goals, via the mass media, create aspirations which are at variance with the realities of impoverished nations and peoples.

POST-COLONIAL LINKAGES

Although most colonial territories have now secured their independence, close economic, political, social and cultural ties remain with former colonial powers. In the last two decades a number of issues have served to sustain the linkages between the centre and the periphery, usually to the detriment of the latter.

The internationalization of capital

Trading companies, such as the East India Company, were an integral part of the colonial process, and in the nineteenth century private investment from European nations went especially to associated colonies. (In the case of Latin America, although there was political independence, there was significant investment from Britain, other European powers and, later,

BOX 2.3 REMNANTS OF EMPIRE, 1994

In 1994 some 30 'colonies' of Europe remained, though they varied in status and size (Figure 2.5). Most are islands too small or remote to be viable as independent states. Britain has 15 dependent territories, of which Hong Kong is due to revert to China in 1997. The remainder have a total population of less than 150,000, and several have less than 500 inhabitants. Britain affords them as much self-government as possible, retaining responsibility for defence and foreign relations, of significance where there are neighbouring countries with territorial claims, as in the case of the Falklands/Malvinas, Gibraltar and Hong Kong.

French territories are much larger, with a total population of 2 million. They are regarded as part of metropolitan France, with their inhabitants having the rights and social security entitlements of French citizens. The other remnants of the colonial period are the Caribbean islands of the Dutch Antilles and Aruba, and Portuguese Macau, which reverts to China in 1999.

Inhabitants of these territories may seek political autonomy, but economic opportunities are limited by size, remoteness and small population. Economic advance might derive from the discovery of new resources, such as oil and fish off the Falklands, by developing tourism, as in the Caribbean and Pacific islands, or in the creation of tax havens or off-shore banking centres, as in the case of the Cayman Islands, but their ties to the metropolitan powers are likely to continue, if only because of their vulnerability on their own.

the USA.) Out of these investments emerged companies with a range of international interests. In the 1950s and 1960s some of these evolved into multinational corporations (MNCs), which sought raw materials, markets and profits on a global basis. They originated in the USA, the United Kingdom and Europe and later, Japan, and tended to concentrate in technically advanced industries such as electronics and pharmaceutical, medium-technology consumer goods such as motor vehicles and televisions, and in mass-production consumer goods such as soft drinks and cigarettes. Their intrusion into the Third World brought capital, technology and employment, but also competition with domestic producers, exploitation of labour and, in some cases, exploitation of laxer pollution controls (Chapter 6).

Oil and the economic crisis of the 1970s

During the 1970s the international economy entered a period of prolonged crisis in which the relationship between the industrialized countries and their former colonies was to suffer lasting, negative effects. The economies of the industrialized countries went into recession, with the economies of the Third World going into crisis, in part due to the reduced market for their exports to the industrialized world on which their economies had by then become very heavily dependent. Only some of the oil-producing countries were able to sustain themselves and, indeed, the escalation in the price of oil on the world market was one of the factors involved in the creation of the crisis in the first place.

The vagaries of market forces were most dramatically felt in the late 1970s, when the Organization of Petroleum Producing Countries (OPEC) recommended a series of oil price rises. Their adverse effect on the economies of the oil-importing developing countries triggered a steep rise in inflation and contributed to a balance of payments crisis in a number of countries, especially in Latin America, which placed a strain on the political solidarity of the Third World. In 1976 OPEC set up a Special Fund for Developing Countries, through which they granted economic assistance to Third World countries, but it fell far short of the increased oil importation costs resulting from OPEC's decisions.

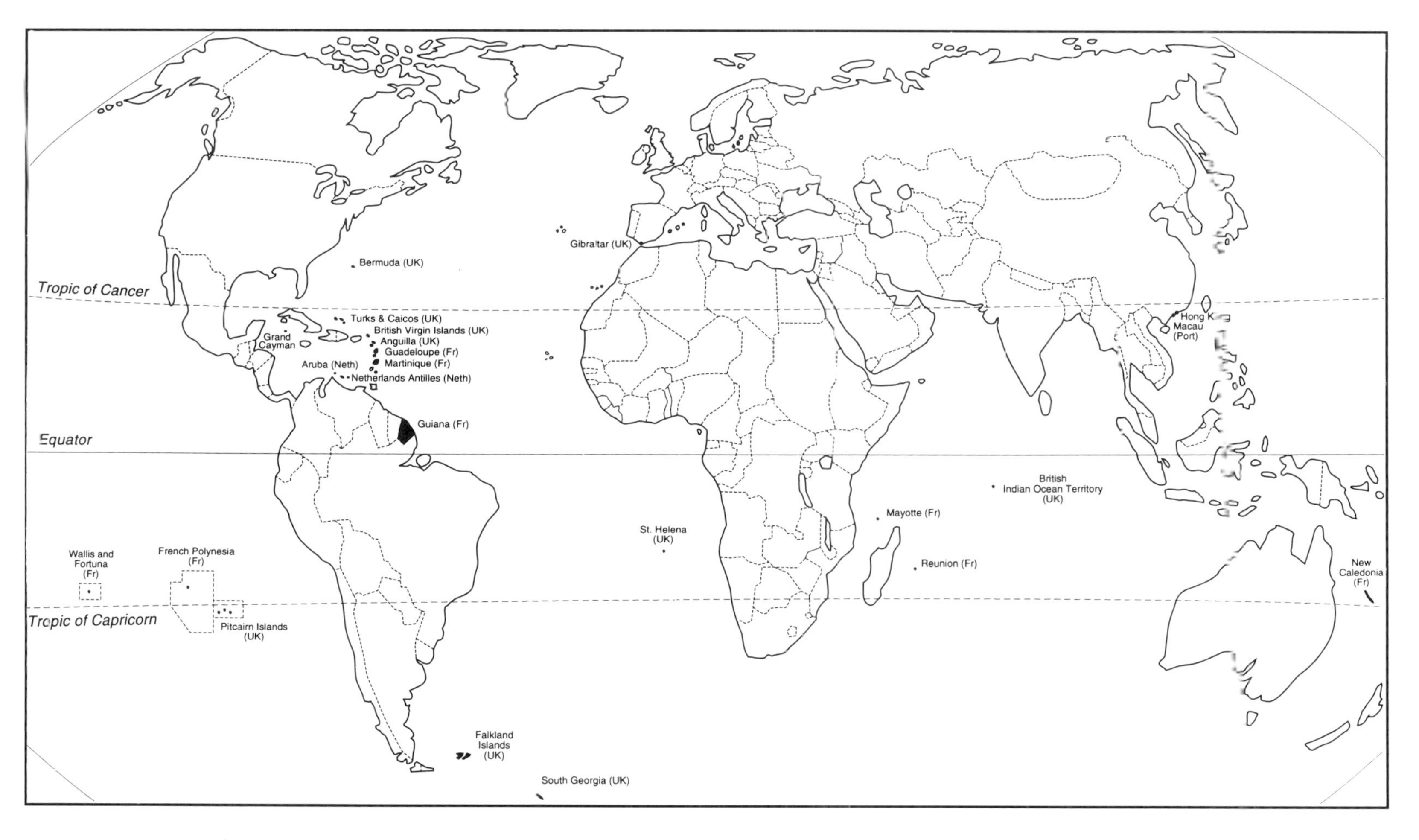

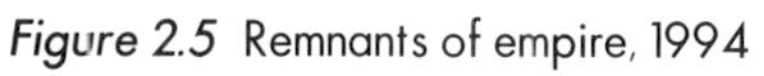

Figure 2.5 Remnants of empire, 1994

The debt crisis

A number of Third World countries became deeply indebted to the international banks, in particular the World Bank and the IMF, in order to cope with the inflationary situation into which their economies were forced in the 1970s. This debt situation worsened during the 1980s, the slow-down in world lending reflecting a weakening of confidence in the global economy. Western industrialized countries, facing recession at home in the early 1990s, reduced their credits to the developing world as its indebtedness grew and repayments faltered.

The main reason for Third World indebtedness continued to be the decline in primary commodity prices on which so many of the debtor nations depended, with African countries particularly badly affected. The World Bank reported that their level of debt doubled in the decade of the 1980s and that debt servicing in 1991 accounted for an average of 21 per cent of the total export earnings of the African states.

A tightening of monetary policy throughout the industrialized world in the early 1990s saw interest rates rise. Then, with the collapse of communism in Eastern Europe, Third World countries faced not only high interest rates, but also fierce competition for credit from borrowers in Russia and Eastern Europe, who benefited from preferential treatment from commercial lending banks because of their more advanced industrial status.

Various strategies were devised for Third World countries to re-negotiate their debt position, the most publicized being the Brady Plan (1989), which aimed to encourage debt-reduction agreements between commercial banks and their Third World debtors. Meanwhile, the World Bank predicted that the 1990s would see a long-term shift away from traditional commercial bank lending to the developing countries. Instead, they would raise money by offering direct foreign participation in their industrial structures.

The West, which during the 1960s had

Plate 2.5 Harbour front, Bridgetown, Barbados This picture captures the complex nature of colonial and post-colonial ties, with the typical English architecture of the cathedral, the modernism of the offices of the Royal Bank of Canada, and tourist craft in the harbour

supported import-substitution programmes, 20 years later criticized those countries still following such strategies. Aid donors made their economic assistance dependent on market reforms. By 1990 the 16 major donor states of the OECD had reallocated most of their economic assistance to those countries implementing reform programmes. Developing countries were now urged to make greater efforts to adopt market-economy structural changes and compete efficiently in the changed international climate of the 1990s. These adjustments have brought a heavy burden to Third World societies, with the 'non-productive sectors' – education, health and welfare – suffering from the lack of funds as large proportions of GNP are used to service their debt. Consequently, the objective of modernization current in earlier ideas of development has become increasingly difficult to sustain in most Third World countries.

The ecological crisis

In the 1980s world attention began to focus on the fragility of the global environment. During 1990 average global temperatures were the highest since records began in the late nineteenth century and some scientists began to link global warming to the 'greenhouse effect'. According to the US-based World Resources Institute, the largest emitters of greenhouse gases were the USA (18 per cent), the former Soviet Union (12 per cent), Brazil (11 per cent), China (7 per cent) and India (4 per cent). Although the largest contributors were in the developed world, the spectacular destruction of the tropical rainforests drew media and popular attention to the Third World.

The convening of the Rio 'Earth Summit' (the UN Conference on Environment and Development, June 1992), attended by most of the world's leaders, placed the issues of climate change, bio-diversity and the protection of the rainforests, on the world agenda. Third World countries criticized the environmental charter signed in Rio de Janeiro, suggesting that the industrialized countries were not doing enough to clean up the environmental pollution they had created themselves, and that they were more inclined to urge environmental constraint on the Third World than to recognize its urgent need to make economic progress. The imposition of such constraints is referred to as 'green conditionality', whereby conditions are imposed on loans and aid programmes, setting maximum and well-monitored pollution levels for new mines and power stations, or requiring the use of 'clean' technologies – even if, in the short term, these are more expensive and less appropriate than local technologies. Developing countries often object to the imposition of such 'green conditionality', and to the linking of environmental issues to trade and debt agreements.

Progress towards environmental improvement remains slow. At a 1995 United Nations conference on the global climate in Berlin, major producers of pollutants and suppliers of fuels expressed opposition to the imposition of limits on 'greenhouse' gas emissions. The United States, the world's largest emitter of carbon, the Organization of Petroleum Exporting Countries, and Australia, a major coal exporter, were all hostile to the introduction of new targets and timetables for reducing emissions. Conversely, the developing countries were reluctant to accept that they should take a larger share of reducing emissions, arguing that the developed world, with one-fifth of the global population but producing four-fifths of greenhouse gases, should take major responsibility.

SUMMARY

The era of European expansion overseas irreversibly affected and altered the non-European world in two fundamental ways. At the international level it created a dependency relationship of the periphery upon the centre, while at the national scale, this same relationship brought about the marginality of internal structures. The effect of the economic relations that evolved was a reflection of the balance of power between a European, and later North American, core and the rest of the world. In consequence of the increasing strength and sophistication of

European industrialization, financial acumen and military power, the peripheral countries came to depend on the First World for their own development. The net effect was the increase in the wealth and power of Europe and, with the exception of the OPEC states, the commensurate decline in the wealth and power of the rest. The control the West exerted over the resources of its colonies was through agricultural and mineral production as well as the direction and **terms of trade**. The sheer discrepancy in global power that was reflected in the evolution of the centre–periphery relationship permitted the continued and renewed penetration of the Third World by capitalist countries, including Japan, in the second half of the twentieth century. The centre–periphery relationship has therefore been perpetuated to the present time and on this hinge most factors that determine the characteristics of the present Third World.

Though the era of formal empire was largely over by 1965, the industrialized Western countries continue to dictate terms of trade and other economic arrangements. Clearly, a force far greater than mere economic or political controls had been generated during the Age of Empire. Apart from economic profit, imperialism had provided Europe with a world it could dominate not only militarily, but also in the technological and intellectual spheres. Western-oriented cultural systems, with their accompanying tastes, fashions and consumer habits, became an intrinsic part of internationally accepted goals. The 'machine age', which had evolved in the West, had unquestioningly come to represent 'modernization' and 'advancement' everywhere – regardless of its consequences. 'Development' policy and planning have, therefore, frequently exacerbated the processes of underdevelopment by treating the symptoms rather than the causes of the problem. Yet progress has continued to be measured against Western models.

This is the background against which the industrialized countries continue to influence and dominate the Third World long after the end of formal empires. Thus, the world economic crisis of the 1970s, from which the industrialized countries began to recover in the 1980s, had a profound impact upon the Third World. The effects of structural adjustment have caused a deep crisis in the Third World with a serious fall in the living standards of people and nations that were already poor.

3

Population

Four-fifths of the global population live in the countries of the Third World. In recent years 95 per cent of the 90 million people born in the world each year have been in Third World countries. These high proportions have come about through persistent high rates of population increase during the second half of the twentieth century, largely the result of declining death rates and continuing high, though now generally falling, birth rates. National rates of natural increase range between 1.5 and 3.5 per cent per annum, rates at which a population will more than double within 40 to 20 years (see Figure 3.1; Table 3.1). In contrast, in the more developed parts of the world natural increase is now mostly below 1.0 per cent per annum, and it would take more than a hundred years for a population to double in size.

This chapter elaborates some of the basic features of the populations of the Third World, in their rates of change (including levels of mortality, fertility and migration), and age and occupational structures, distinguishing these from the population characteristics of the developed world. Since population, in the public mind, is clearly an issue of particular importance for the Third World, one of the main purposes of this chapter is to examine in which respects and to what extent these characteristics of population and population change constitute a problem for Third World countries. It is appropriate that this theme is discussed early in the book for it provides a broad base for the discussion of production (Chapters 4, 5 and 6).

POPULATION GROWTH AND CHARACTERISTICS

Figures for the numbers and characteristics of the population of much of the Third World remain imprecise. There are still countries for which data are incomplete or inaccurate. Censuses are difficult to conduct in countries where levels of literacy are low, and where people may not understand the reasons for questions. They may resent the census as an intrusion on their privacy, especially where the counting of people is associated with attempts to levy taxes. Some data are more difficult than others to collect. Accurate determination of age may be impossible, and women may be reluctant to divulge the numbers of their children, both alive and dead. Even when censuses are taken they may be subject to controversy because of political interference and resulting dissatisfaction with the results. It is known that Nigeria has the largest population of any country in Africa, but it is not certain precisely how large. There was a reasonably accurate census in 1952/53, but the first three post-independence censuses (1962, 1963 and 1973) were controversial and subject to political interference, and were officially abandoned. A national census in 1992, officially accepted in Nigeria and beyond, counted 88.5 million, but the *World Development Report, 1994* (World Bank, 1994a) records a population of 102 million for 1992, and the United Nations estimate for that year was over 120 million. Nigeria is an extreme case of a country with poor population data, but there are many

Crowded street scene, Dhaka, Bangladesh

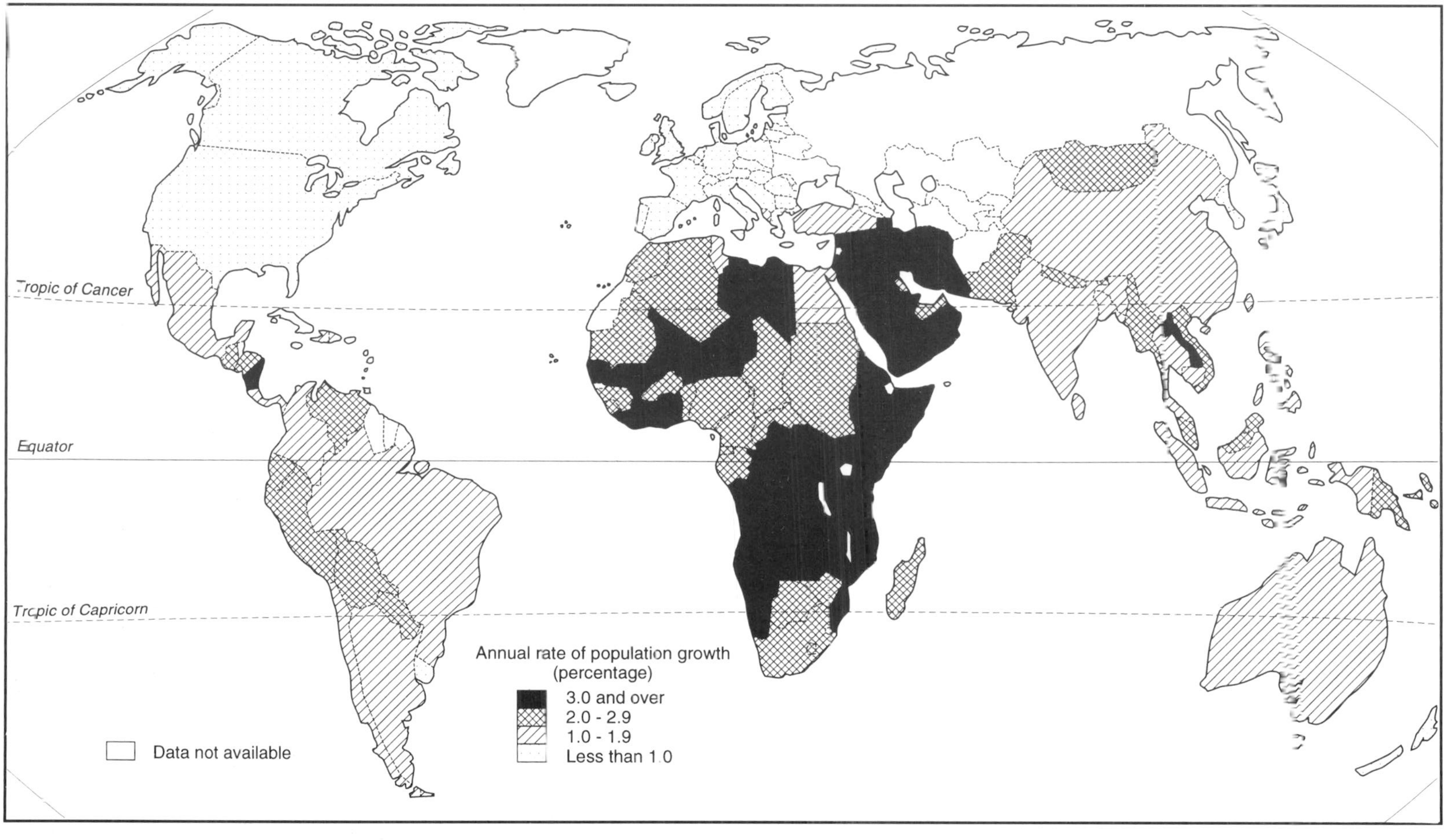

Figure 3.1 Projected annual rate of population growth, 1989–2000
Source of data: World Bank, 1992

Table 3.1 Global and regional population indices

	Population (millions) 1992	*Average annual rate of growth (%) 1980–92*	*Total fertility rate, 1992*	*Expectation of life at birth (years) 1992*	*Birth rate (per 000) 1992*	*Death rate (per 000) 1992*	*Area (000 km²)*	*Density (per km²)*
World	5438	1.7	3.1	66	25	9	111,306	48
Sub-Saharan Africa	543	3.0	6.1	52	44	15	23,066	21
East Asia and Pacific	1689	1.6	2.3	68	21	8	16,369	102
South Asia	1178	2.2	4.0	60	31	10	5,133	224
Eastern Europe and Central Asia	495	1.0	2.2	70	16	10	2,314	212
Middle East and North Africa	256	3.1	4.9	64	34	8	11,015	22
Latin America and Caribbean	452	2.0	3.0	68	26	7	20,507	22
High-income countries	828	0.7	1.7	77	13	9	31,709	26

Source of data: World Bank, 1994a

countries in the Third World that are forced to undertake much of their socio-economic planning without adequate knowledge of the numbers, distribution and age characteristics of their population.

Despite the difficulties over accuracy of data, it is possible to make general statements about population in the Third World. In the past growth was limited through a combination of natural and man-made factors causing high levels of mortality and limited life expectancy. To ensure their survival, human societies had high levels of fertility, producing positive, though normally slow, population growth. There was a high equilibrium balance between fertility and mortality. In some years there were more deaths than births, with 'crisis mortality' in years of famine and disease epidemics, but fertility levels were normally consistently high enough to prevent decline in the longer term. Within the last century, and especially within the last 50 years, in countries of Africa, Asia, Latin America and the Pacific, however, the balance between births and deaths has changed.

Mortality rates have fallen dramatically in recent decades in all age groups, though they remain high for some groups in some countries, particularly for children in Sub-Saharan Africa (where infant mortality rates are typically between 80 and 150 deaths of children aged less than 1 year per 1000 births) where life expectancy was only 51 years in 1991. In the face of such levels of child mortality, measures that result in some control over death and that contribute to the lengthening of life are desirable. Furthermore, communities at local, regional and national levels can take steps to lower mortality – by reducing disease and improving public health – and these can be carried out without specific reference to individuals. Thus mortality rates in the Third World are now fluctuating at relatively low levels, with periodic, short-term, upward trends in limited areas as a result of natural (e.g. drought or plant and animal pest infestation) and man-made (e.g. civil war or inappropriate rural development policies) catastrophes.

In subsistence economies such impacts can be particularly severe. Disease combines with food shortages, wars and political disruption to keep mortality high. At worst, one in four children do not survive until their first birthday, and in some years of 'crisis mortality' there will be absolute population decline. However, with increasing development and with increasing availability of medical care and simple and cheap treatments

Plate 3.1 Health campaign billboard, Niteroi, Brazil The poster, issued by Brazil's Ministry of Health, is part of a campaign against dengue fever. Dengue is spread by mosquitoes, and though not usually fatal, the acute aches and pains it causes give it the alternative name of 'breakbone fever'. The slogan reads 'We are at war. You also must fight against dengue'

such as oral rehydration therapy, overall levels of mortality in the Third World have been consistently falling. In addition, the annual fluctuations in mortality have been significantly reduced as a result of better internal and international food distribution to overcome shortages and associated famines. While living standards for many people in the Third World have not been strikingly improved, some progress has been made in the control of diseases, particularly the prevention and reduction of epidemics. Improvements in public health through better water supply and waste disposal, have also contributed to reduced mortality rates in all age groups (Plate 3.1).

The net effect of lower mortality has been the lengthening of life expectancies at birth (i.e. the number of years a newborn child could expect to live). World Bank data for 1960 estimated mean life expectancies in the low-income countries to be 36 years, in middle-income countries 49 years and 70 years in industrialized countries. In 1992 these life expectancies were 62, 68 and 77 years respectively (see Figure 1.7). Improvement was greatest in the Third World, and especially in the poorest countries. Though life expectancies at birth have risen in all continents, they are still well below the life expectancies of babies born in developed countries (Table 3.1).

With high and annually fluctuating levels of mortality, particularly among infants, high levels of fertility are required to allow populations to reproduce themselves. **Total Fertility Rate** (the most widely used measure of fertility that approximates to the mean number of children each woman will have in her life) is generally much higher in Third World countries than in developed countries (Figure 3.2). Although mortality levels have fallen, fertility levels have not been affected to the same extent, for fertility is more a matter of cultural and personal choice than of levels of development or well-being, though it is linked to these. Fertility levels generally remained fairly static or even rose slightly during the period of greatest improve-

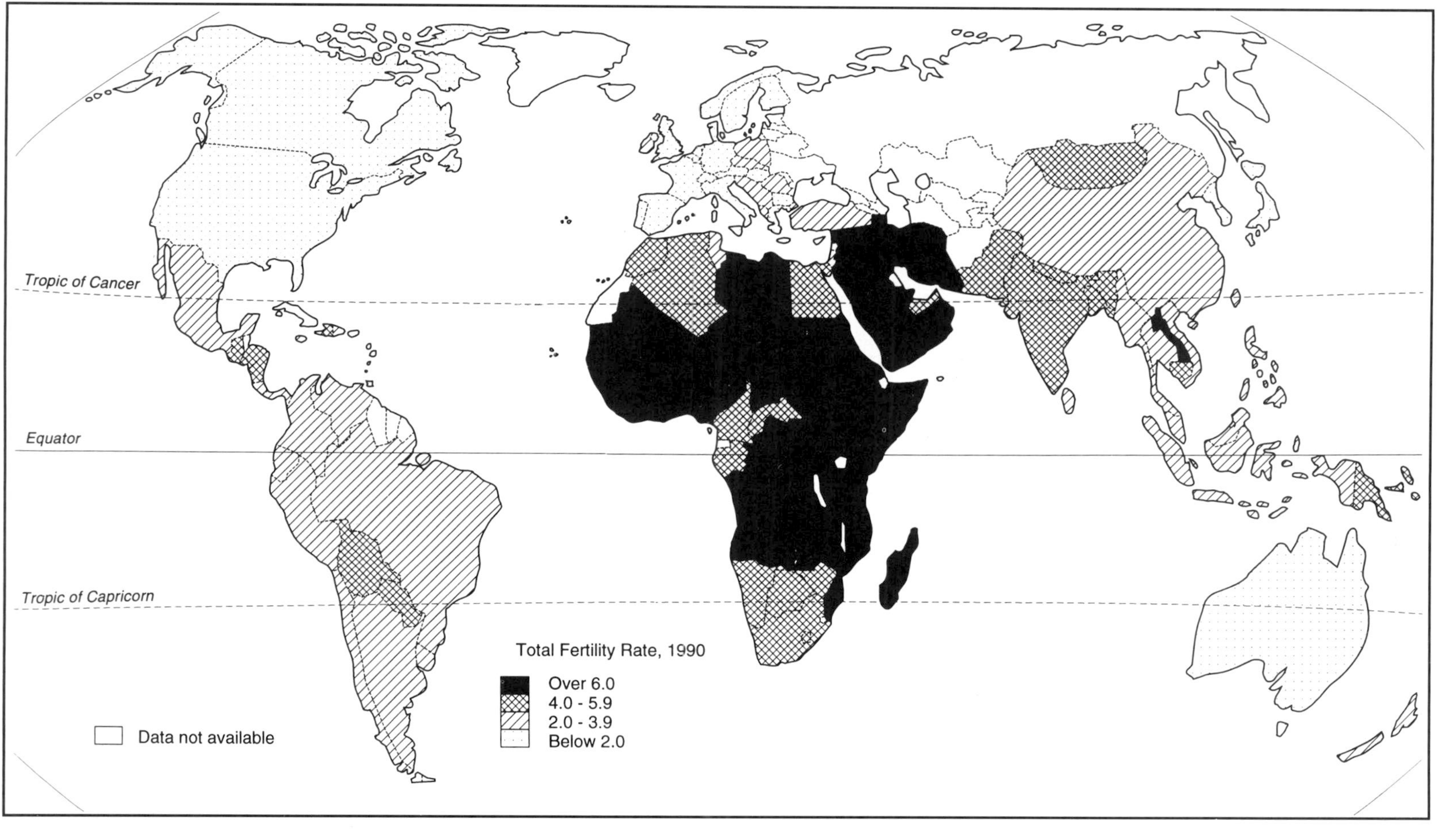

Figure 3.2 Total Fertility Rate, 1990
Source of data: World Bank, 1992

ments in mortality in the 1960s and 1970s. However, the general trend in the Third World has been downward since that time. Fertility remains well above mortality in most regions, and, with more births than deaths in most countries in most years, there is population increase. Where the fall in mortality has been substantial, the gap between births and deaths is large and rates of population growth are highest.

In recent years there has been evidence of a general reduction in fertility, but the rates for Third World populations mostly remain high. Decline has been most marked in China and in several other countries in South and East Asia, and also in Latin America, but has been smaller in Africa and the Middle East. Though high fertility is no longer necessary to maintain growth, it is difficult to change attitudes and practices to effect substantial reductions in fertility that would restore fertility and mortality to an equilibrium. In many societies the virility of men and the fertility of women are confirmed by the birth of large numbers of children. Where people are dependent on low levels of technology, particularly in agriculture, children are economically advantageous. There are more hands to work though there are also more mouths to feed! Boys augment the family labour supply and provide some security for parents in old age, while girls may bring dowries to the families into which they marry, as well as providing domestic and some farm labour. Children are socially prestigious and are valued as economic assets.

Direct or indirect action by governments or communities to influence fertility behaviour is much more difficult to apply than is the case with measures to reduce mortality. Communal measures are more difficult to operate, for the decision whether or not to have children has to be taken by individual men and, most crucially, women. The status of women and the ability of women to control their own fertility, either independently or in consultation with their husbands, is clearly critical. Governments may introduce policies to exhort or even to coerce changes in fertility behaviour, but such policies are difficult to put into practice without the cooperation and consent of individual couples. There must be material and psychological incentives for people to plan and control the size of their families, as well as the means to make this possible through the availability of contraceptives and family planning advice. Achieving fertility reduction in many countries has been in part the result of vigorous direct intervention in family planning activities to persuade women and their husbands of the benefits of such planning.

High population growth in Third World countries is compounded by the effects of mortality reduction on survival of more children into adulthood who then reproduce children in the next generation. Even with constant fertility per woman, the number of births will continuously rise in a population which is experiencing improvements in mortality, and the age structure will become strongly biased to the younger age groups. More than 36 per cent of people in the Third World are less than 15 years of age (compared with only 22 per cent in the developed world), with a further large proportion in the reproductive ages from 15–40 years (Figure 3.3). Age structures are generally pyramidal, with a large base, though, as the age–sex pyramids for Peru and Thailand show, they are sensitive to recent fertility changes with fewer children in each successive age group as fertility falls (Figure 3.4). However, only 4 per cent of the population in the Third World is aged over 65 (compared with 12 per cent in the developed world), though the numbers and proportions of people surviving to old age are rapidly increasing.

While overall there are large and still increasing populations throughout the Third World, there are important variations in rates of population growth between one part and another. Useful generalizations are difficult to make, for with analysis at progressively more detailed scales, modifications and qualifications become necessary. However, one important generalization is the almost universal decline in rates of population growth in recent years, both at a continental scale (Table 3.2(a)) and at a national scale (Table 3.2(b)). Between 1970 and 1990 the total fertility rate fell most in East Asia and Latin America and least in Sub-Saharan Africa. In Sub-Saharan Africa, as elsewhere, there was rapid improvement in mortality rates,

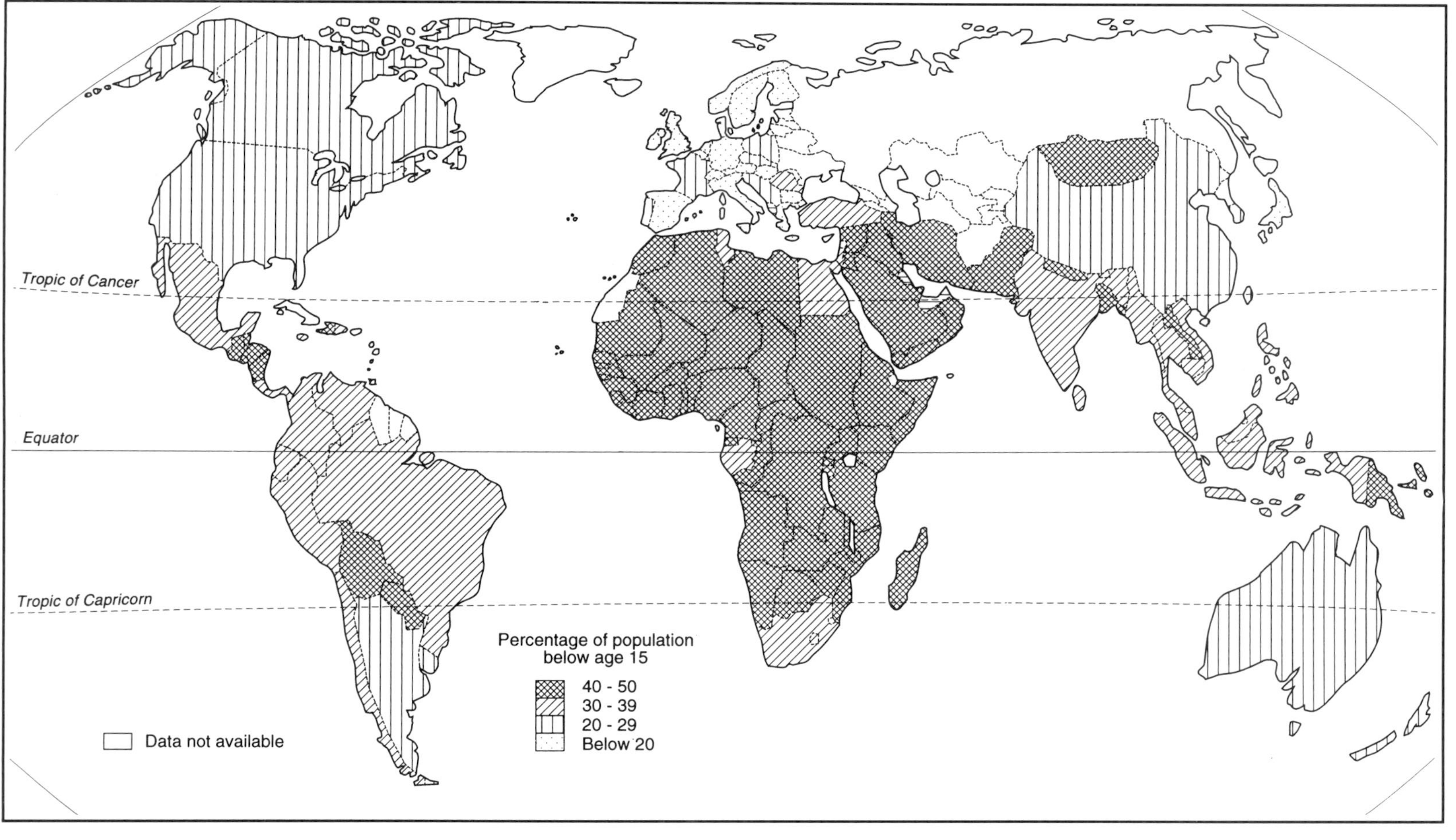

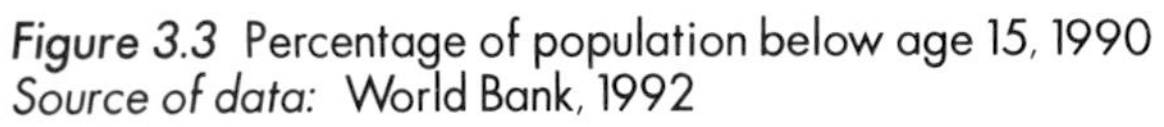

Figure 3.3 Percentage of population below age 15, 1990
Source of data: World Bank, 1992

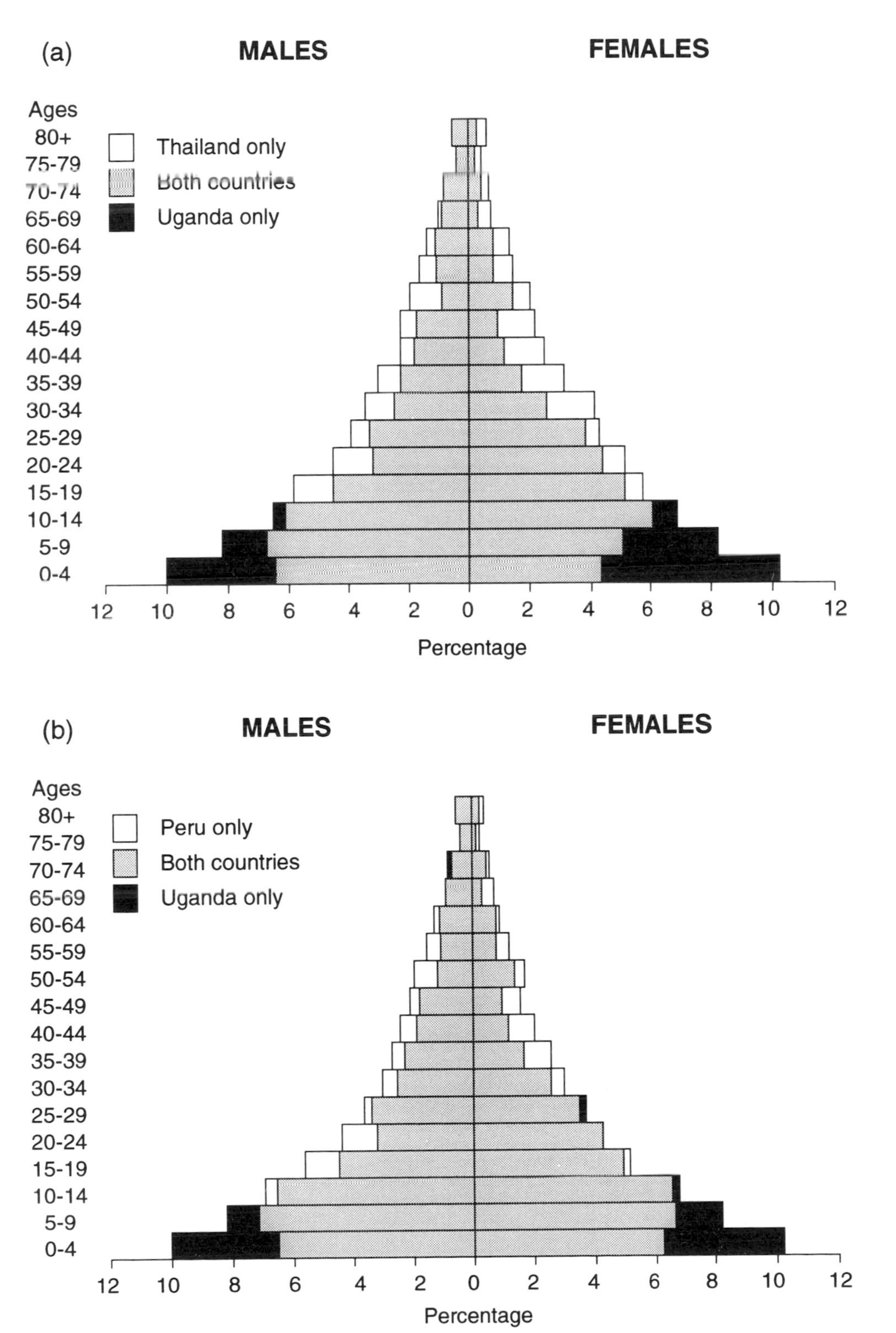

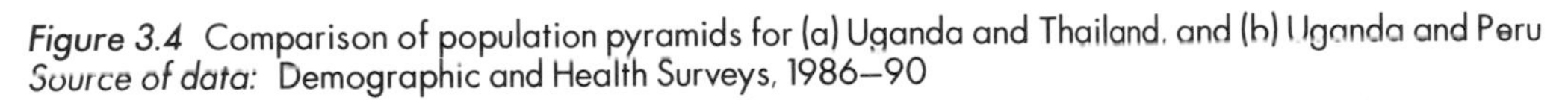
Figure 3.4 Comparison of population pyramids for (a) Uganda and Thailand, and (b) Uganda and Peru
Source of data: Demographic and Health Surveys, 1986–90

Table 3.2(a) Population change by region, 1970–92

	Pop. 1992 (millions)	*TFR 1970*	*TFR 1992*	*% growth per annum 1970–80*	*% growth per annum 1980–92*
Sub-Saharan Africa	543	6.5	6.1	2.8	3.0
East Asia and Pacific	1689	5.7	2.3	1.9	1.6
South Asia	1178	6.0	4.0	2.4	2.2
Middle East and North Africa	253	6.8	4.9	2.8	3.1
Latin America and Caribbean	453	5.2	3.0	2.4	2.0

Source of data: World Bank, 1994a

Table 3.2(b) Population change in the 10 largest Third World countries, 1970–92

	Pop. 1992 (millions)	*TFR 1970*	*TFR 1992*	*% growth per annum 1970–80*	*% growth per annum 1980–92*
China	1162	5.8	2.0	1.8	1.4
India	884	5.8	3.7	2.3	2.1
Indonesia	184	5.5	3.0	2.3	1.8
Brazil	154	4.9	2.8	2.4	2.0
Pakistan	119	7.0	5.6	3.9	3.1
Bangladesh	114	7.0	4.0	2.6	2.3
Nigeria	102	6.9	5.9	2.9	3.0
Mexico	85	6.5	3.2	2.9	2.0
Vietnam	69	6.0	3.8	2.3	2.1
Philippines	64	6.4	4.1	2.5	2.4

Source of data: World Bank, 1994a

so that overall population growth rates rose to over 3 per cent per year (and to almost 4 per cent in a few cases), as they did also in the Middle East and North Africa in this period.

The pattern is given in more detail at a national scale, for the ten largest Third World countries in Table 3.2b, with all except Nigeria having growth rates lower between 1980–92 than in the 1970s. Even in Nigeria, however, the fertility rate was lower in 1990 than it was in 1970, having peaked probably in the late 1980s. Falling fertility and falling rates of population growth are now characteristic of most Third World populations, though the rates of growth are still high by comparison with other parts of the world in the past, and remain highly variable between major world regions (Box 3.1).

POPULATION SIZE, DISTRIBUTION AND DENSITY

At a continental scale there are more people in Asia than in Latin America and in Africa combined. The same order is to be found in population density, which is highest in Asia and lowest in Africa (Table 3.1 and Figure 3.5). Within the continents there are larger areas of high density in Asia than there are in Latin America or in Africa. Particularly in China, the Indian sub-continent and parts of South-East Asia, rural densities above 1000 per square km are common. In Latin America and Africa high densities occur only in a few limited areas, and there are large areas of very low density, for example in the Amazon Basin and in the deserts of Africa.

BOX 3.1 POPULATION GROWTH IN PERU

High rates of population growth, typical of many countries in the Third World, have certainly been evident in Peru, which in the last 50 years has increased in population from 8 million to 26 million. Annual rates of natural increase grew slowly from just over 1 per cent in the late nineteenth century to 1.8 per cent in the 1940s, but the next 20 years saw a massive leap to 3 per cent by the 1960s, when Latin America as a whole had become the world's fastest growing major region. Dire predictions of pending demographic disaster in Peru have been muted more recently as the rate of increase began to decline from its peak in 1960 to 2.6 per cent in 1975–80 and to 2.1 per cent, 1980–92, with an anticipated fall to 1.8 per cent at the end of the century.

In the absence of significant migration to or from Peru, it is the balance of birth and death rates which governs rates of increase. Death rates were initially very high. Mortality from tuberculosis, yellow fever, cholera and other infectious diseases added to the ravages of high infant mortality, gastro-enteritis and malnourishment to generate a death rate of more than 30 per 1000 before 1940, when the average expectation of life at birth was only 36 years. From the 1950s, however, a vast improvement in mortality rates went hand in hand with growth in average income per head and rising GNP. Medical services improved, especially in the capital city, Lima, but changes such as the provision of pure drinking water, more attention to the treatment and disposal of sewage, better personal hygiene and rising standards of education probably had a more widespread effect. Crude death rates fell rapidly to about 15 per 1000 in the early 1960s and to 8 per 1000 in the early 1990s. Average expectation of life at birth rose to 51 years by 1960 and to 65 by 1992.

The trend of birth rates was very different, remaining more or less unchanged at about 45 per 1000 up to the 1960s. A fall in crude birth rates was apparent in the 1970s, and then accelerated from 41 per 1000 in 1970 to 27 per 1000 in 1992, with an expected further fall, since the total fertility rate has fallen from 5.5 children per mother in 1975 to 3.3 in 1992. Family planning and contraception are widely known and understood, though less universally practised. By 1980 three-quarters of women in a representative sample were aware of one or more modern, efficient methods of birth control, and their use is most common, as might be expected, in the cities, especially Lima, and among the better educated. The total population of Peru will undoubtedly continue to grow well into the twenty-first century, given its youthful age structure, but the earlier surge of numbers has begun to subside.

The rapid growth of population and an increasingly youthful age structure have been accompanied by major shifts in the distribution of population. Differential rates of natural increase among the regions have been less important than increased migration and mobility in bringing about redistribution from the harsh conditions of the high Andes to the relatively prosperous coastal zone and to the farming and job opportunities available in the eastern lowlands. Above all, the urban population has grown from 1.5 million in 1940 to 15 million in 1990, Lima alone having a population of over 6 million, some 29 per cent of the national population.

There are great contrasts in population numbers and densities between countries in the Third World. China (over 1162 million) and India (884 million) have about half the population of the Third World and also some of the highest national population densities (Table 3.2(b)). They are followed by Indonesia (184 million) and Brazil (154 million), the largest non-Asian Third World country. In Africa countries with large areas, such as Zaire and Sudan, have relatively small populations, low overall densities and very limited areas with high densities (Figure 3.5). Most countries in Africa and Latin America which have small areas have small populations, though densities in some of them may be high. Rwanda (7.4 million), one of the smaller African countries in area and in population, has the highest density in the continent (over 300 per square km). High figures are also common in small island countries in the Caribbean and the Pacific.

While population densities offer a rough

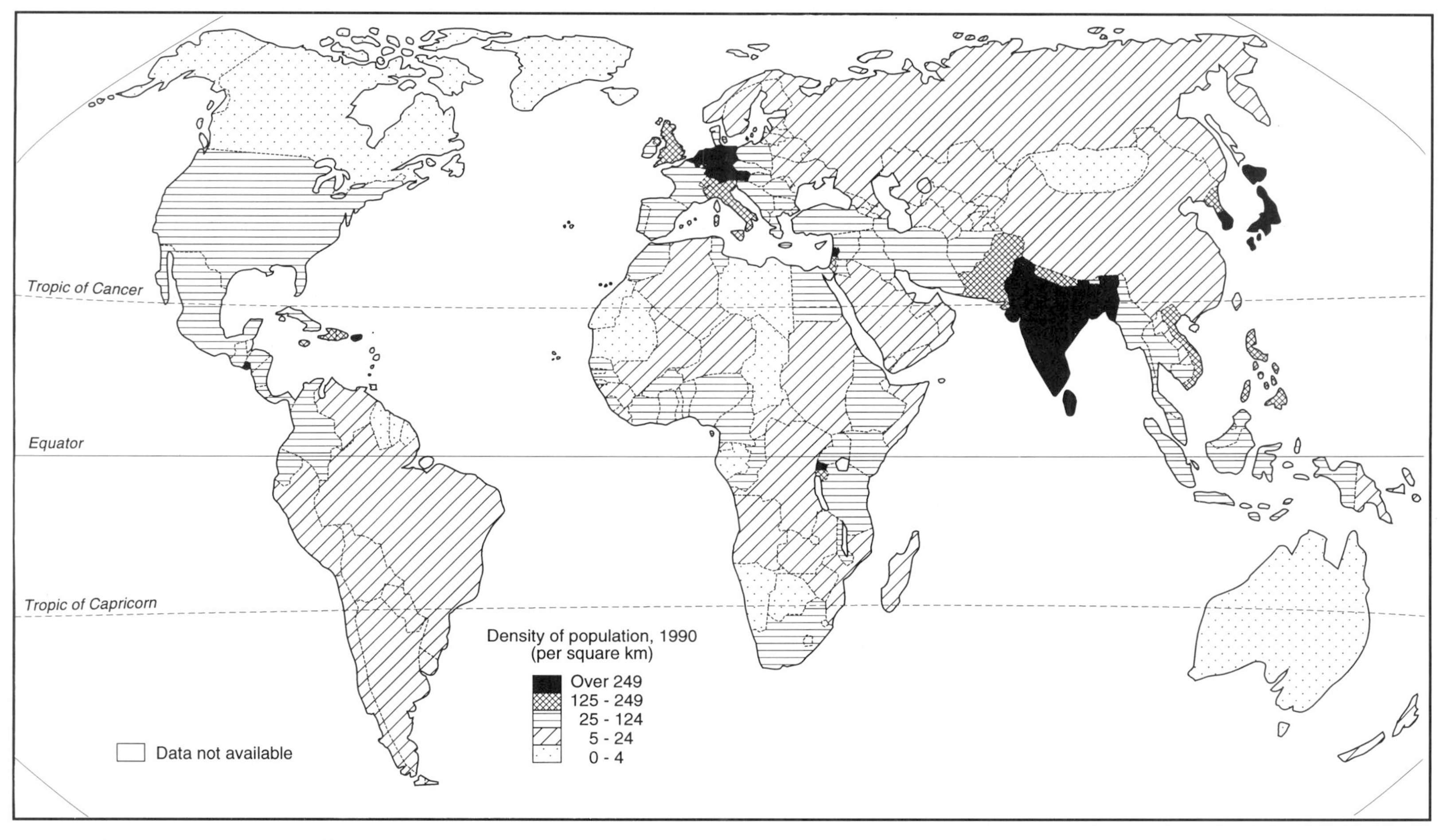

Figure 3.5 Density of population, 1990
Source of data: United Nations, 1992

Plates 3.2 (above) and 3.3 (below) School children, Nepal and Zimbabwe A familiar scene throughout the Third World is of children going to school, on foot or by other transport, as in this case of Nepalese children, on a tractor trailer. Their clean and neatly pressed uniforms often stand in great contrast to their surroundings. The demand for education is reflected in the crowded schoolyard of a secondary school in Zimbabwe

indication of relationships between people and land they must be interpreted with caution. Some of the highest rural densities in the world are in the Nile Valley in Egypt (*c.* 1500 per square km), and are evidence of a remarkable accommodation of people and land that has developed over many millennia. But people there are supported at very low living standards, and it has been estimated that in recent decades about 10 per cent of the best agricultural land has been lost to urban growth. In areas of relatively high density in Asia, Africa and Latin America people may not be well provided for, and even where densities are low conditions may be the same if agricultural systems and practices result in low productivity.

Population size and density are important to Third World countries. A large population provides a large manpower resource, though in most Third World countries there are limited numbers of people with modern education and associated skills. A majority of the population are therefore still dependent on traditional skills, but educational facilities, though limited, have expanded rapidly in recent years (Plates 3.2 and 3.3). In most countries the increase in enrolment rates has more than kept up with the rate of population growth, and the proportion of children enrolled in formal schooling at all levels has risen (Gould, 1993).

Large populations may be a disadvantage in terms of the resources required to meet their needs, especially when numbers are growing rapidly without a corresponding expansion of resources, the most important of which is food. But the rate of food production has outstripped the rate of population growth in most areas, with the exception of Sub-Saharan Africa where per capita food production has been decreasing since the 1980s, even though total production has been rising (see Chapter 4). Countries like Pakistan and Bangladesh, with large populations and limited financial resources, battle with these problems on a large scale, but Green Revolution technology applied since the 1960s has revolutionized their internal food availability, especially of the staple foods, rice and wheat. Countries like Brazil and Nigeria have larger reserves of land and also possess other resources for development (oil in Nigeria and a range of minerals in Brazil). Notwithstanding relatively abundant resources of land and oil, Nigeria, in common with most other African countries, has had problems feeding its population during the last two decades. Its per capita food production has fallen, and the volume of food imports has escalated.

Countries with small populations in large land areas may be at a disadvantage in having too few people to promote economic development. There may be other limiting factors, such as disease. In the past large areas of savanna Africa were unfavourable for human settlement because of infestation with tsetse fly which transmits forms of trypanosomiasis to people (sleeping sickness) and to animals. The spread of disease has also been responsible for the desertion of previously inhabited lands, malaria in parts of the Indian sub-continent and river blindness in river valleys in West Africa. Projects for disease control and eradication are costly to organize and are not always successful. While there has been substantial reduction in river blindness, the global eradication of malaria which seemed possible in the 1950s has not materialized. There has been a resurgence of the disease in the Indian sub-continent and in other parts of the world during the last twenty-five years. Smallpox has been totally eradicated, but now the world in general and many parts of the Third World in particular are facing the threat from AIDS (Box 3.2).

Overpopulation may occur where needs exceed resources, both those in use and those not yet fully developed, though it is difficult to define these circumstances with precision. Relative overpopulation (where there are still resources to be developed) is more common than absolute overpopulation (where the limits of resource exploitation have been reached). Overpopulation is most commonly associated with rural areas and is manifested in high rates of population growth coupled with low productivity, uneven distribution of land, landlessness and underemployment. These are to be found in parts of most Third World countries; for example over 30 per cent of the rural population of Bangladesh are estimated to be without land. Overpopulation is also an increasing feature in urban areas where resources are inadequate to meet needs for

BOX 3.2 HEALTH AND DISEASE IN THE THIRD WORLD

People in the Third World have low levels of health as compared with people in the more developed parts of the world. They suffer from a range of infectious and communicable diseases, and, in more recent times, with increased life expectancy, degenerative diseases are becoming more common. Much of this poor health is associated with poverty and low living standards, but it is difficult to deal adequately with health problems when there are severe restrictions on financial support to provide preventive and curative measures.

A major success was achieved with the global eradication of smallpox in the late 1970s, previously a major killer, but at almost the same time a new disease complex, HIV/AIDS, appeared to add to the health problems of the Third World. It is a disease with global ramifications, with an estimated 12 million people who are HIV+, of whom over 10 million were in Third World countries in 1993. Its impact is greatest in Africa, with 7 million affected, 60 per cent of the global total, with highest prevalence in parts of East and South-Central Africa. The demographic impact of AIDS is in raising mortality rates, especially of the middle and younger age groups, and in Africa it is now the major cause of death in the 20–40 age groups for both males and females. Since these are the working age groups it is already having considerable effects in reducing economic capacity. At present there is no cure and no medical means of protection from the disease; changed behaviour and health education to achieve protection are the only means available to reduce its impact.

Although HIV/AIDS has become a major epidemic and attracts great attention in Third World countries, it is important to remember that there are many other diseases which are common in the Third World, to which very large numbers of people are exposed and suffer from their effects. Among these, malaria presents the greatest threat to health. More than 2000 million people live in areas in which malaria transmission may take place, 500 million of them in areas where the risk is very high. These areas are located almost entirely within the Third World, though in past times malaria transmission was widespread in

Plate 3.4 Baby being weighed at a health clinic in Zimbabwe High infant mortality is one of the indices of Third World poverty, and child health care is therefore an important element in the development process

BOX 3.2 *continued*

both North America and Europe, but has disappeared largely as a consequence of development, a combination of environmental changes and health measures. Malaria remains a major health problem in the Third World because there are ecological factors which favour the breeding of the *anopheles* mosquito which transmits the malaria parasite from one human being to another, and because people in their everyday lives are brought into contact with these mosquito vectors in a variety of ways. There are more than 100 million recorded cases of malaria each year and the actual number much exceeds this. In Sub-Saharan Africa, the most malarious part of the Third World, the number of cases is not known, but it is estimated that at least a million children die from malaria each year. Its effect on the mortality of adults is less, but its debilitating effects on their general health and economic capacity are considerable.

In the 1950s and 1960s it was thought that global malaria eradication could be achieved, using a combination of insecticides to reduce mosquito populations and drugs to kill the parasites. Considerable success was achieved, particularly in the Indian sub-continent, but since then both there and elsewhere there has been a resurgence of malaria. Eradication campaigns were abandoned and the present aim is to control the ravages of the disease through the use of insecticides and drugs. Control is not without problems. Mosquitoes may become resistant to insecticides, and parasites to drugs. A possible but as yet uncertain measure for the future is a vaccine to protect against the disease.

A variety of human activities promote malaria transmission, of which one of the most significant is migration. Infected migrants may take infection to areas otherwise malaria-free. Migrants to areas of new settlement, as in the Amazon Basin and the outer islands of Indonesia (see Box 3.3), are exposed to severe malaria infection. It is difficult to provide effective means of malaria control for mobile people. Problems are exacerbated when people are forced to move as a consequence of environmental catastrophe, war and political disruption.

Much can be and is being achieved in the control and treatment of malaria and other diseases in the Third World through the development of primary health care (PHC). It can be established with relatively low-cost measures which can be carried out by people selected from communities and requiring little training. In PHC more emphasis is placed on preventive than on curative medicine. Much can be achieved through non-medical measures – improvements in water supply, in sewage and waste disposal, in housing, in diets and in food hygiene – such as made great impact on public health in the now developed parts of the world in the second half of the nineteenth century. The goal of the World Health Organization, which plays a major role in promoting primary health care and research into diseases in the Third World, is 'Health for All by the Year 2000'. Though this goal may be unattainable within a limited time-span, there are prospects for continuing reductions in the ill health which afflicts so many people (Plate 3.4).

food, shelter and employment.

Among governments as well as among individuals there are variations in attitudes towards population size. Countries with small populations may wish them to be larger for reasons of prestige and to promote economic development. Such is the case with Libya, which wishes to develop, using its large oil revenues for investment. There are large countries with large populations which, for similar reasons, wish to see them even larger. Conversely there are both small and large countries where the pressures of population on available resources of all kinds are so great, resulting in low social and economic standards for the majority of people, that they make efforts to reduce further growth. Egypt, immediately adjacent to Libya, in contrast has a large rural population and very limited financial resources for investment, and so actively seeks to limit its population growth.

MIGRATION AND POPULATION REDISTRIBUTION

In general the numbers of people in a country are determined by the balance between the numbers who are born and the numbers who die. Migration, the third basic component in population change, involves the movements of people in different ways, for a variety of reasons and with a range of different consequences. Population movements occur throughout the Third World and bring about changes in the size and distribution of population within countries (*internal migration*) and between countries (*international migration*). Some of the movements involve people moving their places of residence permanently from one place to another (*permanent migration*); other movements involve time away from permanent places of residence for long or short periods (days, weeks, seasons or years) followed by return (***circulation***). Circulatory movement is common throughout the Third World, for people have strong attachments to their place of origin and tend to be reluctant to sever permanently their links with them.

The complexity of mobility phenomena is related to cultural, social and economic relationships between areas, and the mix of permanent migration and circulation, between and among rural and urban areas, is related to the level of development in each area and the economic disparities between source and destination of migration flows. It is selective not only by area, with some areas as source areas of migrants and others as destination areas, but also by sex. Men are more likely to become long-distance migrants, often as circular migrants, leaving their women in the source area to look after the family land and to bring up their children. Sex ratios in urban areas, especially in Africa, tend to have a male dominance. Economic pressures are the major factors in influencing people to move, as they are motivated by the prospect of better economic opportunities elsewhere, but social and political factors may also be critical in the migration decision-making process.

The most widely discussed movements in Third World countries are from rural to urban areas. Migrants are 'pushed' from rural areas by low returns from agriculture and shortages of land. They are also 'pulled' to urban areas or other rural destinations by the expectations (not always realized) of a job, of economic opportunity, or of a better life in town. Rural–urban migration and consequent urbanization have been least important for Africa, but are well-developed in Asia and the Middle East where there are strong indigenous urban traditions, and in Latin America where several centuries of colonial influence developed major administrative and commercial centres. The populations of urban places throughout the Third World have mushroomed in recent decades, with growth rates (8–12 per cent per annum) several times the national average, particularly in the major cities (Plate 3.5). Bangkok grew from 2 million to 7 million between 1960 and 1990. Mexico City, one of the fastest growing cities in the world, increased its population by more than 5 million during the last decade, and by the end of this century it and São Paulo in Brazil will have populations of more than 24 million each (see Chapter 7). Such rapid growth

Plate 3.5 Crowded street scene, Kathmandu, Nepal
Pedestrians, cyclists and rickshaws share the road in the shopping area of the old town. Although only about 10 per cent of Nepal's population are urban-dwellers, the urban population grew at almost 8 per cent a year between 1980 and 1992, largely in consequence of in-migration. Notice that in this photo most of the people are young

of urban populations is not only due to rural–urban migration, but also to high rates of natural increase, since people in the young and reproductive age groups are well represented in urban areas. However, in most countries the natural rate of increase is lower in towns and cities than it is in rural areas.

Though large numbers move out from rural areas to urban places, the high rates of natural population growth in the countryside ensure that normally there is not rural population decline in the Third World. Movements of population within rural areas have received comparatively little attention but they are of major importance in redistributing people from more to less densely populated areas. These movements are often spontaneous, but may also be controlled through various forms of planned resettlement. The latter include transmigration in Indonesia, from densely populated Java to the sparsely populated outer islands (Box 3.3), settlement in the Amazon basin in Brazil, and forced redistributions of population, as under the authoritarian regime in Ethiopia in the 1980s (Box 3.4). People may also move temporarily to work in urban places or in rural areas of more advanced development which require additional labour during periods of maximum agricultural activity.

Rural–urban movements and those within rural areas are the consequence of varying degrees of economic pressure and of social need. There are other categories of movement which are the consequence of much greater degrees of pressure exerted by the physical environment and by political factors. In various parts of the Third World environmental refugees have been forced to move by disasters such as droughts, flooding, earthquakes, volcanic eruptions, hurricanes and tidal waves. In the future, global warming may lead to rising sea levels which, in the absence of costly investments in strengthening sea defences, may displace millions of people in vulnerable areas such as Bangladesh and on Pacific islands.

However, strictly speaking refugees are political in origin. In the second half of the twentieth century Third World countries have produced the most refugees, as a result of civil and international conflicts, particularly in Africa and Asia. In Asia these began with the displacement of millions of Muslims and Hindus at the time of the partition of British India in 1947 and with the outflow of Palestinians after the creation of the state of Israel in 1948. They continued with disruptions during the wars and political disruptions in Indo-China, and most recently in Afghanistan and Iraq. In Africa, refugees have become a characteristic feature of migration during the last three decades, particularly in the north-east and south-central parts of the continent (Plate 3.6). In the early 1990s the United Nations High Commissioner for Refugees estimated that there were over 5 million refugees in Africa – more than one in every hundred Africans is a refugee!

International migration has expanded as a consequence of gross economic disparities between less and more developed countries. Some of this is controlled at border crossing points through emigration and immigration policies, but much of it is clandestine and illegal. Some migration takes place between Third World countries, and as their economic differentiation proceeds, this form of movement will become more prominent. The major migration system is of workers to the Gulf oil states from the Indian sub-continent, Sri Lanka, Thailand, and from the Philippines and other Asian Pacific Rim countries, as well as from the Arab world. Many Gulf oil producers (e.g. Kuwait, United Arab Emirates) have populations of which over 50 per cent are foreign-born.

Other international migration flows are from the Third World to the First. Migrants from Mexico and other Central American countries cross the United States border to seek work, and the USA and Canada have replaced Europe as the principal destinations of migrants from Caribbean countries. In some cases there is absolute population loss as a result of emigration. In Guyana, for example, the national population fell from over 750,000 in 1980 to 716,000 in 1990, a massive 5 per cent reduction in a decade. Southern European countries are receiving increasing numbers of migrants particularly from North Africa, across the Mediterranean Sea which, like the US/Mexico border, marks a steep economic cliff between poverty and economic opportunity.

BOX 3.3 TRANSMIGRATION IN INDONESIA

Programmes of planned migration to resettle people have been a feature of economic and social planning in the Third World in the second half of the twentieth century, for example in Amazonia, Tanzania and Sri Lanka. In Indonesia organized resettlement began in 1905 when it was the Netherlands East Indies. In this vast archipelago of more than 13,000 islands, extending from Sumatra in the west to Irian Jaya (the western part of the island of New Guinea) in the east, there are over 180 million people (Figure 9.2). Some islands are among the most densely settled in the world and others are only sparsely inhabited. More than 60 per cent of the national population lives on the island of Java in an area about 7 per cent of the national area, and in some parts density rises to more than 1000 persons per square km. Bali, Madura and Lombok are other small 'inner' islands with dense populations and similar problems of land shortage. Population growth rates in these densely populated islands have been high and problems of matching population to available resources are severe. In contrast, the so-called Outer Islands, with the greater proportion of the land area, have small populations and low densities – in Sumatra about 80 per square km, in Irian Jaya only 6 per square km. They have great potential for agriculture, forestry and mineral developments but have lacked the people to make these possible. Their development would relieve the food deficit in Indonesia and increase rubber, oil palm and other agriculture products for export.

Transmigration, to move people from the inner to the outer islands, was begun by the Dutch colonial administration primarily to meet the need for cheap labour on plantations in Sumatra, and about 200,000 people were moved during the first half of this century. Movement accelerated when Indonesia became independent in the mid-twentieth century, and in the next 25 years more than a million people were moved. The south of Sumatra was a major early focus for migrants and later they were sent to Sulawesi and other islands. Migrants were provided with housing, land and the means to take them through to their harvest. At the end of the 1970s a major programme planned to move half a million households within five years – about 2.5 million people, more than twice the number moved since the start of transmigration. However, the programme was over-ambitious and despite support from the Indonesian government and assistance from the World Bank, it fell behind the projected target. Between 1979 and 1984, 1.5 million people were resettled, and since that time the numbers have fallen sharply.

Problems of moving vast numbers are many and complex. Knowledge of areas to be resettled is often inadequate. Fertile soil has been removed in forest clearance, wild animals have damaged farms, and people have often not been capable of dealing with unfamiliar crops or new tasks in strange environments. Infrastructure – roads, schools, dispensaries – necessary to support settlement is often inadequate, and the health of settlers is poor from malaria and other diseases. In very sparsely populated islands there is the fear that incomers will overwhelm the native populations and destroy traditional ways of life. Differences of religion and language are further complications. From within and from outside Indonesia concern is voiced for the vast areas of tropical forest that are being destroyed as part of the transmigration strategy.

The impact of transmigration on population problems in the densely settled islands of origin has been minimal. In the 1980s the population of Java increased by 18 per cent despite transmigration, and up to a fifth of all migrants have returned to their home islands. But there are positive aspects. Landless migrants from Java have been provided with the means to maintain themselves at comparable and sometimes higher levels than previously. Also, spontaneous migration to the Outer Islands has been stimulated, and the transmigration policy in the future will place less emphasis on official schemes for moving people and more on providing rural infrastructure for development. In this way the government hopes to encourage a more even distribution of population. In addition, the problems of growth of the large population have been tackled by increased family planning programmes, which have been relatively successful, contributing to falling fertility (Table 3.2(b)), by further intensifying agricultural production, by development of Indonesia's abundant resources of oil and gas, and in industrial developments.

Large-scale planned migration of population is difficult to organize and expensive to achieve, even when there are large amounts of land available for settlement. The limited success of the Indonesian experience of transmigration provides an object lesson for other parts of the Third World, where often the land resources to meet needs are less favourable.

BOX 3.4 RESETTLEMENT IN ETHIOPIA

During the last four decades there has been substantial large-scale resettlement in Ethiopia. To bring about this redistribution of population varying degrees of persuasion, force and coercion were used. There were two major phases of resettlement, each associated with the government of the time. From 1950 to 1974 resettlement under the Imperial regime of Haile Selassie was largely spontaneous; about 1 million people settled on land that was largely uninhabited and the area under cultivation in Ethiopia was increased by about 25 per cent. Settlers were mostly subsistence farmers, moving over distances of less than 150 km in families and small groups, with little government involvement. Movements were mainly from the highlands of the Ethiopian plateau to the surrounding escarpments and lowlands where the grasslands and woodlands with lower and less reliable rainfall than on the plateau had been grazed by pastoralists. Few changes were made in traditional methods of cultivation and in crops grown, and these were often poorly adapted to the new environments, creating such problems as extensive soil erosion. Besides spontaneous settlement there were a few government-sponsored settlement schemes for irrigated commercial agriculture, including sugar estates, particularly in the Awash Valley, which attracted large numbers of seasonal migrant labourers from areas of population pressure on the plateau.

The Imperial government was overthrown in 1974 and replaced by a revolutionary socialist regime, committed to raising living standards. It nationalized land holdings, and set up an authority to settle peasants with little or no land to take up unused or under-used land, to alleviate unemployment and to conserve natural resources. Peasant associations were formed to assist in land redistribution. People were also resettled by a commission set up in the early 1980s to deal with the problems associated with severe drought, and its activities were merged with those of the settlement authority. Coercion was a feature of a ten-year plan in this period to move over a million people, but it was overwhelmed by recurring political and environmental crises in the 1980s. Ethnic tension resulting in civil war with Eritrea, together with intensifying drought in northern Ethiopia, forced the resettlement of 600,000 people between 1984 and 1986 with little planning of any sort. They were moved from the higher plateau lands to low-lying lands in the west and south-west, to environments which were alien and to which it was difficult for unsupported peasant farmers to adjust. Settlers suffered from malnutrition, further increasing their susceptibility to other diseases. In the lower-lying lands the traditional cereal crops of the plateau could not be cultivated, and adjustments to new crops were made with difficulty. It took four years or more for settlers to attain self-sufficiency and some resettled pastoral communities were still dependent on food aid after more than ten years.

Traditionally the plateau areas above 2000 m were preferred for settlement. They have good volcanic soils and, perhaps more importantly, are free of malaria, the major health risk in Ethiopia. However, much of the resettlement was in lower altitudes where the disease is endemic, and settlers with no natural immunity suffered high levels of illness and mortality. The risks were increased with irrigation, for this increased the potential for mosquito breeding, and the seasonal movements of migrant labourers contributed to the spread and transmission of malaria. Irrigation schemes also contributed to the spread of bilharzia (a debilitating disease spread by water-borne snails), while in the west and south-west of Ethiopia on the border with Sudan settlers were exposed to the risks of river blindness, sleeping sickness and yellow fever. Such diseases can be reduced by efficient settlement planning and vector control, but these were lacking in these schemes.

Coercion, strange environments, unrealized expectations, exposure to health hazards and limited information available and given to settlers produced psychological stress. Discontent caused some settlers to desert and to return to the places from which they had been moved, particularly in the late 1980s when the drought conditions on the plateau area improved and official coercion was reduced with increasing opposition to the revolutionary regime, which was eventually overthrown in 1991.

The circumstances of foreign migrant labourers are always precarious. They are liable to be expelled or repatriated, particularly if they may have entered the host country illegally, and illegal status means they may be exploited by the payment of low wages, and lack access to social

Plate 3.6 Refugees from the inter-tribal wars in Rwanda crowd onto a bridge on the Rwanda–Zaire border, 1994

security provisions. In West Africa in the 1970s and 1980s foreign workers became the scapegoats for deteriorating economic conditions, and there were major expulsions from many countries, notably from Nigeria in 1983 and 1985 following the downturn in its oil-based economy. Migrants may also find themselves in war zones or in civil disruption, as were migrant labourers in Kuwait at the time of the Gulf War in 1991.

FACTORS IN POPULATION CONTROL

Family planning and control have figured more importantly than any other element in discussions concerning population in the Third World in recent decades – with various and often vigorous reactions. Planning and control have been considered only in terms of limiting or reducing population, whereas they should be seen more objectively as the means whereby the social and economic potential of people may be fulfilled both nationally and individually. The planning and control of population are very emotive issues about which it is difficult to be objective, and reactions are influenced by political, religious, cultural, social and economic factors.

Problems of population on a global scale, and in the Third World in particular, have become a major concern since the 1950s. Sufficient data have become available through national and international organizations to make fully evident the magnitude and implications of population growth. These organizations include national census offices and institutes for population studies, with international involvement within the United Nations (e.g. the Population Division of the Economic and Social Council) and its special agencies (e.g. World Health Organization, Food and Agricultural Organization, World Food Programme, UNICEF). In the 1970s the UN Fund for Population Activities (UNFPA) was set up specifically as a focus for work on population: to provide support for better and more coordinated investigations and to seek ways and means of alleviating and possibly solving the

"She's got a point there—does she take the pill before or after what would have been a meal time?"

Paradoxes raised by policies aimed at controlling population

wide-ranging problems which populations present. It was felt that pressures of population in many parts of the Third World could be reduced if political, social and economic structures were modified, but cultural, religious and political differences do not make consensus between states on population matters easy: for more than two decades after it was set up the World Health Organization was not allowed any direct involvement in aspects of population control.

In retrospect some of the approaches applied to population planning and control in the 1950s through to the 1970s were crude and lacking in insight. They were paternalistic and patronizing on the part of the developed world towards the Third World, and to a large extent were responsible for the belief that the former wished to control the numbers of people in the latter. These approaches were possible through important advances in contraceptive methods, particularly the contraceptive pill and intra-uterine devices (IUD), which were more efficient than any contraceptive aids previously available. It was thought that these new methods could be applied on a large scale among people with little or no knowledge of what they involved or implied. It was not appreciated that methods which might be acceptable in some parts of the world would not be acceptable in others. In the Third World they have been interpreted by many as an attempt by the more developed countries to limit numbers and reduce the threat that these might otherwise present. In addition, many religions and cultures are strongly pronatalist. They have blessed and encouraged large numbers of children, viewing with outright disapproval some of the means available for family planning.

Government and non-government bodies, particularly in the USA, have been responsible for the development of many of the approaches to family planning and control, and for making the new means of contraception available. Considerable investigation of couples' knowledge of, attitudes to, and practices in family planning have been undertaken in Third World countries, and programmes have been designed to make family planning measures more widely available. These programmes had varying effects and degrees of success. Some of the most significant fertility declines have been achieved in relatively small populations where it has been possible to develop motivation for the acceptance of family planning and to provide the means. This has been particularly effective in small islands like Mauritius, and in rapidly modernizing and highly controlled states, such as Hong Kong, Taiwan and Singapore.

However, successes have also been achieved in large countries, including Indonesia, Thailand and Bangladesh. The greatest achievement on a large scale has been in China. There contemporary official authority has combined with a traditional acceptance of authority to achieve remarkable reductions in fertility rates in a very short time through contraception and abortion, without assistance from outside the country. China, the most populous country in the world, has developed on socialist principles since the 1940s and has major programmes for planning and controlling its number of people.

More than 70 per cent of couples in the reproductive age groups practise family planning; and the overall rate of natural increase has fallen to 1.6 per cent per annum, while for the central provinces with large populations it is below 1.0 per cent. In China on a large scale and in Singapore on a small scale more authoritarian measures have worked towards limiting their populations.

India has had variable success in its measures to control population growth. Attempts to spread the adoption of family planning have ranged from encouragement to coercion. Coercion was contemplated in the 1970s involving legislation which would have made sterilization compulsory after the birth of three children to a couple, or four if the first three were of the same sex. They were strongly resisted and reaction against them was a major factor in the downfall of the government of Mrs Indira Gandhi in 1977. Most recent evidence, from the 1991 census, is that there has been a slowing in the annual population growth rate to 2 per cent. In Mexico, a country with a medium-sized population which had been growing rapidly, an important reduction in fertility was achieved in a few years as a result of a major family planning programme. The annual rate of increase has fallen from 3.4 to 2.4 per cent, but even with this reduction the population can still double in size in 25 years.

What has become clear is that fertility reductions are possible where there are positive and direct interventions by governments in fertility change. Indirect methods – improved incomes, expansion of women's education – also have an effect, and the greatest successes are achieved where both direct and indirect approaches are developed. The largest falls in fertility are in those countries with improvements in the overall quality of life for the majority of the population, and where governments have simultaneously developed vigorous family planning programmes.

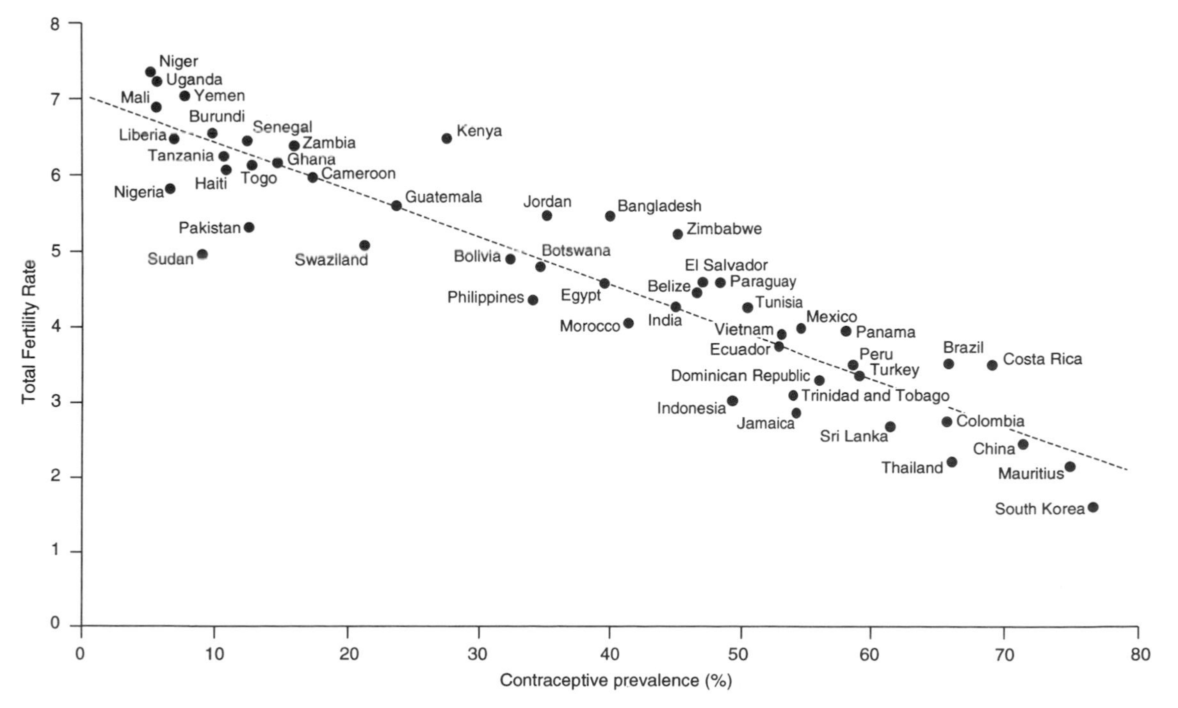

Figure 3.6 Fertility and contraceptive prevalence, late 1980s. Those countries with high fertility tend to also have low rates of contraceptive use, and as contraceptive use rates rise so levels of fertility fall. The relationship described in the figure suggests its general validity for all regions of the Third World, and offers a strong justification for promoting family planning campaigns
Source of data: Robey *et al.*, 1992

The one critical objective in many countries has been to raise the use of contraception, as measured by the **contraceptive prevalence rate (CPR)**, for it has been shown that there is a strong relationship at an international level between the total fertility rate and the CPR (Figure 3.6). This relationship has been the basis of many population policies developed by governments in recent years, and a basic assumption of international agencies in making loans for fertility reduction. A strong thrust of policy has been to enhance the availability of modern contraceptives and of information about their use. These are targeted particularly at women, for example in maternal and child health programmes, where the link between fertility and childhood survival can be made explicit. This is a supply-side approach – improving the supply of contraceptives ensures their more widespread and more effective use. However, problems remain on the demand side – many women and their husbands have neither the wish to restrict family size nor, even if they wished to, the money to buy contraceptives even if they were locally available. Even so, in Sub-Saharan Africa, where demand constraints have been most prominent, some governments (in Kenya, Botswana, Zimbabwe) have developed supply-led family planning programmes with some success. In Kenya the CPR rose from 7 per cent in 1979, when the TFR was at nearly 8, to 27 per cent in 1989, with a fall of 13 per cent in TFR to 6.7. While all the fall cannot be attributed to the increasing use of modern contraceptives (in Kenya mostly the pill and IUDs), it certainly was a factor in achieving a higher demand for their use by women, but men are much more resistant to the use of condoms (Plate 3.7).

Plate 3.7 Family planning advertisement in Kathmandu, Nepal 'Panther' condoms are brashly promoted by a neon hoarding

To maintain a population at replacement level, with a balance between the numbers who die and the numbers born, requires 'zero growth'. Very few Third World countries currently approach this, in contrast to the developed world where all countries now have low growth rates (below 1 per cent per annum), and in some populations are falling. The processes by which developed countries have progressed through a demographic transition from high mortality and fertility to low or zero growth after mortality and fertility decline seem to suggest a generally applicable model that can be applied to Third World countries. However, the factors influencing the stages in the transition are difficult to identify and evaluate. Furthermore, there are considerable differences of opinion about the causes of the transition even in Europe and North America. The demographic transition in the developed countries can be studied only retrospectively, and the more it is studied in detail the more it becomes evident that it was not uniform in character. The recent and ongoing transitions occurring in the Third World take varied forms, and do not seem to have had a common sequence or causation. However, a general downward trend in both fertility and mortality is apparent, even though the pace of change seems to be highly variable from country to country.

Economic development is undoubtedly an important factor in reducing fertility; improved economic status and higher standards of living encourage people to have fewer children. Third World countries which have substantially reduced their population growth have also experienced economic development, with increased incomes for considerable proportions of their peoples. Extending economic benefit, especially to the poor, seems most important in changing attitudes and practices. Economic development can also bring improvements in public health, financing curative and preventive medicine on a large scale. As a result, mortality rates decline, but most importantly there is reduction in infant mortality through the better care of mothers and children. Parents are then more likely to accept that large numbers of children are not necessary to ensure that sufficient survive to provide for family needs. In Africa, substantial reductions in mortality, and especially infant and child mortality, in the last two decades have taken place despite disappointing, if not negative, economic change. While rising incomes improve child health through improved nutrition and other service provision, clearly rising national or household incomes are not a necessary requirement for mortality decline if there are also parallel health improvements.

Economic development also allows levels of education to be raised; improvements in literacy and progressively higher standards of attainment are influential in reducing family size. They are particularly important for mothers, and raising the enrolment rate of girls in schools has been seen throughout the Third World to be a major strategy to strengthen the demand for smaller families. Women gain new perceptions of their own economic opportunities and appreciate that one factor in realizing these is to have fewer dependent children. Better educated women are better able to understand the methods and techniques of birth control, and can appreciate that it is not sufficient to follow the exhortation seen on a wall in India some years ago to 'Practise family planning this weekend'. However, in many Third World countries women have traditionally occupied positions of social and economic inferiority to men. Advances in education have come more slowly to women and they lag behind seriously in literacy and in other respects. This situation needs to be changed radically and rapidly, not only for greater acceptance of family planning but also for women's rights to equality. It has to be seen as part of a broader set of measures enhancing the status of women, by legislation and through practice, for example in improving women's access to paid employment.

Clearly a great many factors operate in complex relationships to determine the control of population numbers. Understanding why control is necessary is essential for the widespread acceptance and practice of family planning. Couples cannot be forced to accept and to practise family limitation by whatever means. Greater attention has to be given to these matters as they are determined and interpreted within the Third World. More needs to be known of the attitudes and feelings of countries, ethnic groups, families and of men and women separately towards these matters. Suggestions of direction or dictation from the more developed parts of the world on matters so delicate and personal are likely to be rejected. Nevertheless, according to the International Planned Parenthood Federation, between 200 million and 300 million couples in the Third World now use modern means of contraception, but before the end of the century there will be 900 million couples of child-bearing ages who will require access to these.

VIEWPOINTS ON POPULATION

Until two decades ago the main perspective on Third World populations was towards the reduction of numbers to lessen the threat of population out-stripping resources, a threat first enunciated by the Rev. Thomas Malthus at the end of the eighteenth century. He and neo-Malthusians who have developed his ideas in recent years have argued that populations need to control their growth to achieve an equilibrium with resources available, notably food. Adjustment to equilibrium must come about either as a result of 'positive checks' (increasing mortality) or through 'preventive checks' (reducing fertility). Clearly the former is morally unacceptable and ought to be avoided;

hence the emphasis on the introduction of fertility control through family planning and on the need for this on a very large scale. This Malthusian approach is based on the presumption that population growth is a, or even *the*, prime cause of poverty.

Within the last two decades the views of Third World countries on population and its related problems have become much more widely known. This is partly due to their increasing contribution to debates on a range of global issues, focused in a series of international meetings organized by the UN concerned with the environment, habitat and women in particular and with population in general. World Population Conferences were held under United Nations auspices at Bucharest in 1974, Mexico City in 1984 and Cairo in 1994, with countries represented at a political level. A consistent argument at all three conferences has been that the main problem is not so much numbers of people but their poverty, the need to alleviate this and the means to do so. Population is identified as a symptom rather than a cause of the problem. By alleviating poverty and attaining higher standards of living for a greater number of people, the prospects for reducing numbers through lower fertility would be greatly enhanced: 'Development is the best contraceptive!'

A majority of Third World governments supported the revised UN World Plan of Action on Population at the 1994 Cairo Conference. This offers new directions of thinking and action on population that link direct and indirect policies. Countries that wish to reduce fertility are urged to implement development programmes and educational and health strategies which, since they contribute to higher standards of living, have a decisive but indirect impact upon demographic trends, including fertility. They are also urged to intervene directly through promoting developments which could increase the demand for fertility reduction, including reduction in infant and child mortality through improved nutrition and care, greater involvement of women in socio-economic development, protection for children and the aged, and wider educational opportunities. Specific family planning programmes need to be integrated with these developments. International cooperation is called for to give priority to assisting national efforts in order that these programmes and strategies be carried into effect.

The complexities of the population issues have been brought to the fore with the realization that there is not one population problem but many. Third World countries with specific population policies feel that these various problems have been recognized and that the perspective should be widened to include other matters besides control over numbers. The extension of control through family planning remains of prime importance, though with the need for there to be greater concern for human rights and family welfare in the strategies employed.

The more developed parts of the world are expected to continue supporting population programmes, but in the wider context of 'aid for development' rather than in the previously narrower context of 'aid to limit population'. What undoubtedly is needed is a major redistribution of resources on a global scale and this was elaborated in the work of the Brandt Commission (1980) which proposed a programme for a major transfer of resources from richer to poorer countries to contribute to breaking 'the vicious circle between poverty and high birth rates'. But action to achieve such a transfer is difficult to effect and the political will in developed countries to do so is weak. It is difficult to be sanguine about the possibilities of solving the population problems of the Third World in the foreseeable future.

It is encouraging that the relationships between environment, resources and people are now, and especially since the UNCED Rio Conference in 1992, receiving more attention in all parts of the world than they have in the past. In the run-up to the conference Third World countries wanted the population question to be given low priority on the agenda, on the grounds that population growth was to be seen as a symptom not a cause of local environmental deterioration. By contrast, First World governments wanted population placed high on the agenda, since it could then be seen as an important contributory factor in the global and local environmental crises and tackled as such. In practice both views prevailed, and population was considered by all as one of several factors that required direct consideration by governments, with international support for sensitive population policies set in a broad development context.

Population issues are also central to global economic development. At the heart of the need for economic development is the issue of jobs and employment creation in the Third World. With rapid population growth over several decades the numbers of people, both male and female, in the economically active age groups (aged 15–60 years) are large and growing. They need to be able to sustain a livelihood for themselves and their families, either in rural areas or, often after migration, in urban areas. Where fertility limitation programmes have been successful the number of job seekers will be falling. As a result of the relative success in fertility reduction in China compared with India from the mid-1970s, the number of new job seekers in the 1990s is now less in China than it is in India, even though the total population is nearly 300 million more. However, the style and substance of recent employment creation in rural and urban areas in the Third World has tended to emphasize technological change and increasing capital investment (see Chapters 5, 6 and 7 below), so that the number of jobs per unit of output has been falling.

The jobs question is not simply a matter of the numbers and the distribution of new employment opportunities, though these are clearly important, but also the structure of these opportunities – who the jobs are for: skilled or unskilled, educated or uneducated, and particularly it affects the gender aspects of the employment question (Plate 3.8). Women, in the past less likely to be in formal or even informal employment, are now seeking and being offered a more direct share in the labour market. Figure 3.7 identifies the proportion of women recorded in the national labour force in 1992, a proportion that is relatively high in the developed countries and in Latin America, but conspicuously low in Middle Eastern countries, reflecting cultural as well as economic differences. The high levels recorded in some parts of Africa

Plate 3.8 Women street traders, Oaxaca, Mexico Women of a range of ages sell small quantities of flowers and vegetables in a street market

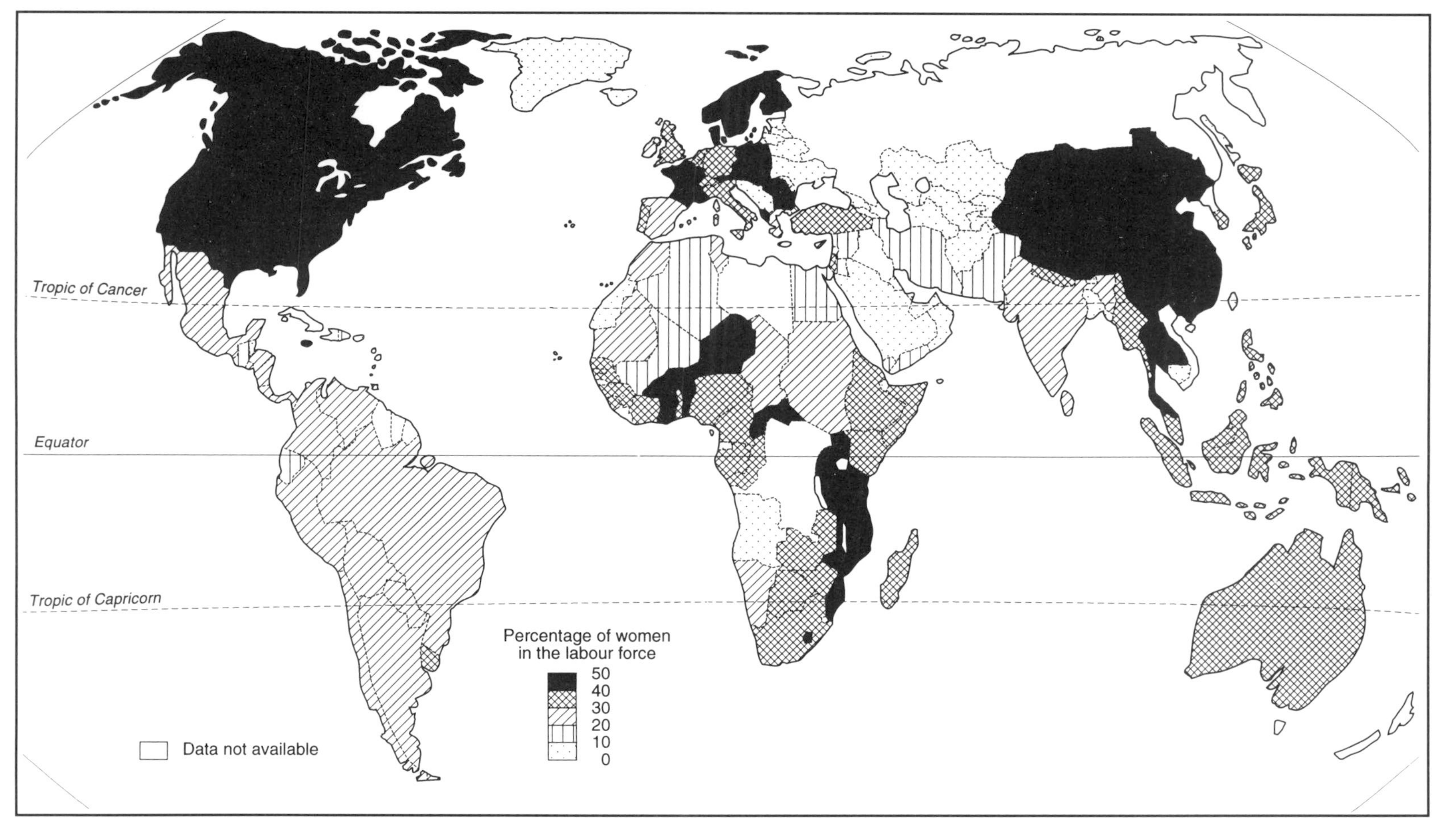

Figure 3.7 Women in the labour force, 1992
Source of data: World Bank, 1994a

reflect the traditional role of women in agriculture, together with the impact of male migration, leaving women behind to carry out farm tasks (see Box 5.1). Everywhere, however, the proportion of female employment has been growing and is likely to continue to grow as the economic status of women is enhanced. More women have the educational qualifications and the wish to work outside the home, and these are enhanced where family size is smaller and less of women's time is of necessity devoted to domestic and family responsibilities. The growth of women's employment has been particularly important in the new electronics industries in the Asian NICs, and women are strongly represented in the urban informal sector throughout the Third World, especially in domestic service. With a greater concern for increasing the numbers of women employed and their status within the labour market, countries need to develop even more rapidly than if the gender proportions were to remain unchanged.

Better appreciation of the problems of population in these wider contexts links with greater direct concern for and understanding of the questions of population growth and distribution, which in turn may lead to greater pressure upon national and international authorities for action. The problems of population in the Third World are a global issue. They must remain so. While we write of achieving balance between population and resources we must not lose sight of the fact that people are the greatest of all resources. The numbers of people in Third World countries, their characteristics and qualities, their distribution and movements are of crucial importance. Without people other resources have no meaning.

SUMMARY

Population is a Third World issue. There are high rates of population growth, though these have been falling in most parts of the Third World – most rapidly in Asia, where there is rapid economic development coupled with vigorous involvement of government in family planning programmes; least rapidly in Africa, where there are still low levels of development that allow children to be economic assets to the family from an early age. Governments have not normally been politically committed to family planning and do not command the resources or the moral authority to significantly affect fertility behaviour. But population cannot be the primary cause of increasing poverty and environmental degradation. There are certainly economic and environmental benefits arising from slowing the rate of population growth – more control of local environmental pressures, fewer job seekers, fewer children to be educated and to be provided with health care – but lower rates of population growth will not in themselves greatly stimulate development. Measures to facilitate fertility reduction need, in particular, to be sensitive to the needs and wishes of women. Better educational and employment opportunities for women and generally enhancing their economic and social status have been shown to be associated with fertility reduction throughout the Third World.

Population redistribution through migration is closely related to development, relieving population pressure in some overpopulated regions and providing hands to work in new economic activities in rural and urban areas. Since migrants tend to be better educated and more responsive to economic opportunities, migration systems at the national and international level reflect economic differentials between source and destination of the migrants. They also reflect the economic relationships between source and destination that contribute to sustaining and even widening these differentials. Migration and redistribution are symptoms of the patterns and processes of economic and environmental pressures within the Third World.

At a global scale, however, this discussion of population has also shed light on the broader questions of the relationships between the First World and the Third World. The conceptualization of population growth as necessarily a 'problem', a constraint on development that needs immediate and direct attention by government, is a view that is most prevalent in developed countries. Governments and individuals in Third World countries are much more likely to view population change as an opportunity rather than a threat, to emphasize the positive and longer-term effects of changing population structures and distributions and rising skill levels – not so much more mouths to feed as more hands to work or even more minds to be applied to development – rather than the negative and, it is hoped, rather shorter-term effects of recent population growth on the declining per capita availability of income, food and jobs.

4

Farming Systems and Agricultural Production

Agriculture is still the basic concern of most people in the Third World, whether as landowners, tenants, smallholders, peasant farmers producing largely for themselves, or as landless labour. It is crucial to the prospects of development. Some countries have made great strides towards industrial development; some have been enriched by the possession of mineral resources, especially oil, with strategic value on the world market. But most are heavily dependent on agriculture not only for their basic food supplies but also for exports with which to finance essential imports. Many farming families consume the greater part of the food they produce, but as towns and cities have grown, production for the urban market has become of increasing importance. In terms of national aggregates, food production in many countries is as sufficient to feed growing numbers as it was 30 or 40 years ago. However, this is by no means true of all countries, and in most of the developing world, poverty still denies many people access to enough to eat. Hunger and famine in Ethiopia, the Sudan and Somalia are but the most recent poignant examples of how close so many rural people and so many rural production systems are to the very margins of survival.

The predominance of agriculture and the rural sector in the Third World as a whole may be seen in Table 4.1 by the fact that 61 per cent of the working population is still employed in agriculture, a proportion that has fallen from 66 per cent in 1970. There are, however, very considerable variations from one area to another. Employment in Sub-Saharan Africa, the Far East, South and South-East Asia was strongly agricultural in 1970 and there has been

Table 4.1 Agriculture in the Third World, 1965–90

	Employment in agriculture as % of total		*Rural population % of total*		*Share of agriculture in total GDP*		*Food and raw materials as % of exports*	
	1970	*1990*	*1965*	*1990*	*1965*	*1990*	*1965*	*1990*
Sub-Saharan Africa	77	72	86	71	40	32	70	29
Far East	67	64	81	50	37	21	48	18
South and SE Asia	70	66	82	74	44	33	57	24
M East and N Afr.	61	40	65	49	20	—	24	12
Latin America	41	27	47	29	16	10	48	29
Third World	66	61	76	56	29	17	42	20

Sources of data: FAO, 1988, 78–9; World Bank, 1992, 222, 248

◀ Women harvesting rice, Vietnam

only a relatively small drop to 1990 in percentage terms, though the total numbers in agriculture have grown substantially over the period. In Latin America and the Middle East, economies are not so dependent on agricultural employment, even less than they were in 1970. In both of these regions, the *total numbers* employed in agriculture have not increased greatly (by only 8 per cent in the Middle East and by 12 per cent in Latin America between 1970 and 1990). These regional differences broadly correspond to differences in GNP per head (see Figure 1.2). In general, the poorer the country, the greater its dependence on agriculture, with relatively few opportunities for other forms of employment. But the differences within the Third World are less striking than the contrast with the developed world, where agriculture occupies on average only 6 per cent of the total employed, falling to as little as 2.5 per cent in the USA and 2.1 per cent in the UK.

A second major feature of Third World agriculture as a whole is the low productivity of labour. Crop yields may be of the greatest importance where there is pressure of population on resources, and peasant farmers sometimes show considerable expertise in coaxing high yields from small plots of land with little equipment, but only at the cost of intensive labour. In general, it is the productivity of agricultural labour that determines farm incomes, and in this respect the contrast between the Third World and advanced economies is very striking. Table 4.2 shows, for a sample of countries, the value added in agriculture per head of the population employed in agriculture, and may be taken as an approximate guide to the contrasts, not only with the developed world, but also within the Third World.

It is very clear that the productivity of labour in agriculture corresponds quite closely with average values of GNP per head in Latin America; only Ecuador and Bolivia, among the countries listed, have a GNP per head of less than US $1,000. All of the African countries in the table fall below this level, and in the Asian list only Turkey, the Philippines and Thailand have a GNP per head over $1000.

Many reasons may account for the low productivity of agricultural labour. The major factors may be grouped under two headings: lack of capital and the lack of land. Lack of adequate land may be a result of the pressure of population, leading to the fragmentation of farms, or it may be the result of encroachment by large estates, but small farms are also the result of the simple fact that there are severe limits to the amount of land that peasant households can cultivate with the tools and labour at their disposal. Thus low productivity is also the result of a lack of capital for the purchase of better implements and seed, or farm animals, fertilizers, storage facilities, pest control, irrigation and water control (Plates 4.1 and 4.2). Low productivity implies a low standard of living and the impossibility of accumulating savings and therefore of making investments in improvements. It often implies malnutrition, which itself limits productivity. This situation

Table 4.2 Productivity of labour in agriculture, 1990. Value added in agriculture per head of economically active population in agriculture 1990 (current US $)

Developed world		*Latin America*		*Asia*		*Africa*	
USA	29,369*	Argentina	10,143	Turkey	1,492	Morocco	1,424
Australia	25,247*	Venezuela	3,491	Philippines	948	Egypt	1,002
France	23,028	Brazil	3,127	Indonesia	680	Nigeria	485
UK	18,102	Mexico	2,300	Sri Lanka	592	Kenya	295
		Ecuador	1,451	Pakistan	541	Zimbabwe	275
		Bolivia	1,162	Thailand	533	Ethiopia	157
				Bangladesh	393	Tanzania	146

Note: *Data for 1989
Source of data: World Bank, 1992, 224

Simple highland agriculture

Plate 4.1 Reaping barley, Lake Titicaca, Peru In the high Andes men and women collaborate in harvesting a meagre crop of barley by hand

Plate 4.2 Men winnowing wheat, Nepal Men use the simple device of beating wheat stalks against flat stones to separate grain and straw. In the background can be seen terraces without boundary walls

has long been identified as 'the vicious circle of poverty' from which it is so difficult to escape.

A third major characteristic of Third World agriculture results from the changing balance of urban and rural sectors, and the rapid growth of towns and cities. In the past rural communities lived more or less self-contained lives, remote from urban concerns and producing largely for themselves, but the growth of towns has created a market for both the produce of the countryside and for displaced rural labour. Urbanization has, of necessity, created an irreversible move towards the commercialization of subsistence farming. Although the data may be uneven because of variations from one country to another in the ways in which rural and urban populations are defined, the reduction in the share of rural population has been quite striking. Among the major regions, it is only in Latin America, however, that the trend has gone so far as to reduce the rural population (29 per cent) well below a half of the total numbers (see Table 4.1). Nevertheless, in spite of the rapid growth of towns and cities, most people still live in the countryside, often quite remote from basic urban facilities.

A fourth major feature has always been the reliance of many Third World countries on exports of foodstuffs and raw materials of agricultural origin. It is this fact that has been seen by many authorities as one of the basic reasons for slow and halting progress in development. Prices for agricultural produce on world markets have generally tended to be vulnerable to great fluctuations and to have risen far less than the prices of manufactured imports. Exports of agricultural products are still fundamental to the Third World, though less so now than was formerly the case. From 1965 to 1990 they have fallen from 42 per cent to only 20 per cent of total Third World exports, but again there are considerable differences from one major region to another, and even more striking differences among individual countries. Although oil has been largely responsible for the emancipation of the Middle East and North Africa from agricultural dependence, and in some parts of Latin America, the Far East and South-East Asia the expansion of trade in manufactured goods has reduced dependence on agriculture, for many countries agriculture remains almost their sole source of export earnings, and must not only provide the revenue to pay for imported goods, but also for the interest on foreign loans.

Over half of the countries in the developing world relied on agricultural products for 80 per cent or more of their exports in 1965, but by 1990 this number had fallen to about a quarter (see Figure 1.9). It is true that there has been a substantial rise in the exports of manufactured goods from many other countries besides the Newly Industrialized Countries of the Far East and Brazil, but in many cases the raw materials for manufactured exports are still derived from local agriculture. Value is added to agricultural produce by processing, which puts the final products, such as canned and frozen foods or semi-finished raw materials, into an industrial category in the export statistics, but does not reduce the importance of agriculture as the basic generator of economic wealth. To a certain extent, the same comment may be made of the share of agriculture in GDP, which has also declined substantially from 29 to 17 per cent between 1965 and 1990 in the developing world as a whole, and is now as little as 10 per

This Christian Aid cartoon contrasts Third World and First World food issues, with commodities important in the First World originating in the Third, but not easily available there

cent in Latin America, but it is still responsible for generating a third of the output of Sub-Saharan Africa and South Asia.

THE VOLUME AND COMPOSITION OF AGRICULTURAL PRODUCTION

Information about food and agricultural production is published annually by the Food and Agricultural Organization (FAO). It is based on returns to FAO from each country which are analysed in various ways to estimate the physical output of crops and livestock, weighted by local price differentials and combined to produce indices of agricultural and food production from year to year. It should be stressed, however, that national returns, especially from the poorer Third World countries, are not always accurate, not least because governments cannot afford to maintain elaborate data collection services, so that production figures, areas cultivated and yields are based on very general estimates. Moreover, definitions of different categories may vary and sometimes change from one year to another.

Figure 4.1 shows the volume of agricultural production by major sectors of production and for five major world divisions as well as for the Third World as a whole. Three-quarters of the volume of agricultural production is of staple foodstuffs providing carbohydrates and vegetable protein in the form of cereals, root crops and pulses. Only in Latin America is their share significantly lower (at 50 per cent), with a higher proportion of fruits, nuts, oil crops, sugar, coffee and other specialized crops. In general, this is indicative of a deeper commercialization of production in Latin America. It is also notable that animal products make up some 18 per cent of Latin America's volume of production, reflecting not only a contribution to international trade, as for example from Argentina and Brazil, but also a higher standard of living than in other parts of the Third World, where animal products generally make up no more than about 10 per cent of the volume of production.

There are, as is to be expected, great variations from one region to another. Table 4.3 gives information about the areas harvested of specific crops as percentages of the total land use in the crops listed. The areas in permanent tree crops such as coffee, rubber, etc., are not available, and the relatively small areas in green vegetables and fruits are omitted. The regional breakdown is more detailed than that which is given for the volume of production.

The choice of specific cereals and root crops is obviously adjusted to environmental conditions. Wheat is the major staple of most temperate and sub-tropical regions – in northern China and north India, Pakistan, the Middle East and North Africa (with barley as the other major staple), and is important in Mexico and the southern parts of Latin America. Rice is the major staple in more humid, tropical regions such as South and South-East Asia and the Far East, and is of local importance in Latin America and Africa. Maize, originally domesticated in the New World, is of particular importance in Latin America, but has also spread widely wherever climatic conditions permit. Millets and sorghum, with drought-resistant qualities, are basic subsistence crops in areas of highly seasonal and often highly variable rainfall in South Asia and the Middle East and in the savanna lands of West Africa and the Sudan. Among root crops, cassava (manioc) is adapted to hot and humid climates and is the major staple in parts of the Amazon basin, and in Central and West Africa (where yams share primacy of place among the root crops), and throughout the Pacific.

The choice of food crops reflects many other factors besides adaptation to environmental conditions of climate, soils and water supply. Food preferences change slowly, and agricultural practices associated with specific crops may involve long-standing skills and customs. Culture, religion, culinary traditions, habit and custom are all deeply bound up with diet and therefore with the choice of crops or livestock. Rice has a symbolic value in much of the Far East, just as maize does in Mexico. Pork and wine are anathema to followers of Islam and the cow is still sacred in parts of India. The culinary traditions of China, India or Mexico, no less than those of Western Europe or North America, reflect and also reinforce agricultural

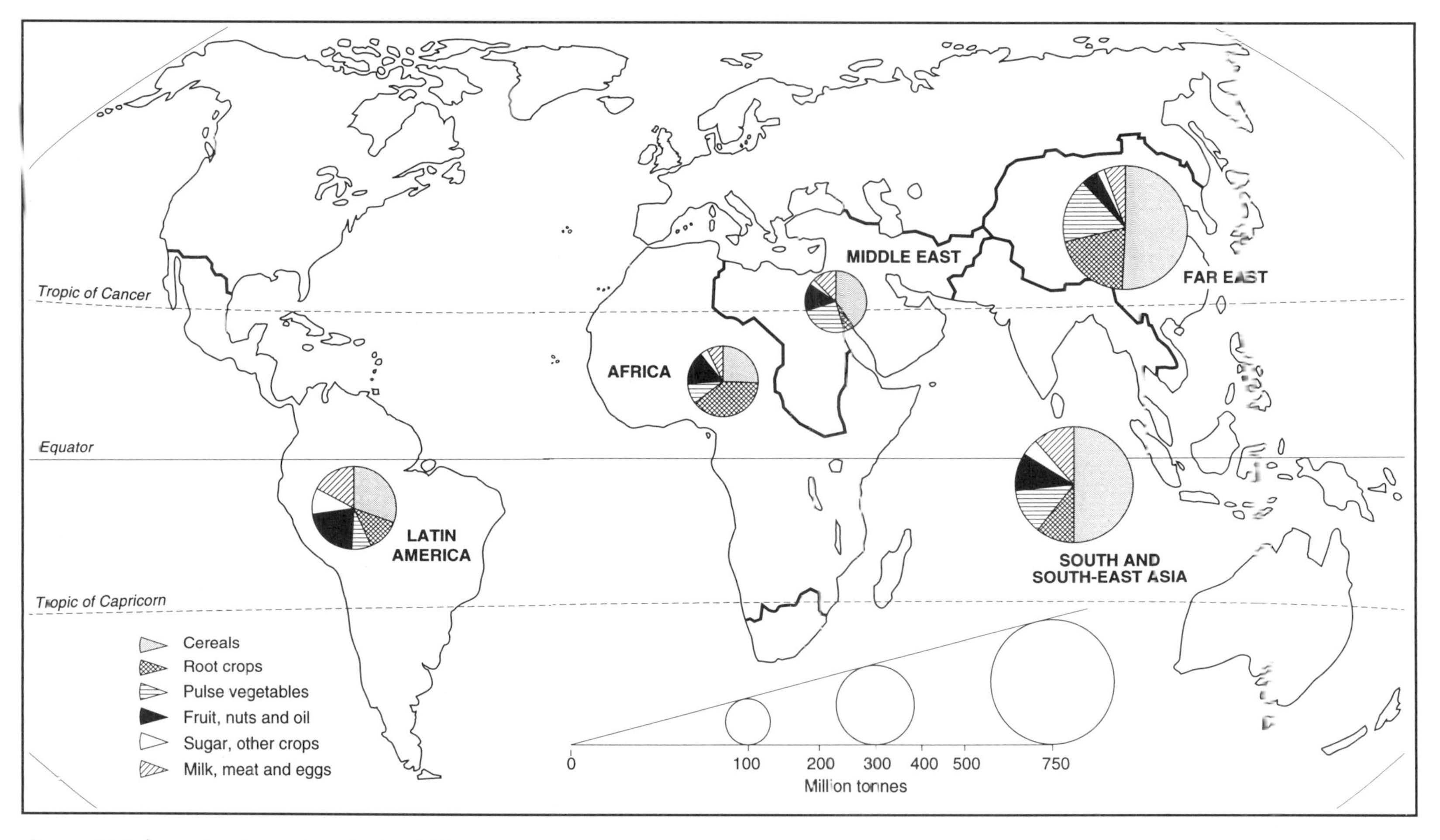

Figure 4.1 Volume of agricultural production, 1988
Source of data: FAO, 1988

Table 4.3 Percentages of total area harvested in specific crops, 1988

Region	*Total million ha*	*Wheat*	*Rice*	*Maize*	*Other cereals*	*Roots*	*Pulse and soya*	*Oilseeds*	*Cotton*	*Sugar*
Africa										
N and Sahel	26.7	19.0	2.3	5.8	45.3	1.7	12.5	8.8	3.9	0.8
W and Central	51.0	0.7	5.6	7.3	48.1	18.1	9.1	7.8	2.6	0.7
E and S	30.0	2.4	6.6	36.9	17.0	14.0	12.0	5.7	4.3	1.2
Asia										
Middle E	39.9	54.7	2.1	3.1	26.1	1.0	7.8	2.8	2.3	0.1
South	183.7	16.6	28.8	3.7	18.5	0.9	14.0	9.7	5.5	2.3
SE	60.1	0.2	61.4	15.9	0.9	7.7	6.5	4.7	0.5	2.2
Far East	163.9	21.8	25.9	15.1	7.3	7.1	10.0	7.8	4.1	0.8
Latin America										
Southern S America	19.3	27.8	1.1	13.5	9.9	1.3	25.1	16.8	2.5	1.6
Eastern	51.3	7.0	11.7	26.8	0.9	4.5	34.0	1.1	5.8	8.2
Andean	8.2	3.2	14.1	29.8	12.4	13.3	10.1	3.0	5.6	8.6
C America and Caribbean	20.2	4.6	3.3	43.6	13.1	2.6	16.0	3.0	17	12.1
Third World	620.3	16.0	22.7	13.3	16.6	5.6	13.6	5.5	4.0	2.7

Note: 'Other cereals' are mainly barley in North Africa and the Middle East, and mainly millets and sorghum in the rest of Africa and South Asia.
Source of data: FAO, 1988, 105–9

preferences. Such linkages are strongest, of course, where production is for subsistence or for local markets, as it often is in the Third World.

Patterns of food preferences, and the farming associated with them, have been regarded as quite resistant to change, but this resistance can be easily exaggerated. In the New World Indian farmers quickly adopted European crops and livestock where there were clear advantages or where there was an easy fit with traditional methods. American crops, especially maize, manioc and potatoes were widely adopted in the Old World. In many parts of tropical Africa, maize and manioc (cassava) form a staple of many diets. Today, one very important factor making for change is the tendency to adopt foods and diets which are given prestige by their association with North American or Western European usage, and which are strongly promoted in the interests of transnational corporations. A prestige consumption of wheat in the form of bread or pasta in place of traditional diet, processed meats and dairy products, may often lead to imports of food from the industrial world at the expense of local production. Cigarettes, soft drinks and imported processed beef products symbolize a seductive association and are heavily promoted by advertising.

Among other factors influencing the choice of crops and affecting attitudes towards change, the pressure of population may put a premium on crops of high nutritional value which yield well in terms of calories per hectare; rice is often preferred to cassava on such grounds, for example, even though it requires much more labour. Strategies of crop production may need to trade off risks due to climatic variability against maximizing the production of food in a good year. Thus a balance may need to be struck between high yielding crops such as rice, which may be liable to fail if rainfall is late or insufficient in a given year, and crops such as millets or cassava which are lower yielding, but can withstand greater climatic variation. Finally, and of rapidly increasing importance, are the commercial pressures exerted through price structures, access to national or international markets, international food aid, government policies and the introduction of new crops or varieties of crops.

Most of the cereals, root crops, pulses and vegetables are for domestic consumption within the Third World countries themselves and often by the farmers and in the villages which produce them, but some crops are more strongly associated with national and international commerce. Soya beans are a commercial crop of substantial importance in some of the wealthier Third World countries. Brazil is the second largest world producer of soya beans after the USA and is ahead of China, where soya is an important and traditional subsistence crop. Argentina, where soya beans make up some 25 per cent of the volume of production, is the fourth largest producer. Oilseeds, cotton and sugar enter into commerce far more than most of the cereals and pulses. Sunflower and linseed (for vegetable oils) make up some 10 per cent of Argentina's agricultural output; the Sudan has a high proportion of its arable land devoted to sesame seeds, and both countries export vegetable oils. Although India, Brazil, Pakistan and China are the leading Third World producers of cotton, it is widely grown in Africa and Latin America both to supply local textile industries and the international market. Sugar dominates the land use of the Caribbean region, where it occupies about a quarter of land in crops, and the region is, of course, of great importance in the international sugar trade.

A significant proportion of cultivated land is devoted to tree and shrub crops, especially in the humid tropics. Best known is the highly commercialized plantation production of such crops as rubber, tea, coffee, cocoa and bananas. Although, in general, they occupy a relatively small proportion of the land under cultivation and of the volume of production, they are of much greater value per tonne. Coffee occupies less than 4 per cent of cultivated land in Brazil; cocoa less than 6 per cent in West Africa; and tea only about 14 per cent in Sri Lanka. Production of plantation products for export tends to be highly localized where there is easy access to ports, railways and good roads, and in some small countries may dominate the pattern of land use, as sugar does in Cuba. While the output of such crops has a significance extending far beyond the proportion of land they take up, it is important to recognize that, as with the production of cereals, a large part of the output from tree and shrub crops goes towards domestic consumption within the Third World. Bananas, for example, are commercially important export crops in Central America, parts of the Caribbean and in Colombia and Ecuador, but, as plantains used in cooking, they are also important in the internal consumption patterns of tropical Africa and in much of Latin America. Other fruits, such as mangoes, avocados, papayas (pawpaws) and many others provide important elements of diet and are an integral part of village agriculture.

The livestock industries tend to be characterized by high value per tonne, but a low share of the total volume of output and an even lower output per hectare of cultivated land. There are obvious problems in summing together the volume of livestock products such as meat, hides, wool, milk, eggs, and there are equally great problems in attempting to gauge the importance of livestock in agricultural production by adding together the numbers of sheep and cattle or pigs. This is also a sphere in which national or FAO estimates of production are probably the least reliable. However, an attempt is made in Figure 4.2 to show the distribution of 'livestock units' in relation to the population engaged in agriculture by using well-established equivalents, which must, of necessity, be very approximate measures. It is clear that livestock is most important in Latin America and least important in the Far East and in Africa. In spite of those parts of China such as Tibet and Inner Mongolia where livestock is the basis of subsistence and the Chinese predilection for pork and poultry, the Far East makes a poor showing in the volume of production and in the number of livestock units. In Africa arid and semi-arid regions can often be used only for extensive grazing, and tsetse fly precludes livestock production from large areas. But for a very large number of people in Africa and in Monsoon Asia either poverty or excessively small holdings mean that most of the protein consumed must be of vegetable origin. Indeed, the consumption of meat, milk and milk products is clearly related to levels of income. The contrast is very clear between the volume of meat, milk and egg production in the developed world, which amounts to some 28 per cent of the total volume of agricultural production, and the general level in the developing world, which is only 10 per cent, rising to higher values in Latin America and the Middle East.

FARMING SYSTEMS AND THE ENVIRONMENT

The outline of agricultural production which emerges from the FAO statistical series gives very little insight into the ways in which Third World farming households organize their productive activities with the modest means at their disposal. It is difficult to make sensible generalizations, for there are great variations from one area to another or from one farm to another. Complex realities have to be simplified and classified, and this can be done in various ways:

- by relating agriculture to environmental types, such as the farming of tropical forests, savanna lands or arid and semi-arid regions;
- in a regional and cultural context, contrasting India with China or Latin America;
- through the scale of landholding from the small peasant farm to the plantation, the ranch and the large estate;
- by way of the apparent, but often misleading, dichotomy between subsistence and commercial farming;
- or along a gradient from 'traditional' to 'modern', depending on the degree to which modern technologies are used.

Some of these issues will be discussed later in this chapter and in Chapter 5, but it may be useful at this point, bearing in mind that some 80 per cent of the farm families of the Third World may be regarded as small-scale peasant farmers, to look more closely at some typical farming patterns and their relationships to environmental conditions. Traditional farming systems are often quite finely tuned to achieve an adequate productive level from local resources, while at the same time maintaining an ecological equilibrium with the environment. Under modern conditions, it is often this equilibrium that comes most severely under pressure.

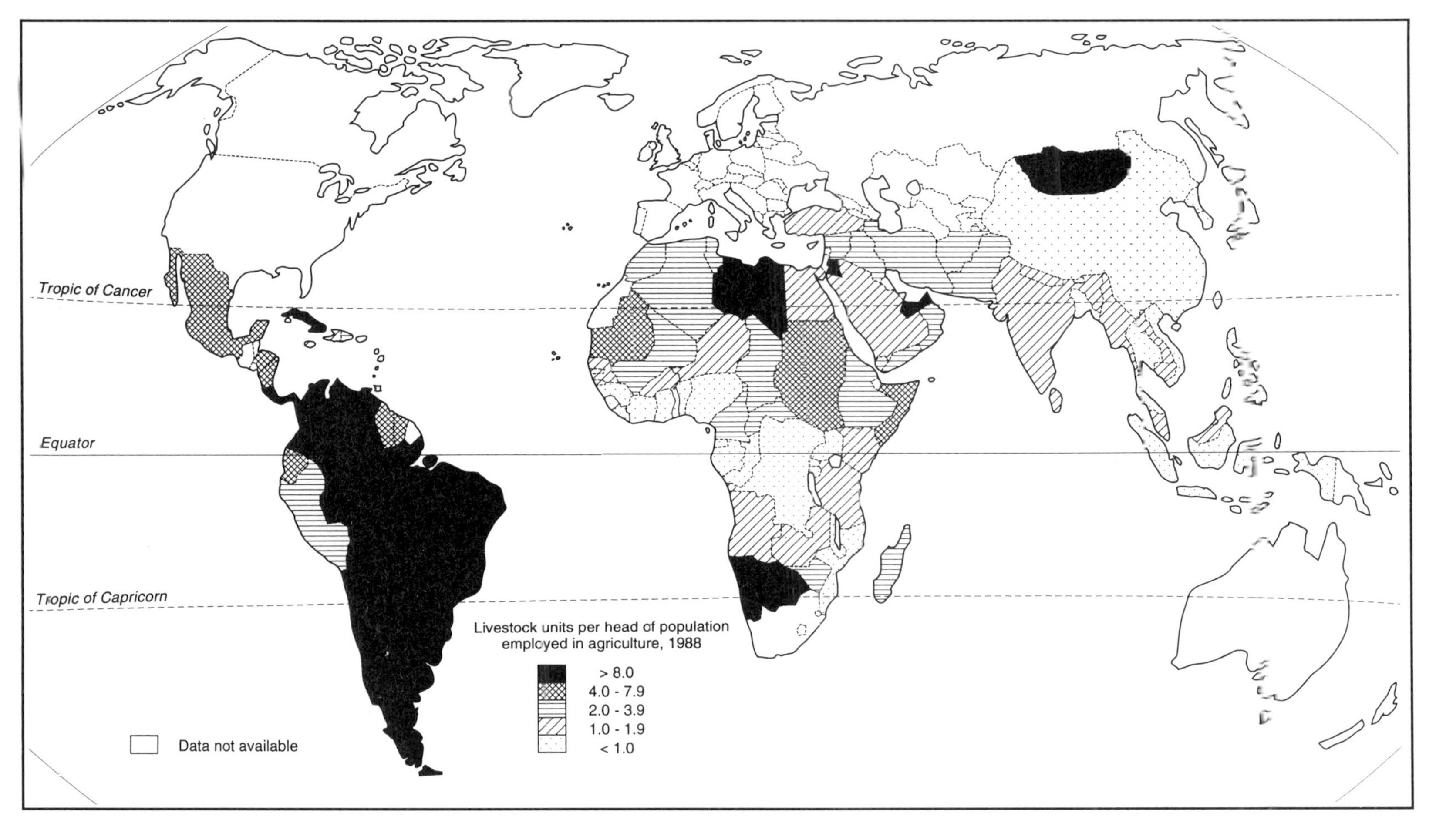

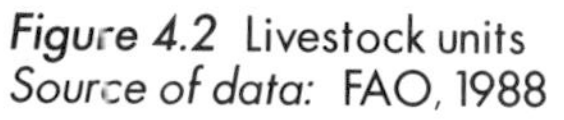

Figure 4.2 Livestock units
Source of data: FAO, 1988

Shifting agriculture and bush fallowing

Shifting agriculture in various forms occurs mainly in tropical rainforest and savanna grasslands. It is sometimes known as 'swidden' farming or 'slash and burn', but there are many local names: *conuco* in Venezuela, *roça* in Brazil, *chitimene* in Zambia, *milpa* in Central America and *ladang* in Indonesia, for example. The essential feature is the cultivation for a few years of clearings made in the forest which are then allowed to revert to secondary vegetation for a time while new clearings are successively opened up. The ratio between the period of cropping and the length of the fallow period during which vegetation is allowed to regenerate varies a great deal. In the rainforests of Central Africa one or two years of crops may be followed by 8 to 15 or more years of fallow; in semi-deciduous and dry forest zones 2 to 4 years of crops can be followed by 6 to 12 or more years of fallow. There is, however, a great deal of regional and local variation, ranging from *ad hoc* and irregular movement of individual farmers from one clearing to another to complex arrangements of clearing and fallow in long period cycles which may last up to 45 years. Long-period shifting cultivation was formerly widespread in West and Central Africa, and in parts of South-East Asia, Central America and the Amazon basin.

Shortages of land and pressures of population have led in many areas to change, and particularly to shorter lengths of fallow. So great is the variation that various types of shifting cultivation have been identified; forest fallow implies a lapse of 10 years or more between periods of cultivation; a shorter fallow of 3 to 4 years, too short for the establishment of shrubs or trees is sometimes known as grass fallow, and the same term may be applied to long fallows in drier areas. Bush fallow generally implies a fallow period of less than 10 years. Long-term shifting agriculture has effectively been replaced by bush fallow over most of West Africa and the original rainforest is practically destroyed. No more than remnants of former shifting cultivation remain in Latin America, and in parts of South-East Asia (e.g. in some states of Malaysia) it is illegal.

Clearings are almost invariably made by hand methods in much of Africa, though chainsaws are increasingly being used in Latin America; the larger tree stumps are left, and branches and forest debris are normally burnt, especially where there is a drier season. In Zambia woody material is gathered from other areas to burn on the clearing and thus increase the ash content. Cut vegetation is sometimes used as a mulch, as in western Colombia. Crops may be planted directly in the ash of burnt-over land or the ground prepared by hoe and digging stick. The use of the plough is relatively rare, especially in the African humid tropics.

It has frequently been observed that the system is adapted to a forest ecology and the fragility of tropical soils. Plant nutrients are concentrated in the vegetation rather than in the soils from which nutrients tend to be rapidly leached by heavy rainfall in combination with high temperatures. Clearing of the forest breaks the normal cycle of plant nutrients. The ash resulting from the burning of cut vegetation reduces soil acidity and makes some minerals immediately available for crop production, but fertility falls off rapidly and so do yields, so that in the third year of cultivation they may be only a half of what they were in the first year. The decline in fertility and crop yields is usually the main reason for the abandonment of a clearing and the move to a new one, but it has been noted (in Venezuela, for example) that the growth of weeds may make the opening of a new clearing less laborious than continued cultivation of the old one.

Cropping systems are often complex, and the initial impression may be one of chaos. Intercropping of different species is normal, and sites are chosen with a precise eye for differences in levels or drainage conditions. The variety of crops in a small area fulfils a number of functions: it ensures a varied diet and a phasing through the year of both harvests and labour demands; it helps to reduce pests and diseases; and it makes possible a complete ground cover at various levels of height so as to reduce the impact of rainfall (to some extent emulating the layered structure of natural vegetation) and thus reduce the risk of soil erosion. As well as temporary clearings, there may be intensively cultivated gardens near permanent settlements in which fruits, tree crops and specially valued crops may be grown with the help of manuring by household wastes.

Once considered as wasteful and backward, shifting agriculture with long fallow sufficient for the regeneration of forest has more recently been seen as an adequate response to the problem of maintaining soil fertility and conserving the resources of the environment in the humid tropics. Under optimum conditions it yields both subsistence and cash crops while allowing the regeneration of forest and the recovery of soil fertility. Furthermore, it is a system that provides adequate subsistence for a low investment of labour. Research in widely dispersed areas suggests labour inputs of no more than 120 to 200 days a year per hectare of cultivated crops. In contrast, traditional rice agriculture in China, without mechanization, used to require up to 500 days per hectare per year, yielding a lower quantity per man hour of grain or grain-equivalent. But as long-period forest fallows are reduced to bush fallow or grass fallow, so the need for labour increases with greater attention to cultivation of soils, weeding, the application of green manures, etc.

Shifting agriculture has shown considerable flexibility. New World crops such as cassava, maize, peppers, papaya and many others have long been incorporated into 'traditional' African agriculture, for example. Crops for subsistence are usually combined with production for sale, sometimes of vegetables and fruit for urban markets, sometimes of export crops such as cocoa, coffee, rubber and palm oil. Ghana's cocoa and Nigeria's success in palm oil were built on this combination of tree crops for cash and bush fallow or shifting cultivation for subsistence crops. To some extent a current equivalent is the combination of bush fallow in warm, humid montane areas of Bolivia, Peru and Colombia with the illicit cultivation of coca, from which the international drug trade is supplied, and which brings illegal wealth to peasants in remote areas.

However, in a changing world, shifting cultivation and bush fallow face two difficult problems, one associated with the growth of population and the other with modernization. Increase in population implies a progressive shortening of the fallow period. The cycle of recuperation by the growth of secondary vegetation becomes less complete, leading to a long-term fall in yields unless there are compensating additions of fertilizer or new seed strains. However, since cultivated land occupies a greater proportion of the available land, total production is normally increased, at a greater cost of labour and with declining yields per cropped hectare. A low-level equilibrium of soil fertility may be reached which can maintain a greater total output from the same amount of land under longer fallows, and can therefore support a modest increase in the density of population.

The modernization of farming under systems of shifting cultivation meets other obstacles. Clearing is usually done by cutting and burning, leaving tree stumps in the ground. Full mechanized clearing is expensive and far beyond the reach of most peasant farmers. Cultivation of cleared land is usually by hoe or digging stick and the presence of tree stumps and roots may preclude the use of the plough, even if peasant families were able to use draft animals. Weeding of intercropped fields and planting must be done mainly by hand. There are thus clear limits to the area that can be cultivated by a peasant family – normally no more than 2 or 3 hectares in much of Africa. Artificial fertilizers are expensive and less effective than in temperate climates. The intensification of production by the use of mechanized clearing and the plough is difficult, expensive and potentially dangerous from an ecological point of view.

Wet-rice farming

In contrast to shifting cultivation, wet-rice farming has for many centuries been able to support very high densities of population in humid tropical and sub-tropical areas, particularly in Monsoon Asia. It is associated with landscapes in which the hand of man is dominant: carefully levelled plots, surrounded by low water-retaining banks or bunds, banks of terraced slopes and, above all, detailed and unremitting attention to the control of water. With few exceptions, it is an agriculture conducted by

Plate 4.3 Rice and wheat terraces, northern Pakistan The small, level plots are bounded by low ridges to retain water and slightly stepped, so that water may flow from one level to the next. In the micro-scale of such agriculture, a vegetable is grown along the dividing ridges

peasant farmers at a relatively low level of technology combined with intensive labour and a high level of traditional skills (Plate 4.3).

The basic outlines are well known. In South-East Asia, for example, land is flooded, then cultivated by ploughs drawn by oxen or water buffaloes. Soils are puddled and the surface rendered smooth, ready for planting. Seedling rice may be grown in nurseries and transplanted by hand into the wet soil. Later the fields are drained and the harvest taken, most commonly by hand. One rice crop a year is the normal practice, but much depends on the availability of water, the length of the growing season and the amount of labour available. Three crops in two years have been noted in Sumatra, where rainfall is abundant throughout the year. Two rice crops a year, characteristic of southern China, require careful scheduling of labour and good water supply as well as abundant labour. Even three crops a year can be grown in highly favoured areas with rapidly maturing varieties and very intensive cultivation, e.g. in Java, Indonesia. In most areas, however, double cropping is achieved with other crops than rice, planted in rotation in order to maximize returns from the land. Throughout Monsoon Asia there is also a multiplicity of rice varieties, many of them quite local in their distribution, but adjusted to very varied conditions, particularly in relation to the length of the growing season or the depth of water in flooded fields.

Multiple cropping, in the sense that more than one crop a year is produced, is normal in rice-growing areas, rice alternating with other cereal crops, root crops such as sweet potatoes, beans or vegetables. In Taiwan, for example, intensive rotation is geared to seasonal weather and available irrigation water in a three-year system; rice is grown with benefit of summer rains and seasonal irrigation, and is followed by unirrigated sweet potatoes and wheat during the winter and spring. In the second summer, the same land is devoted to green manure crops, groundnuts and soya, relying on rainfall rather than irrigation water, followed by sugar for 15 months.

Intercropping, involving the growth of more than one crop at the same time, is also indicative of the intensity of land use. A second crop may be planted before the previous crop ripens; quick-growing crops such as green vegetables may be planted at the same time as crops that mature more slowly so as to extend the harvest period more evenly through the year. There are, therefore, many variations in cropping systems to make the best use of land and labour in the light of seasonal variations in climate and water supply.

The farming structure as a whole depends on the maintenance of fertility at an adequate level to support the heavy demands made on it by multiple cropping. The preoccupation of rice farmers in south and central China with the need to maintain soil fertility by recycling back to the land all possible wastes of vegetable and animal origin has long been well known. Vegetable wastes are composted; in some areas vegetable wastes are fed to pigs, raised both for their meat and their manure; human excreta are used (with consequent health risks); and green crops are raised for the sole purpose of ploughing them back into the ground to increase the organic content of the soil. There is a 'migration of fertility' towards the neighbourhood of towns where wastes from foodstuffs, animals and people are used to enhance the productivity of the land. Upland vegetation may be raided to make composts for the benefit of irrigated land, and the contrast in China between carefully attended wet-rice land and the neglect and erosion of the uplands is also evident.

Even where such painstaking husbandry is not practised, however, wet-rice cultivation maintains fertility at a high enough level to permit permanent cultivation without fallow. Flood waters and irrigation bring silt and soil nutrients to the cultivated fields; blue-green algae on flooded rice fields fix nitrogen from the air; and soil leaching is prevented by the impermeability of puddled, worked soils tending towards neutral pH values. Yet it is also evident that rice yields respond well to the application of organic and mineral fertilizers.

The labour requirements of rice cultivation are much higher than those for shifting cultivation even though draught animals are used. Moreover, the returns from an increase in the intensity of labour (weeding, transplanting, application of manures, detailed attention to irrigation and drainage) tend to be higher than in other forms of tropical agriculture. The point at which additional labour produces little or no return in yields occurs at a very high intensity of labour. Thus, rice cultivation can both feed and employ high densities of rural population, as it does in much of India, south and central China, Java, the Tonkin delta and other parts of South-East Asia. In addition, wet rice cultivation can be, and often is, supplemented by cultivation of nearby uplands for annual or permanent crops. For example, on peasant holdings in Malaysia, rice is often combined with rubber; in China with tea cultivation; and in Java with coffee. Indeed, the relative security of rice cultivation (in spite of occasional disastrous flooding or the failure of monsoon rains) permits farmers to grow speculative or income-producing crops on adjacent lands.

Finally, wet-rice cultivation is flexible in that output can be greatly increased under modern conditions. Improved strains of rice, simple forms of mechanization, and increased inputs of mineral or organic fertilizer promise increased yields without completely overturning the social fabric built around agricultural operations. Above all, the provision of irrigation seems to offer the greatest scope for a sustained increase in output, though the planning of such schemes is generally outside the control of the farmers themselves.

Given its advantages, it is not surprising that rice cultivation has spread widely from its original heartland in Monsoon Asia. In Africa it has long been a basic crop in Madagascar, but because wet-rice requires much more labour than bush-fallow farming, it is not greatly favoured in this land-rich but labour deficient continent, even though there have been major rice 'experiments' in West Africa – notably in Nigeria, Sierra Leone, Côte d'Ivoire and in the great bend of the River Niger in Mali and Niger. In coastal West Africa, from Guinea to Côte d'Ivoire, upland rice, which has labour inputs more akin to those for traditional cereals, is common. In Latin America wet-rice cultivation has been largely commercial rather than for subsistence, and has become well established in Brazil, Colombia, Peru, Ecuador and Guyana.

Montane agriculture

A major characteristic of traditional peasant farming is its use of local environments so as to take advantage of different qualities of slope, aspect, soils or local climates (Plate 4.4). A wide range of products secures variety in diet, provides raw materials for building, fuel and craftwork, and also affords some insurance against the loss of a particular crop. But it is in truly mountainous country that there is the clearest attention to the use of different ecological niches defined by altitude, slope, aspect and water supply. Patterns of exploitation and the institutional arrangements adopted to exploit the maximum range of environmental resources differ with respect to scale. Such systems are characteristic of mountain areas in the Himalayas, Afghanistan or East Africa, but they are perhaps best developed in the Andes, particularly where strong relief over small distances brings together highly contrasted ecological zones. No more than a few kilometres may separate the harsh conditions of an alpine tundra climate from coastal desert or tropical rainforest. Thus, a wide range of ecological niches may be found within the territory of a single community or even an individual farmer. On a larger scale, high-altitude communities aim to produce surpluses of potatoes, other root crops and animal products such as meat and wool which are exchanged for maize, wheat, beans or other surplus crops produced by valley settlements. Such traditional exchange systems may be conducted by barter at a convenient market or point of exchange, but simple barter was generally replaced many years ago in Latin America by conventional money prices which were not necessarily related closely to prevailing prices in the 'national', urban markets. Dislocation of these local systems has equally often followed the integration of formerly isolated regions into national markets.

Plate 4.4 Montane agriculture, northern Pakistan The Karakoram Mountains offer a hazardous environment because of seismic activity, slope instability and low rainfall. Nonetheless the area is exploited by tapping glacial meltwater to provide irrigation for the valley bottoms and terraced lower slopes, to grow fieldcrops and trees for fruit, building material and fuel. Note the erosion along the valley slopes and the bare land above the line of cultivation

Before the Spanish conquest of Peru, highland communities quite often controlled outlying settlements at other altitudinal levels in order to provide themselves with a varied range of products including maize, beans, fruits and vegetables. One such product was the leaf of the coca shrub, the use of which was carefully controlled in the Inca Empire, but which has now become one of the important bases of peasant farming in remote areas of the eastern Andes. In modern times highland peasant farmers and even communities acting as a whole have acquired land and colonized the humid montane forests on the eastern slopes of the Andes at 2500 metres or lower, maintaining links with the highlands through their family networks. Production and exchange are organized on commercial lines rather than by barter, but it is a complementary system of exchange which is more or less internal to the zone of contact between altitudinal zones and is independent of the large-scale commerce by which products such as coffee, bananas and beef are exported from the eastern slopes of the Andes to urban and international markets (Box 4.1).

Livestock

In the Third World as a whole, livestock play a subsidiary role in relation to crop production. It could be said, in general terms, that animal products tend to be more costly in terms of the calories and protein they yield and in terms of the resources they use than the products of arable farming such as rice, wheat or beans. Many Third World families can afford meat only as a luxury. Nevertheless, there are very considerable contrasts from one area to another. Much of Latin America, the Middle East, interior Central Asia and North Africa have very strong traditions of stock-raising, but livestock play a subordinate role in the diet of China, South and South-East Asia and much of tropical Africa. Three kinds of situation can be seen:

- farming systems in which livestock are the dominant basis of a distinctive way of life and in which arable farming has little place;
- farming systems in which animal power is indispensable for the working of arable land;
- situations in which livestock are quite unimportant compared with arable production.

The arid and semi-arid lands of the Old World are the classic territory of the nomadic pastoralists, traditionally pursuing a way of life dominated by the needs of their animals, and almost constantly on the move seeking the pastures which spring up after the rains, and sometimes following a complex schedule of movement to take advantage of seasonal and highly localized rainfall in different places (Plate 4.7). Long-distance transhumance between summer pastures in the mountains and winter pastures in the lowland plains took communities over hundreds of miles in the annual cycle. Mobility is the dominant motif of a way of life in which movement is constant so that possessions have had to be portable. Animal products provided the raw materials – wool, hides, sinews, hair and bone – for the original 'mobile homes', furnishings, tools and clothes as well as food. The Maasai of East Africa are one such group of specialized pastoralists, traditionally living on the blood and milk of their cattle,

BOX 4.1 ALTITUDINAL AGRICULTURE IN THE ANDES

The Tarma Valley in central Peru (Figure 4.3), focused on a small town of that name at an altitude of 3000 metres, contains level land on the valley floor and moderately sloping land, carefully terraced from before the Spanish conquest in the sixteenth century, and traditionally the location for intensive, irrigated production of maize and beans, the nitrogen-fixing qualities of leguminous plants making possible continuous cultivation (Plate 4.6). Irrigated alfalfa or lucerne, introduced by the Spaniards, later helped to supply fodder for the mule trains operating along the length of the valley and beyond, and also helped to maintain fertility for the production of maize and wheat (also introduced by the Spaniards). On the steep slopes of the valley sides, and on moderate slopes above, barley could be grown up to 3900 metres, and root crops, especially potatoes, up to 4200 metres, with a characteristic cycle of temporary cropping for a few years followed by a long grass-fallow period. The third element in the pattern was the grazing of common pasture land at these and even higher altitudes, with sheep (introduced by the Spaniards) tending to replace the indigenous herds of llama. Individual peasant farmers in the villages around Tarma normally had access, therefore, to irrigated valley-bottom land, temporary plots for rain-fed production of temperate crops and to common pastures – in all, a varied system of mixed farming. Improved roads have increased accessibility to external markets, and farming has become even more intensive. Highly valued irrigated land is now frequently used for the production of fresh vegetables and flowers for the Lima market, and fertility is maintained at a high level by the careful use of animal manures, sometimes brought down from the hills.

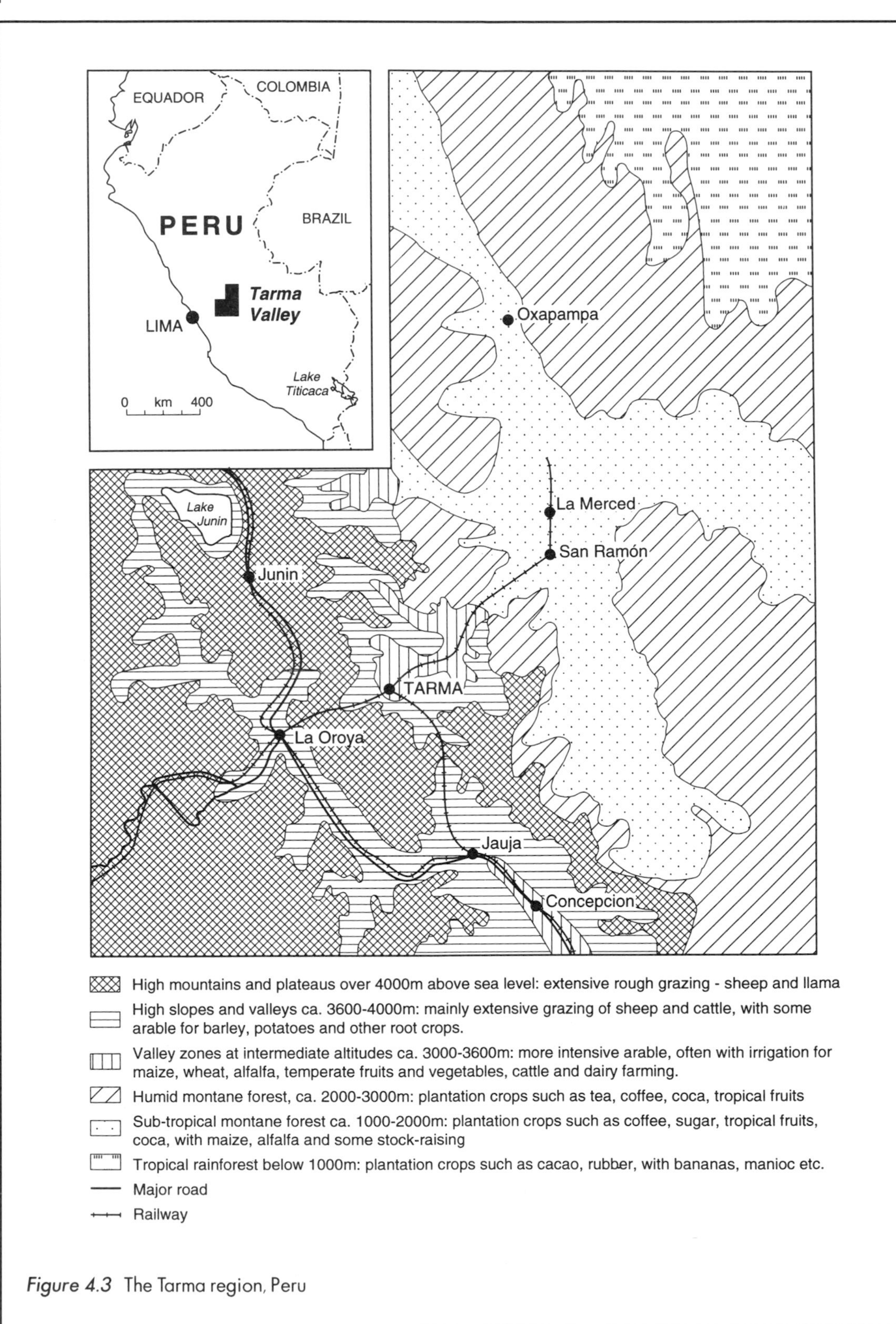

Figure 4.3 The Tarma region, Peru

Mountain terracing

Plate 4.5 Agricultural terracing, Nepal Steep Himalayan hillsides, with intrinsically unstable slopes, require careful management for sustainable agriculture. On these slopes near Kathmandu the upper terraces are wider, and not flat but sloping, to permit water to run off. These are typically used to grow wheat and barley. The lower terraces are flat, with retaining walls, within which wet rice is commonly cultivated

Plate 4.6 Agricultural terracing, Tarma Valley, Peru In this case, on gentler, lower slopes in the Andes, the terraces are wider and less regular

Plate 4.7 Sheep and nomad, Thuburbo Maius, Tunisia As well as seasonal migration to grazing and water, pastoralists also follow more local movements to sustain their flocks. In this case a shepherd is moving his flock around available grazing in an arable area

Plate 4.8 Livestock at a well, Sudan In drier areas access to water is essential, but excessive trampling by the animals severely erodes the land around the wells, and they remove most of the local vegetation

at once the basis of their way of life and the target of their aspirations, trading or bartering with agricultural peoples for the grain they needed. Specialized pastoralists and farmers, each with distinctive ways of life, came together in a kind of symbiosis on the basis of such exchanges.

Pastoral nomadism is almost everywhere under threat. Traditional mobility is undermined by modern preoccupations with the accumulation of material possessions. The provision of health services and education for a mobile population is difficult. Government policies of social control tend to be antipathetic to the mobility of pastoral people. The demarcation of states, provinces, administrative districts and the establishment of rights to land have constricted and circumscribed the freedom of access to seasonal or ephemeral pastures. The pastures themselves are threatened by over-grazing, not least in the African Sahel as a result of droughts, but also more generally as a result of increasing herd sizes with the use of modern veterinary medicines that reduce herd mortality (Plate 4.8). Moves to improve pastures or breeds of stock generally imply fencing or controls over the free movement of animals. Even the Maasai in Kenya and Tanzania have taken up farming to ensure their food supply.

Livestock farming in Latin America has a very different character from its Old World counterparts. Cattle, introduced by the Spanish in the sixteenth century, flourished and ran wild in the natural grasslands of Latin America, with no serious indigenous competitors or predators. They spread widely from Mexico to Argentina wherever there were natural grasslands to be occupied. Medieval Iberian institutions were adapted to become the basis of North American style cattle ranching. Vast pastoral estates were accumulated in Mexico and elsewhere (becoming targets for land reform in the twentieth century). In Brazil and on the eastern slopes of the Andes, from Colombia to Bolivia, forests have been cleared and their rich bio-diversity destroyed, in order to make way for the monotony and low productivity of grassland for stock-raising. In much of Latin America cattlemen enjoyed, or claimed, a macho reputation and a status above that of the cultivator.

Perhaps the most successful pastoral farming in the Third World as a whole was achieved in the Pampas of Argentina. The gaucho was the Argentine cowboy, rounding up half-wild cattle for meat and hides, but with more intensive settlement, stock-rearing was revolutionized when soft and palatable European grasses invaded and superseded coarse indigenous varieties, and improved breeds of cattle were also supported by cultivated alfalfa. Argentina supplied the British meat market and by 1930 had achieved a degree of wealth and development which rivalled that of Australia or New Zealand.

Livestock play an important role in a range of Third World environments:

BOX 4.2 HIGH ALTITUDE PASTORALISM IN THE ANDES

Two very different structures of pastoral farming come together in the High Andes. The high *puna* of southern Peru is plateau country over 4200 metres above sea-level, dominated by coarse tussock grasses and the tundra-like vegetation that gives the region its name. It is a harsh land, well above the upper limit of arable cultivation. Indian communities on the *puna* raise flocks of llama and alpaca, but their pastoral economy has long been linked symbiotically with agricultural communities at lower altitudes, with which they exchange wool, meat and hides for maize, potatoes and other goods. The whole was a large-scale subsistence unit in which exchange was regulated by a barter system or by conventional money prices. With the expansion of an international market in the nineteenth century for alpaca wool, highly valued for its fine, long staple, the raising of alpaca was strongly stimulated. Revenue from the sale of llama meat and its inferior wool by the traditional mechanisms of the rural market is channelled into domestic needs, notably the purchase of flour, potatoes, sugar and kerosene or other necessities. On the other hand, the trade in alpaca wool was and is handled by itinerant merchants rather than by the traditional mechanisms of the rural market.

- in the arid and semi-arid lands;
- in the natural grasslands, especially the savannas;
- in mountainous areas (Box 4.2).

They also have a more profoundly integrated role with arable and crop farming wherever the plough is used. In Asia and in much of Latin America, as in Europe, arable farming is synonymous with the use of the plough, which in turn has traditionally implied draught animals such as horses, mules, oxen or water buffalo. The maintenance of draught animals is thus essential for crop production, so that grazing, fodder crops and weed-fallow are integrated with crop production in a carefully structured system which also provides animal manure for arable land, as well as meat, wood, hides and milk. Draught animals are as important to farming in South and South-East Asia and the Middle East as they once were in Europe (Plate 4.9). Animal power, needed for ploughing, was also important in many other farming activities, but here again, there is change, not wholly for the better. Tractors increasingly replace animal power, requiring imported inputs of oil, fuel or spare parts and thus increasing the dependence of farming on external markets and sources of supply. Pastoral systems become increasingly marginalized (Box 4.3).

In some regions of the Third World livestock are relatively unimportant. In much of Africa south of the Sahara, hoe cultivation is the norm rather than the plough, and shifting cultivation or bush fallow is adapted to hoe cultivation, so that the plough has made relatively little headway. Stock-raising has also been inhibited in parts of West and Central Africa by the presence of the tsetse fly, carrying with it the disease of trypanosomiasis or nagana in cattle. It is evident from Figure 4.2 that these areas have far fewer large stock than in the semi-arid pastures of the Sahel, but there is a thriving use of small stock – chickens, guinea fowl, ducks, etc.

Other farming systems

As exemplified by the discussion in this section, it is evident that traditional farming systems have adapted very sensitively to the opportunities offered by the environment as well as to the limitations it has set. These are no more than examples and there are many other characteristic patterns of farming such as the rain-fed crop-fallow systems of north China, northern India and Pakistan. Many forms of irrigation, ancient and modern, show skill and ingenuity in their use of water and soil. It is precisely because of the complexity of adjustments to environment and to human needs that drastic modern innovation carries such great dangers as well as potential benefits.

BOX 4.2 *continued*

It is revealing, however, that the proceeds of alpaca sales, conducted by the men, are regarded as supplementary cash that can be spent by them, not on household goods, but on fiestas, drink and personal expenditures of a similar kind.

In contrast to the communal grazing of alpaca and llama on the open *puna*, sheep-raising has been associated much more strongly with the evolution of large estates, though Indian peasant farmers also took readily to the raising of sheep even as early as the sixteenth century. From the mid-nineteenth century, international trade in wool began to bring cash incomes to those in a position to raise sheep on a large scale. Unused and communal lands were expropriated and incorporated into privately owned haciendas devoted to the raising of sheep, and sometimes of cattle as well. Indian herders and their flocks were tied to the estates and required to care for the owners' flocks as well as their own. The drive to improve breeds of sheep required the separation of improved stock from traditional breeds, the fencing of pastures, and thus the eviction of Indian herders and their flocks, leading to rural crises of employment and landownership which have not yet been resolved satisfactorily even after the land reforms of the early 1970s.

Plate 4.9 Camel and horse ploughing, El Djem, Tunisia Traditional agriculture in the Third World continues to make much use of draught animals. Here a camel and a horse are used with simple iron ploughs to prepare a small field for a crop

BOX 4.3 MARGINAL PASTORALISM IN AFRICA

Settled agriculture and state bureaucracies have increasingly marginalized pastoral peoples. The apparent anarchy of their way of life makes them unacceptable to the modern state. In many areas they are subject to policies which erode their traditional freedoms of movement and persuade them into permanent agricultural settlement. But by its very nature, pastoralism is associated with drought-related environments and such areas are in any case marginal for any other form of farming activity. Are pastoralists a threat to such environments or are they part of an ecological adjustment process which preserves such areas from more permanent degradation? The most significant area of pastoralist activity in the Third World is Africa. Approximately a third of the continent is scrub savanna and permanent grassland where the 500 mm or less of rainfall per annum is extremely seasonal and highly unpredictable. Unsuitable for agriculture, nomadic herding represents the only potential human activity. Around 10 million herders of cattle, camels, sheep and goats occupy a vast zone of savanna which stretches from Senegambia eastward to Somalia and then in an arc southwards into Kenya and Tanzania. Here, wet and dry years come in phases sometimes extending over several years, as they did in East Africa in the early 1960s, when herd sizes were depleted by more than 50 per cent.

Over millennia herdsmen have contrived their own ways for dealing with these situations. One strategy deployed by pastoral nomads has been to encroach onto the lands of settled farmers in areas which are normally wetter. In the past conflict between people practising such different ways of life has often been settled to the advantage of pastoralists, with frequent disputes over grazing territories leading to the development of a group of young men whose initiation to adult life involved armed conflict. With the development of a warrior cohort, the subjugation of surrounding agricultural peoples became a way of life for

BOX 4.3 *continued*

many such pastoral groups. Fortunately for the farmers, such opportunities are today more restricted. It therefore becomes ever more imperative to provide some acceptable alternative for pastoral people during lean years. Attempts by planners to improve the situation of herders in such environments have sometimes created more problems than they solved. Sinking boreholes may produce gluts of animals in one location and create the circumstances for total destruction of plant life. Similarly, increases in human population produced by improved medical facilities or improved opportunities for marketing stock causes increases in animal populations and such imbalances can exacerbate environmental deterioration.

Scientific research in this area is difficult because of the unpredictability of rainfall and range management strategies have been deployed with varied success. Attempts are now being made to understand the subtle interactions of a delicate set of inter-relationships. Recent work by Mortimore (1989) has shown that pastoralists form an important part of the ecological chain linking man, plants, animals and soil. Each herd animal contributes to a complex ecology by selective grazing, with a finely balanced set of activities which relates to both rainfall periodicities and to its unpredictable distribution. Nomadic herders clearly understand these relationships and adjust their movements and their herd sizes accordingly. A natural inclination by pastoral agronomists to try to limit herd sizes in good years may not be wise. It seems that increasing herd size to meet the stocking potential of pastures in such wet years is probably a better policy than allowing vegetation to waste by persistently understocking and risking total loss of regenerated savanna through bush fires. The characteristic booms and slumps in numbers of animals, which reflect ecological cycles in an uncontrolled environment, replicates the natural order and may be the best strategy, providing some reasonable alternative is provided to tide people over the lean years without destroying a way of life which embodies deep environmental knowledge. This is a problem which remains intractable, especially in a part of the world where political instability often undermines any such policy. The most common cause of changes in population densities and of environmental disasters has not been the bad practices of pastoral nomads but the product of modern armed struggles and the subsequent movements of refugees. It is in this way that the unacceptable human cost of drought turns crisis into total catastrophe.

RECENT TRENDS IN FOOD AND OTHER AGRICULTURAL PRODUCTION

Agricultural systems are rarely static and under modern conditions of population growth, rapid urbanization and technological change, Third World farming faces difficult challenges to its expansion. This section sets out to explore how agriculture has responded to these challenges. There has in fact been an overall expansion in agricultural production in the Third World which is far greater than would have seemed possible thirty years ago. Production has doubled in volume since 1960; a rate of growth in the developing world of 3.3 per cent per year in the 1970s fell slightly in the 1980s to 3.0 per cent (2.6 if China is excluded), far greater than the rate of 0.5 per cent per year recorded for the developed world. There are major regional variations in rates of growth. In Figure 4.4 indices of agricultural production show the contrast between impressive growth of some 80 per cent from 1970 to 1988 in the Far East (mainly China) and a modest growth in Africa of some 35 per cent over the same period. Except for Africa, rates of growth in the 1970s were similar in the rest of the Third World, but began to diverge after 1981, with Latin America lagging slightly behind continued expansion in South and South-East Asia and the Middle East.

Yet what is of the greatest importance in the developing world is the ratio between rates of increase in agricultural production and rates of growth of population. Published figures suggest that this ratio has, overall, been positive since 1970. Per capita food production in the developing world rose by about 14 per cent during

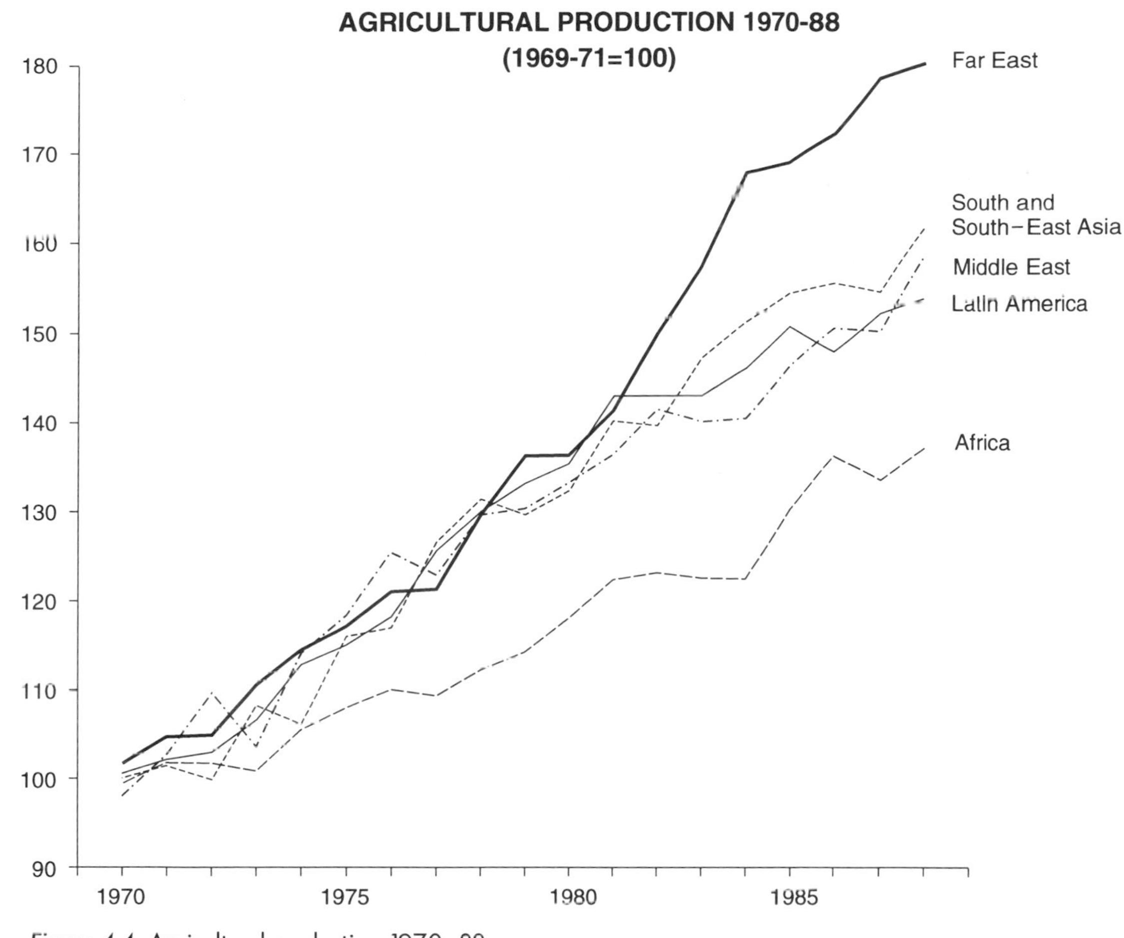

Figure 4.4 Agricultural production, 1970–88
Note: Middle East includes Libya, Egypt and Sudan, which are excluded from Africa
Source of data: FAO, 1988

the 1980s, but with substantial regional variation (Dyson, 1994). Based on FAO statistics, Figure 4.5 shows indices of food production per head in 1988, with 1969–71 as the base figure of 100. Only in South and East Asia, and especially in the Far East, has food production significantly outpaced population growth. In Latin America food production per head reached a plateau about 1981 and has experienced a slight fall since then, entirely due to large areas of export wheat land in Argentina being taken out of production due to low world prices. In the Middle East per capita food output rose in the 1970s but fell slightly during the 1980s. It is only in Africa that it can truly be said that there is a fundamental crisis in food production which has signally failed to increase with its explosive population growth since 1970. The published figures suggest that there has been a continuous decline since the mid-1970s, more slowly in the later period, but by 1988 the deficit was some 16 per cent compared with 1970.

It must be remembered, however, that the levels in the official sources are averages based on national returns which are often incomplete or approximate estimates, and certainly not accurate to within 1 or 2 per cent. There is a great deal of variation from one country to another and from year to year. Figure 4.6 demonstrates how widespread is the tendency for food production to fall short of population growth, even though sub-continental averages may show a positive margin. For example, over

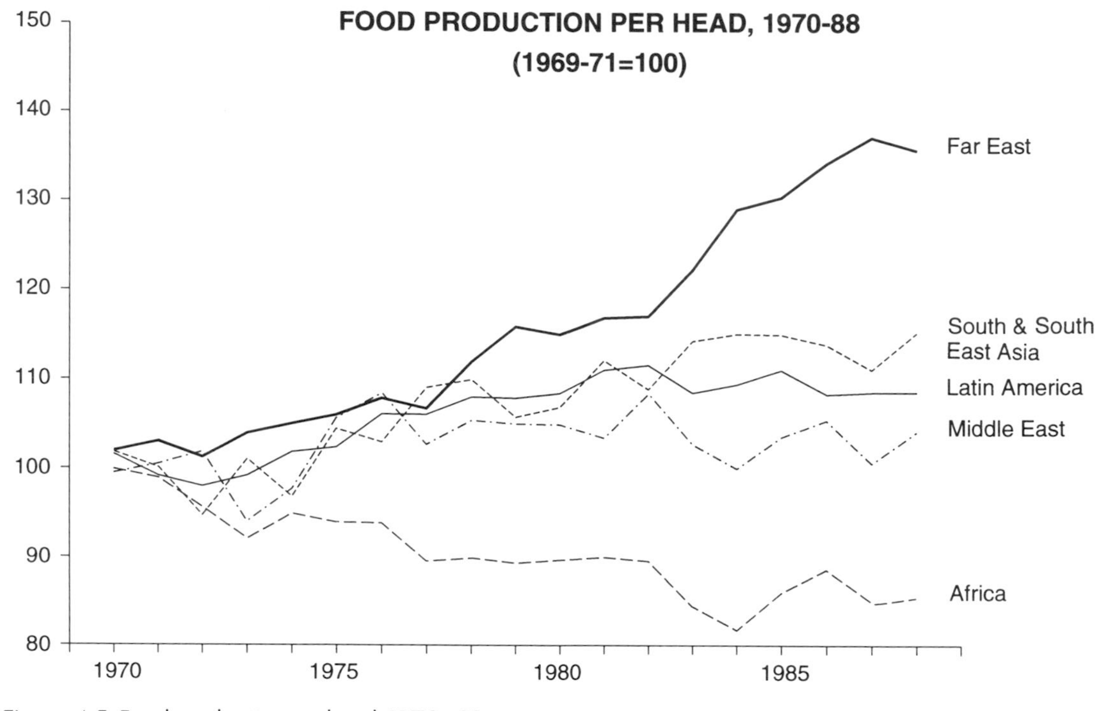

Figure 4.5 Food production per head, 1970–88
Note: Middle East includes Libya, Egypt and Sudan, which are excluded from Africa
Source of data: FAO, 1988

the decade to 1990, only 35 countries in the developing world produced more food per head than ten years previously, but 55 produced less; of these, 31 were in Africa and 13 in Latin America.

Small changes in the average ratio of food production to population growth, even if they are accurately measured, are no sure guide to the incidence of food shortages. Given that there are often gross inequalities in the distribution of income, low or declining average food supplies imply that many of the poor must suffer from malnutrition and hunger. Many more are so near the margin of subsistence that political failure or climatic fluctuations from year to year – drought, flooding, the late or non-arrival of monsoon rainfall and other weather situations – may bring disaster.

The question of whether food production in the Third World is or is not increasing faster than the growth of population tells us little about whether people are, or could be, adequately fed. A range of issues are involved, such as food wastages or losses in storage and between producer and consumer, the efficiency of transport, and above all, internal inequalities in the distribution of income or access to land. Poverty, not the lack of food supplies, is most often the cause of hunger and malnutrition. For many peasant farmers poverty and the lack of food are directly associated with the fact that there is insufficient land to farm. Increasingly, however, poverty and hunger are associated with the urban poor or the rural landless population.

In many Third World countries, essential foodstuffs must regularly be imported. In the Third World as a whole, expenditure on food imports as a share of total merchandise imports amounted to 15 per cent in 1965, declining to 11 per cent in 1990. Latin America maintained its food imports at 12 per cent of the total, and in Africa they slightly increased from 15 to 16 per cent. In some countries the volume of cereal imports has increased a great deal. In Egypt grain imports more than doubled between 1974 and 1990, when they were equivalent to about 80 per cent of internal production. Dependence

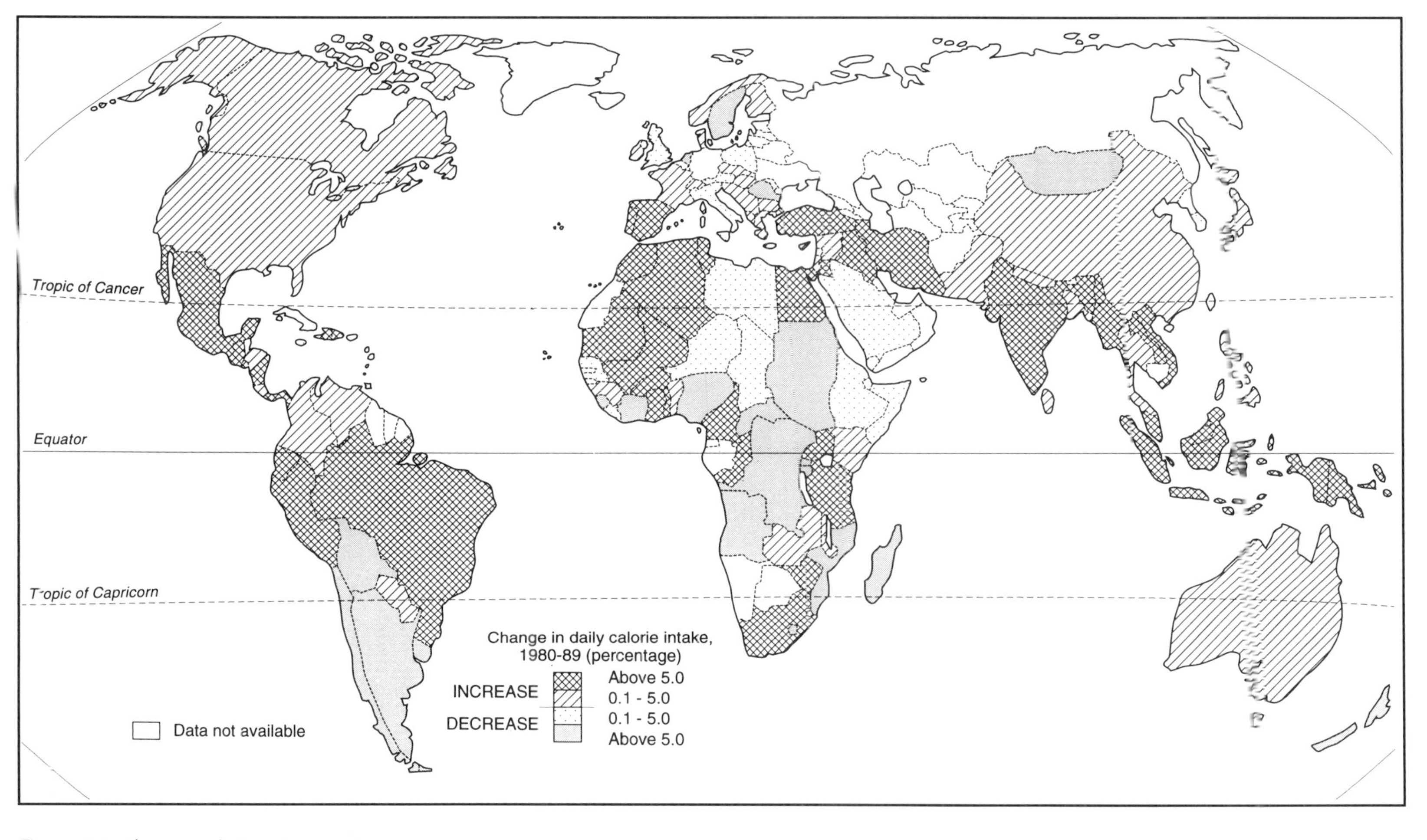

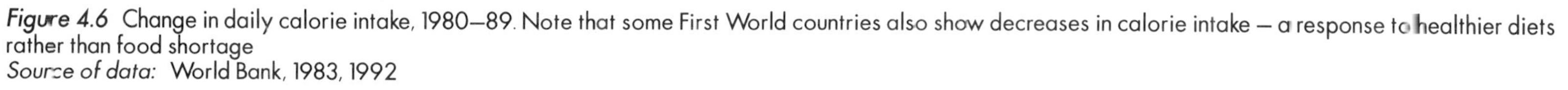
Figure 4.6 Change in daily calorie intake, 1980–89. Note that some First World countries also show decreases in calorie intake – a response to healthier diets rather than food shortage
Source of data: World Bank, 1983, 1992

Table 4.4 Indices of cereal production in the Third World, 1988 (1969/71 = 100)

	Third World	*Africa*	*Latin America*	*Middle East*	*South and SE Asia*	*Far East*
Cereals						
production	168	147	155	177	164	178
yields	155	133	140	138	151	183
area harvested	108	109	111	129	109	97
Wheat						
production	224	145	165	192	212	295
yields	191	174	139	168	160	257
area harvested	118	83	118	114	132	115
Rice						
production	160	159	177	103	164	157
yields	146	109	141	107	145	157
area harvested	110	145	125	96	113	100
Maize						
production	181	153	149	173	185	226
yields	153	125	139	145	143	181
area harvested	118	122	107	119	130	125
Millets						
production	94	146	30	85	84	63
yields	109	138	106	38	104	147
area harvested	86	106	28	120	80	60
Sorghum						
production	94	141	155	202	128	73
yields	130	135	133	100	140	194
area harvested	98	104	117	203	91	38

Source of data: FAO, 1988, 105–9

on grain imports in Mexico increased from 2.9 million tonnes in 1974 to 7.6 million in 1990. In Latin America, cereal imports were equal to 21 per cent of cereal production, and 13 per cent in Sub-Saharan Africa. Such figures should not be taken as any more than a very approximate guide to the failure of agriculture to feed growing populations. Rapidly growing cities, especially when they are also ports, may be more accessible for imports of cheap foreign grain than for the domestic product. Especially among urban consumers, imported foods such as wheat in the form of bread or pasta, dairy products and processed meats, are given prestige by their association with North American or European usage. Cereal imports may reflect a shift towards more profitable cash crops for export. It is also evident that massive imports of grain from the developed world, either in trade or as aid have sometimes served to distort traditional farming and marketing systems by undercutting local production.

The general issue of food production in relation to population growth, important though it is, should not be allowed to eclipse the very real progress that has been made towards the expansion of agricultural production. It is necessary to turn to the question of how far this expansion has resulted from increases in the amount of land cultivated and how far it is a result of more intensive land use and increased yields.

In Table 4.4 the growth of production between 1969/71 and 1988 is shown for the major cereal crops, together with figures showing the increase over the period in yields and in the area harvested. In the Third World most of the 68 per cent gain in cereal production since 1970 has come from greater yields, which have increased by 55 per cent in the area harvested.

In the Far East, the area harvested has actually declined, and the biggest increase was in the Middle East (29 per cent). In South and South-East Asia cereal yields are half as great again as they were twenty years ago. Third World wheat production has more than doubled, while maize and rice have increased by 50 per cent. The greater part of this increase in wheat, maize and rice has been brought about as a result of greater yields, though the increase in area harvested, especially in wheat and maize, has been above the average for cereals as a whole. This may be attributed partly to a move away from millets and sorghum and partly to a real increase in the area under cultivation.

Over a longer period, between 1950 and 1980, it is thought that 60 per cent of the increase in the developing world's output of cereals could be attributed to higher yields, and 40 per cent to the expansion of area cultivated. This figure conceals a very important shift during the period as a whole. In the 1950s the expansion of the area harvested was responsible for 82 per cent of the increase, and increased yields only 18 per cent. In the 1960s the balance was reversed as increased yields accounted for 75 per cent of the total increase as against 25 per cent due to increased area. This trend continued into the 1970s, when the corresponding figures were 84 per cent and 16 per cent, and through into the 1990s. Grigg concludes that 'in the 1980s yield increases accounted for nearly all the extra output in both the developed and the developing world' (Grigg, 1993, 88).

Increases in area harvested stem from a variety of factors, not least the expansion of double cropping in favoured areas, but also from changes in the ways statistics are collected and published. It is also, and most obviously, a result of gains made in the extent of arable land, either as a result of conversion from pasture, as often used to be the case in the developed world, or from the expansion of irrigated land or the clearing of forests. It is therefore helpful to look at changes over the last twenty years in the pattern of land use in the Third World before examining the increases in yields associated with the Green Revolution.

THE EXPANSION OF AGRICULTURAL LAND

The broad pattern of land use in the Third World is shown in Figure 4.7. What is immediately striking is that in many areas the proportion of land in arable farming is quite low. Only in South Asia is as much as half the land in arable farming, and in Africa less than 10 per cent. Much of the arable land, perhaps a half in the Third World as a whole, and very much more in Africa, lies fallow in any one year. Grazing land for animals is usually unimproved, often arid, and rarely as productive as permanent pastures in the humid climates of Western Europe or eastern North America. Pastures take up about a quarter or more of available land area, especially in southern South America and southern Africa, but it is scarce in South and South-East Asia, where livestock farming is relatively uncommon.

The expansion of cultivated land in the Third World which is apparent from the data on cereal production outlined above can be set into the context of changes in land use which are summarized by major regions in Figure 4.8. There has been a substantial growth in the extent of farmland. Between 1972 and 1987, 77 million hectares were added to the area of farmland; 41 million hectares of arable, or about 6 per cent since 1972; 26 million hectares of pastures and 10 million hectares (some 15 per cent) in permanent crops such as coffee, tea, cacao, bananas, rubber or oil palm. It should be added that over this period there has been a considerable expansion of irrigated land, some of which represents an increase in the potential of existing arable land, but much of which must be a net gain from pasture or from hitherto unused land. Overall, irrigated land increased by 34 million hectares (26 per cent) from 1972 to 1987. Considerable emphasis was given to the expansion of irrigation in the agricultural planning of the 1960s and 1970s, involving major schemes, but also the proliferation of simple tube wells and small-scale projects, especially in Asia.

Much of this gain has been at the expense of

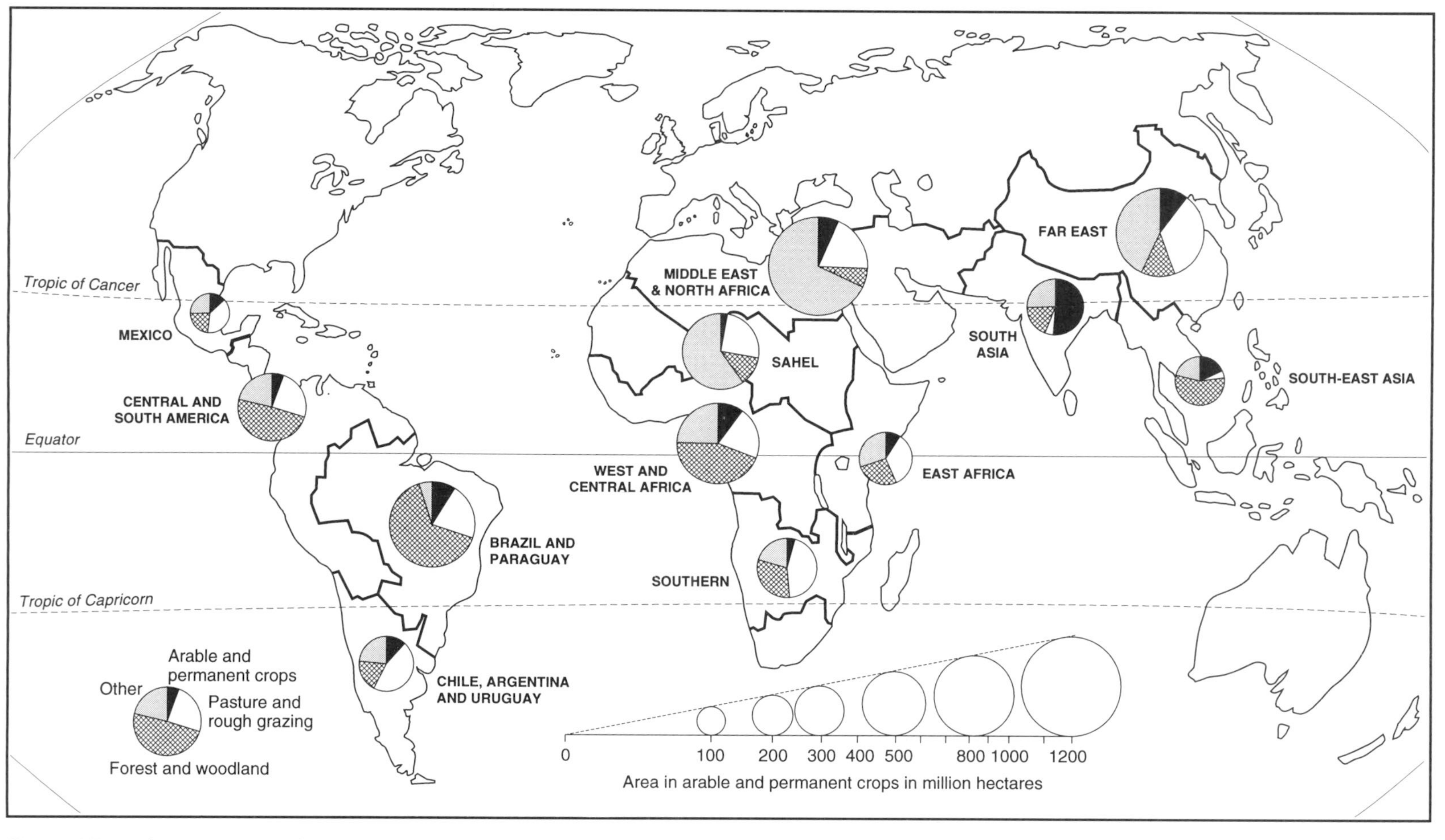

Figure 4.7 Land use patterns in the Third World, 1988
Source of data: FAO, 1988

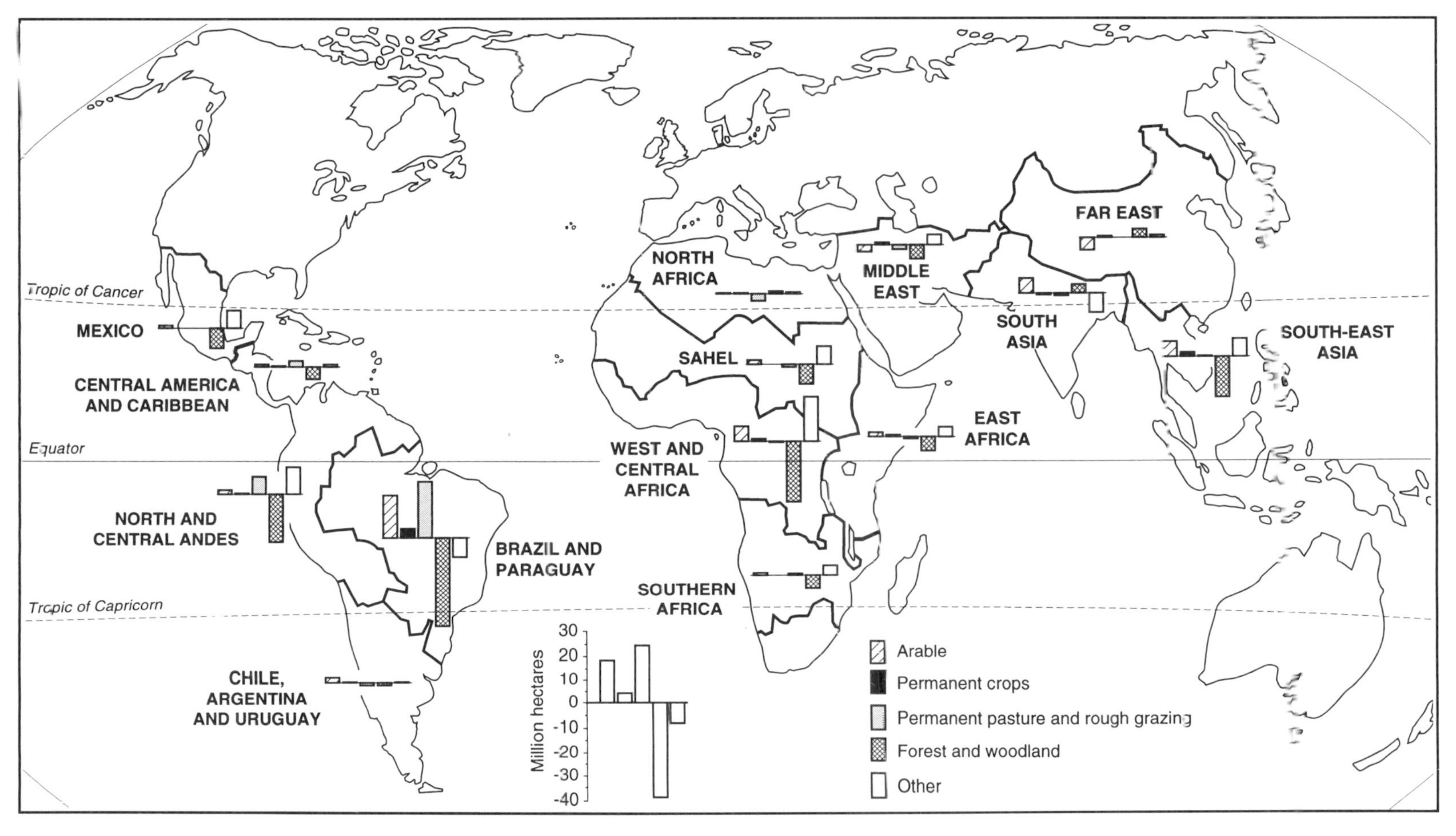

Figure 4.8 Changes in regional land use, 1972–87
Source of data: FAO, 1988

forest and woodland. No less than 130 million hectares of forest and woodland were lost between 1972 and 1987, amounting in total to an area about the size of Germany, France and Spain put together. Most of this loss has been from rainforests of West and Central Africa (26 million hectares), the Amazon basin (39 million hectares from Brazil alone), and from South-East Asia (17 million). The other major category of land use to show an increase has been the amount of unclassified, unused and useless land, which has increased by 52 million hectares, much of it in the arid and semi-arid lands of North Africa and the Sahel which have been seriously affected by processes of desertification.

Until the 1950s, increases in agricultural production in the Third World came largely from the expansion of agricultural settlement into new lands. Agricultural colonization on a large scale was, of course, responsible in the nineteenth and early twentieth centuries for the settlement of much of Canada and Australasia. It motivated the frontier movement across the USA, the expansion of Russian settlement in Siberia, and also the settlement and colonization of the temperate lands of Argentina and Chile. Expansion on a smaller scale occurred in many of the tropical colonies to produce export crops to satisfy the demands of the industrial countries or to provide land for growing populations. In much of Africa, Latin America and South-East Asia there was still land for the taking even after the frontier of settlement had closed in the temperate lands, but there had never been a great deal of new land for colonization in South Asia or the Far East.

The expansion of agricultural land has slowed down in the last 30 years. Since 1965 it has been no more than 1.3 per cent per year even in Latin America, and averages only 0.5 per cent in the Third World as a whole. In part this slackening of settlement is due simply to the fact that good, accessible land has already been occupied and what remains tends to be of marginal quality, relatively inaccessible to national or international markets, or requires heavy investment in irrigation, drainage or clearing before it can be effectively used. Furthermore, the relative shift in the terms of trade between agricultural and industrial products has tended to make colonization less profitable than it was in the late nineteenth century; migration to the towns is frequently seen to be more desirable than a hard and risky existence on the pioneer fringe. In some countries the climate of opinion is unfavourable to the uncontrolled expansion of large estates in new territory, and public policy often requires that new frontier settlement should be undertaken by medium- or small-scale farmers who often lack the necessary capital.

Nevertheless, although much of the modern pattern of agricultural colonization is underwritten and planned by governments, 'spontaneous' colonization still occurs where suitable land is available. In Latin America most new settlement in the western margins of the Amazon basin and in southern Venezuela has taken place as a result of the piecemeal infiltration of peasant farmers. In Peru, for example, the expansion of settlement down from the Andean ranges to the humid montane forests has been accomplished with a close and continuing integration between the old highland settlements and the new clearings. Kinship networks link the two as trading links are established. Much of this kind of settlement is simply done by occupation on a squatter basis, and although pioneer settlers may be evicted by nominal owners, as they have been in eastern Bolivia, they live in hopes of official recognition of their right to occupy the land they cultivate.

A controversial process of settlement has often followed the building of roads in South America east of the Andes. Road building is accompanied on the one hand by a flurry of ownership claims and land speculation in the offices of the capital city, and on the other hand by a drift of squatter settlements along the line of the road itself. Squatters clear and cultivate the land, only to be evicted if the legal owner sees the possibility of profitable exploitation after the land has been cleared at no cost to himself. The squatter must either move on, or remain as tenant or wage-paid labourer. As Alistair Hennessy has pointed out (Hennessy, 1978), the Latin American frontier is far from being the nursery of democratic freedoms, independence and rugged individualism which were the qualities claimed by Frederick Jackson Turner for the American frontier in the USA. The characteristics of the recent frontier in Latin America are more often those of poverty, refuge, instability, lawlessness and illegal drug traffic.

Since the 1960s attitudes towards agricultural expansion to new land have been changing. In the semi-arid lands of the Sahel the unwise extension of arable farming and the over-grazing of poor scrub and seasonal grasslands have exacerbated the deterioration of the natural habitat and have assisted in the process described as desertification. Expansion of agricultural land in the humid tropics involves the destruction of rainforests and of other natural vegetation on a large scale, and the processes by which the demand for food has induced the replacement of true shifting cultivation by bush fallow has had dramatic effects on the tropical forest. Almost all the rainforest of West Africa and much of what there once was in Central Africa has disappeared. Rainforests in Amazonia and in parts of South-East Asia are severely threatened. Concern has been expressed for many years about the potential destruction of plant and animal species and the reduction of biological and genetic potential which this implies (Box 4.4). Since the 1930s fears have been expressed about the loss of soil and the higher incidence of flooding as a result of the destruction of natural ecosystems by unwise cultivation. Massive soil erosion followed the expansion of coffee plantations in Brazil in the nineteenth century with resulting flooding and silting in the valleys. There is also concern that the destruction of forests adds its quota to other sources of atmospheric carbon dioxide stemming from the high rates of consumption of fossil fuel in the developed world. There are many unknowns about the growth and implications of the greenhouse effects deriving from the expansion of carbon dioxide in the atmosphere, but it is important to recognize that awareness of the need for conservation transcends national boundaries and that it is greater in the developed world than among the vast majority of people who live in the Third World.

PRODUCTION PER HECTARE

It is clear from Table 4.4 that since 1970 the yields of major subsistence crops in the Third World have increased more rapidly than the area cultivated. Yields, normally measured as kilograms per hectare, may be taken as a measure of the productivity of land. Yet it is important to recognize that productivity of land can be measured in other ways; by the value of output per hectare or by the nutritional output per hectare, for example, volume of output (i.e. yields per hectare) makes it impossible to compare directly the productivity of land in different crops. For example, potatoes yield far more per hectare than wheat. The value of output per hectare is a measure of the yields, in the light of the valuation put by the market on different kinds of crop or product. One of the reasons for the low productivity of land in the Third World is that farmers remote from urban markets and reliant on their own production must concentrate their efforts on crops necessary for their own subsistence, which may be highly productive in terms of nutritional value, but which command low prices on the national or international markets. For example, it has been observed that in Venezuela, productivity per hectare (by value of output) on peasant farms created by agrarian reform was only three-quarters of the national average, but low values per hectare resulted to a large extent from the concentration of the agrarian reform sector on staple subsistence crops of low market value (maize, beans and rice) rather than unduly low yields in terms of volume of output.

Physical yields of specific crops are, nevertheless, low in the Third World as a whole and the contrast with the developed world is very striking. The average yield of wheat in the Third World (excluding China) is only 70 per cent of that of the developed world market economies; of maize only 34 per cent; and of rice 44 per cent. Such overall figures conceal very considerable and significant variations. Low yields may be associated with environmental factors, the lack of capital, the use of poor seed, lack of fertilizers and low technology, and they are sometimes the result of long continued cultivation of poor, exhausted and eroded soils.

BOX 4.4 CLEARANCE OF THE FOREST

Much concern has been expressed over the period since 1970 about the loss of the tropical rainforest to agriculture and other uses. In fact clearance of woodland has been an integral part of agricultural activity since man first became a cultivator and pastoralist, as natural vegetation was cleared to provide land for crops and animals. Large areas of Europe and North America were cleared in prehistoric and historic times to provide for the advance of the frontier of settlement and agriculture. The 'advance into the rainforest' is thus merely a logical continuation of this process, and though the particular concern is with the loss of forests in Amazonia, rainforest clearance has long been a facet of peasant agricultural advance in Africa and Asia.

However, in the late twentieth century, such advance is regarded with greater scepticism and hostility, arising from a growing interest in 'Green' issues and concern about the place of deforestation and forest burning as *one* factor contributing to 'global warming' and possible climate change. The scale of change is difficult to ascertain. Some countries have better surveys than others; even modern remote sensing techniques cannot always distinguish between original (primary) and regrown (secondary) forest; estimates may include very different types of woodland; and while 'green' groups may exaggerate the scale and rate of loss, governments may seek to underestimate the areas supposedly removed. Estimates for the late 1980s suggest that there may be some 24.5 million square kilometres of rainforest, with about 45 per cent in Latin America, 37 per cent in Africa and 16 per cent in Asia and the Pacific.

These forests are encroached upon in many ways – for fuel-wood and commercial logging, by shifting cultivators, agricultural colonization schemes and agribusinesses, and cattle ranching. They may also be affected by large-scale development projects such as mining and the flooding caused by large hydroelectric schemes. In addition to the direct loss of clearance, deforestation may give rise to a diminution of bio-diversity, the loss of raw materials for industry, agriculture and medicine, as well as damage to the local environment in terms of erosion and silting, and possible impact on the wider climatic environment.

Deforestation in Amazonia has prompted particular concern in recent years. An estimate prepared for the 1992 Rio conference on the environment suggested that 800,000 square kilometres had been deforested, half of that in Brazil. However, that figure might be set against a *total* area for the Amazon basin of 7.35 million square kilometres, of which 67 per cent is in Brazil, and the reminder in Bolivia, Peru, Ecuador, Colombia, Venezuela and the Guianas.

'Clearing the forest' is seen as an integral part of the colonization process. Coming from non-rainforest environments settlers see land clearance as essential to establishing their farms, encouraged by government officials anxious to promote 'modern' agricultural and ranching activities. From their perspective clearing and burning the forest is a rational decision, but an important consideration is not merely the loss of the forest, but that some tracts which are cleared may not be used most effectively, are quickly abandoned or never put to use at all. There are also other consequences in terms of impacts on the rivers from domestic pollution by settlers and from mines and lumber mills, and conflicts with surviving native peoples. Though forest clearing in Brazil has caused the most controversy, the advance by Andean countries into their eastern foothills and river headwaters has caused severe erosion, with consequent short-term land use, flooding and silting.

It is also important to note that this process of deforestation is not unique to Latin America. It was estimated that by 1988 *all* of the primary rainforest in India, Bangladesh and Sri Lanka had been lost, and half that of Thailand and the Philippines. If colonization has been portrayed as a major threat, it should also be recognized that there are very large losses of forest to commercial logging, particularly in Africa and South-East Asia. Japan alone accounts for 40 per cent of the world's tropical timber trade, mainly from Indonesia, and Sarawak and Sabah in Malaysia.

Nevertheless, there are exceptions. Wheat yields in Mexico (4111 kg/ha) and Egypt (4747 kg/ha) were more than three times those of Canada (1212 kg/ha) in 1988. Environmental factors are very different, of course, but in Canada the productivity of capital and labour is much higher and much more important than the productivity of land.

Average yields in the Third World may fall far short of what has been achieved in Europe or Japan, but they have improved greatly in many countries. Between 1965 and 1989 cereal yields in the developing world as a whole increased by 90 per cent. In South Asia and the Far East they have doubled; in Latin America they have increased by two-thirds, and even in Africa they have improved by about a quarter. Average yields of root crops have done less well, improving by about a quarter since 1965. Two factors of importance in sustaining greater output have been a rapid increase in the amount of land irrigated and the consumption of fertilizers. However, even more important has been the creation and introduction of improved plant and animal strains. Such improvement, by empirical methods, has been part of Western agricultural tradition since the eighteenth century, but the major modern breakthrough occurred in the early 1960s, through the work of Norman Borlaug on developing higher yielding varieties (HYV) of rice and wheat. By 1970, the so-called Green Revolution was being heralded as the means by which poor farmers could secure their subsistence and also develop a surplus for sale to the burgeoning towns. Its greatest impact as a philosophy of transformation has been on farming in the Indian subcontinent, China and South-East Asia, where production of basic cereal crops has increased dramatically. Table 4.5 shows how, from small beginnings in 1966/67, the percentage of cropped area in HYV cereals in India grew rapidly to 1984–85, particularly in wheat and rice.

More recent data are hard to obtain directly, as HYVs have become the norm, but their continuing spread can be indirectly deduced from rapid growth in fertilizer use. In 1950 4.0 and 0.5 kilograms of chemical fertilizer per hectare were applied in China and India respectively; the equivalent amounts in 1988 were 262 and 65 kg per hectare.

'Whoever could make two ears of corn ... to grow upon a spot of ground where only one grew before would ... do more essential service to his country than the whole race of politicians put together.'—*Gulliver's Travels*

This cartoon is based on a quotation from Jonathan Swift's *Gulliver's Travels*, first published in 1726, which anticipated the significance of the 'Green Revolution' of the late twentieth century

Table 4.5 Share of heavy yielding varieties in major cereals in India, 1970–85

	% cropped area in HYV cereals					
	Rice	*Wheat*	*Sorghum*	*Millet*	*Maize*	*5 main cereals*
1970–71	14.9	35.5	4.6	8.9	7.9	15.1
1980–81	45.4	72.3	21.1	31.3	26.8	42.8
1984–85	56.9	82.9	32.5	49.3	35.5	53.3

Source of data: Lipton and Longhurst, 1989, 34

The introduction of new varieties has not been without problems. There are many indigenous varieties of rice in Asia, selected by traditional methods to suit local conditions. They may have qualities such as short stalks to prevent lodging, long stalks to withstand flooding; they may be quick-maturing or have some immunity to local pests and diseases. Most of them secure reasonable yields under varied conditions of local weathers and soils rather than high yields under optimum conditions. HYV rice, first associated with the International Rice Research Institute in the Philippines, was deliberately bred for specific qualities using modern techniques.

Early HYVs of rice yielded well under optimum conditions, which meant careful water control (i.e. irrigation), heavy application of fertilizers and pesticides. Newer and better tasting varieties have since been bred for other qualities and plant-breeding institutions offer a continuing response to specific needs – such as rice which will withstand moisture shortage or show resistance to specific pests or diseases.

The introduction of new varieties of wheat from Mexico and the USA paralleled the improvement of rice, but, translated to Asian conditions, also implied irrigation, fertilizers and pesticides to fulfil their potential. Hybrid maize has also been successful in raising yields, though less dramatically, and although there has been progress in HYV millets and sorghum, they have been less advantageous than wheat or rice and slower to spread. Progress with root crops has been less outstanding, and it is to the misfortune of tropical Africa that so much of its staple production is in the form of root crops such as yams and manioc. Even in Monsoon Asia, where adoption of HYVs of rice and wheat has been rapid in irrigated areas, progress has been slow where cultivation depends on direct rainfall or on flooding during the monsoon, as in much of Bangladesh.

Modern varieties have made possible a vast increase in food output, but there have been other and less desirable consequences. The introduction and spread of new varieties has often been associated with growing inequalities in the rural sector. It was most often the larger-scale farmers who were able first to take advantage of the assistance and subsidies offered for the introduction of the new crops. They were also among those who felt able to shoulder the risks involved in changing traditional practices. Larger-scale farmers could also afford the additional inputs of irrigation water, pesticides and fertilizers required by the new techniques. In countries such as India and Pakistan, where landlord and tenant systems operate, poor tenants were often evicted to make way for the new techniques, and thus joined the army of landless labour. Small farmers who owned their own land, and were dependent on high interest loans for raising capital, often found the new technology too expensive. Competition among landowners for land to increase acreages for mechanization raised the price of land, tempting small landowners to sell. Eventually they may be squeezed out altogether and forced to seek employment as wage labourers or to migrate to the towns to swell the numbers of urban unemployed.

The diffusion of the so-called Green Revolution has followed a rapid course in the 1970s and 1980s in favourable areas well endowed with water, good land and urban markets. In such areas it has spread down to the level of many small farmers who have reaped some benefit from higher productive capacity. However, landless rural labour has failed to gain any advantage and in many areas has suffered from the lack of employment as mechanization has taken place. The Green Revolution has meant little or nothing to farmers in less favoured upland areas, areas remote from urban markets and areas in which subsistence crops are still the mainstay of the farm or village.

There are other problems associated with such innovations. High technology agriculture increases dependence on external supplies of fuel, fertilizer and pesticides. Large petrochemical companies provide all three support elements so that the economies of all but the largest developing countries become increasingly dependent on the international vicissitudes of the oil industry and its suppliers. The world economy is not yet sufficiently well organized to cushion the developed world, let alone Third World countries, from the effects of inflation problems and variable oil prices. In ecological terms, the impact of high technology and sophisticated agronomy is perhaps even more insidious. Most local strains of cereals which new varieties replace have been bred by trial and error over many generations. There is concern about the narrowing of the genetic pool which the loss of traditional varieties may bring about. Old varieties may yield fewer calories but they often have a higher protein content and are more disease-resistant. New varieties have generally required pesticides of which increasing amounts are likely to be required as pathogens acquire immunity. It is sometimes argued that this vicious spiral towards high-cost vulnerability could only be halted by the reappraisal and incorporation of locally resistant plant species and local peasant knowledge.

SUMMARY

The agricultural patterns of the Third World are variable and highly complex. They vary greatly in the scale of operations, in productivity, and in their combinations of subsistence and cash crops and livestock. These are often agricultural systems of great subtlety in the adjustments of farming practice to the natural environment and to national economic conditions. Agriculture is not only the source of food supplies for the generally more impoverished populations of the countryside, as well as for the expanding urban centres, but it must also provide exports to support the balance of payments, and provide raw materials for processing and manufacturing. Agriculture remains crucial to the livelihoods of the majority of people in the Third World and in recent years increases in output and general conditions in rural areas have been given high priority in development plans. However, the raising of agricultural productivity, whether in bringing new land into production or increasing the yield from existing land, has proved to be a very difficult task. Introducing innovations in agriculture may have high costs, and also favour the relatively less-disadvantaged members of the rural community. Innovations to increase productivity and advances into new land to increase output may also carry environmental risks.

5

Agrarian Structures and Rural Development

The previous chapter was concerned with the character and trends of agricultural production in the Third World, and with the ways in which agricultural systems in contrasting circumstances are the products of a long process of adjustment to frequently hazardous environments and frequently marginal economic circumstances. In spite of continuing problems of malnutrition, and powerful media images of the impact of famine, Chapter 4 demonstrates that in the modern world of growing population and an increasing pace of relentless commercialization, Third World farmers have responded by raising their output, partly by increasing the land under cultivation or the number of livestock produced, but also by substantial increases in the intensity of farming and the level of yields. Advances in agricultural production are required to feed growing populations, and in many countries, especially in Africa, higher production has not meant any relaxation of the constraints of hunger, malnutrition and poverty; in many areas the situation has become worse. There is thus a continuing need to increase the output of agricultural produce and the availability of food within the Third World.

An important influence on the way in which agricultural output may be increased and modified is the way in which land is controlled. Forms of landholding, patterns of tenure, and the size of farming units have major bearing on the possibilities for agricultural innovation.

PATTERNS OF LANDHOLDING

Ownership of agricultural land has long been characterized by inequality, and not only in the Third World. In Brazil, 53 per cent of all agricultural units are less than 10 hectares in area, but they occupy only 2.6 per cent of the agricultural land, while 0.9 per cent of units are over 1000 hectares in size and occupy some 44 per cent of the agricultural land. Other Latin American countries show similar characteristics, with a dichotomy between a smallholding peasantry and massive estates, and similar patterns of inequality are evident in Asia and Africa. In Indonesia pressure on the land, especially in Java, creates many tiny holdings, with half less than 0.4 hectares and three-quarters less than 1 hectare. At the other extreme, large estates devoted to the production of palm oil, rubber and other tropical crops occur, mainly in the more sparsely populated Outer Islands, and especially in Sumatra. In Africa landholding patterns often reflect colonial histories of white settlement superimposed on traditional African farming. Throughout the Third World there is a profound contrast between the many, who own only little land (as well as those who are landless), and extensive farm areas which are controlled by a few.

Large holdings of land have been accumulated in various ways and with different points of departure in time and space. In some cases they involved the withdrawal of land from communal usage and the subordination of a peasant population. They have also been

◀ Woman harvesting sugar cane, India

created in virgin territory, requiring the import of labour from elsewhere, and representing new additions to the area of cultivated land. In Latin America, cattle introduced by the Spaniards multiplied on natural grasslands and savannas. Control given to *conquistadores* over these herds of semi-wild cattle for their meat and hides led gradually to control over the land they grazed and the formation of large estates, ranches and *haciendas*. Their modern counterparts are the cattle estates in the Amazon basin created by the destruction of the rainforest in order to satisfy the apparently insatiable and world-wide demand for hamburgers! Nineteenth-century demand for tropical crops also led to the creation of large landholdings, particularly plantations, to produce coffee, sugar and cotton.

Conventional thought has long assumed that policies to encourage the adoption of modern agricultural techniques should be focused on the medium or large-scale estate. Economies of scale are possible; credit is more easily acquired; and the scale of resources available implies a minimization of the risks which plague the small farmer. Such holdings can and do introduce new agricultural techniques and more efficient methods of production, including mechanization. In South-East Asia it has been the larger farmers who have enthusiastically embraced the Green Revolution with its heavy demands on careful managerial control of inputs and capital investment, and they have reaped high rewards as a result. In Mexico the expansion of agricultural production in the north is associated with the innovative capacity of large or medium farmers, and although small farmers (*ejidatorios*) have willingly adopted the cultivation of high yielding hybrid maize, they have been more reluctant than medium farmers to adopt the cultivation of sorghum, for which there is little subsistence demand. In general, it has been the large-scale producers who have seemed most ready to adopt technical innovations in the production of export crops to compete on the world market.

Yet it is by no means always true that large estates are productive and efficient in their use of either land or labour. There are many examples in Latin America and Africa of large estates engaged in extensive and low-productivity pastoralism, and employing very few labourers or herdsmen, side by side with overpopulated and intensively used small farms. In economic terms the marginal productivity of land, labour and capital is anything but equal. It can be argued that the high profitability of extensive pastoral estates is not compatible with the social function that land should perform in providing employment and food for overpopulated rural areas.

Nevertheless, owners of large estates in some parts of Latin America, for example, have followed a thoroughly rational policy of investing capital where it can secure the highest return. The market is king, but this has often meant that capital accumulated by extensive farming on large estates, or as rents from tenant farmers, has not been invested in agricultural improvement, but rather in urban real estate and property development, the shares of multinational companies, or a *pied-à-terre* in Miami or Paris. Savings accumulated in the countryside of the Peruvian sierra, the Kenya Highlands or the Ganges plain, are channelled into the national urban and industrial sector or to the international banks.

Large estates and plantations, controlled by individuals or large (and often foreign) corporations represent one extreme of landholding patterns in the Third World. At the other is the survival of what are described as 'traditional' farming systems. In these usufruct, or use-right, arrangements prevail, where land is held by a group, such as a lineage, family or tribe, whose members share the use of land and cultivate it or raise stock mainly for subsistence. Such 'traditional' structures were common in the Third World before the arrival of the Europeans, and are generally characterized by subsistence agriculture, using simple technology and with little participation in processes of commodity exchange. Such 'pre-capitalist' systems survive in parts of Latin America, Africa and Asia, but most have been modified by colonialism and contact with contemporary market forces. The continuing use of the term 'traditional', however, reflects a perception that these are the most conservative elements in Third World agriculture, and those most resistant to change.

For many farmers in the Third World, access to land is determined by some form of tenancy arrangement with a landlord. Many of the large landholdings are not directly farmed by their owners, but parcelled out into tenant farms operated under a variety of conditions. It is under some of these forms of tenancy that the cycle of low productivity, low returns and poverty reaches its lowest point.

In pre-industrial Europe the subordination of peasant farmers by powerful groups and the creation of a landlord class dedicated to the extraction of rents, tribute and taxation was a common feature. Such feudal structures were gradually eroded or overthrown in Western Europe, or destroyed by revolution in Eastern Europe and in parts of the Far East. However, in much of the Middle East, Latin America, the Indian sub-continent and South-East Asia, the hierarchy of landlord and money-lender, merchant, trader and tenant farmer still remains strong, and in some ways the currents of economic development have strengthened it further. The kind of secure long-term tenancy with fixed money rents and compensation for improvements which is common in Europe and North America is rare in the Third World, and a variety of tenurial systems exist, generally weighted against the interest of the tenant.

Labour-service tenancies tend to be associated with relatively poor farming techniques on large estates, particularly in South-East Asia and the Middle East. For the landowner the labour-service tenancy represents a means of securing income from the land with a minimum of expenditure. Tenants supply labour for the cultivation of the landowner's crops and the care of his stock in return for the right to cultivate a plot of land or to graze a few stock. Sums paid out in wages tend to be minimal. Tenants must focus their attention on subsistence production, and they are tied to the landlord's enterprise in such a way that improvement of agricultural methods becomes very difficult.

In share-cropping tenancies the essential feature is that the harvest is shared in some fixed proportion between landlord and tenant. In some cases the landlord provides the land and no more; in others, he may provide seed, implements and fertilizers or credit. The share-cropper has a greater scope for decision-making than the labour-service tenant, but only within the narrow limits imposed by the landlord, who frequently dictates what crops should be grown and how they are to be cultivated. Share-cropping has often been associated with commercial crops, and particularly those for which methods of cultivation to produce a standard product are relatively straightforward. Cotton cultivation in nineteenth-century Peru and rice in modern coastal Ecuador were based on this system. Under a share-cropping system the landowner is able to extract an income without directly participating in production himself, but by selling his share of the crop and often by buying his tenant's crop at a low price for subsequent sale with his own share. He may also gain from charging his tenants higher rates of interest on loans he provides than on loans he takes out from the banks. The middleman role of the landlord under share-cropping systems is therefore important and may come to dominate his activities.

'Modernization' of agriculture has begun to lead to the decline of these and similar tenurial systems in the Third World. The adoption of improved varieties of crops and stock, more intensive use of land and mechanization have frequently induced landowners to evict labour-service tenants and share-croppers in order to work the land or raise stock under their own direct management. Former tenants are thus reduced to landless labour or to unemployment.

PEASANT FARMING AND RURAL SOCIETY

Peasant farmers are commonly seen, not always correctly, as the most characteristic group in Third World rural society. Who, then, are the peasants? They have been defined in various ways, but the essential features of definition seem to be:

- a peasant has access to land as a means of production – he or she may own it or be simply a tenant, but not a landless labourer;
- the land is worked mainly by the labour of the household;

- the land is the primary source of income or subsistence;
- the peasant participates in a subordinate status in a wider society, usually the nation state.

Peasant farming is thus not necessarily associated with subsistence farming only, but with subsistence and/or commercial production. It may be, and often is, strongly oriented to the market. Nor is it necessarily associated with low technology, though this is often the case in the Third World. Finally, while peasant farmers must by definition have access to land, it does not follow that they have access to enough land to support themselves at a satisfactory level of living in the light of local expectations. It is often this fact, and the lack of alternative economic opportunities, that is responsible for low productivity, and not an intrinsic lack of ability among peasant farmers.

Given the variety of environmental resources and methods of production, it is clearly impossible to suggest any fixed size for the average peasant farm or even an upper limit. In almost all parts of the Third World, except for some countries in Latin America and parts of the Middle East, median farm sizes are less than 5 hectares and in some countries the median figure may be less than 2 hectares. To some extent these figures relate to environmental factors. For example, in countries in the semi-arid regions of Africa such as Chad, Mali, Niger, Sudan and Botswana, median values are less than 5 hectares, but in the African humid tropics less than 2 hectares. In areas of intensive rice production in Asia, such as Bangladesh, India and Indonesia, holdings tend to be very small, often below 1 hectare. Pressures of population may also be relevant to the small size of holding, and in much of Africa the smallness of holdings often reflects the difficulties of cultivating any more than a few hectares by traditional methods of clearing and hoe cultivation.

Small-scale peasant farming is a predominant element in Third World agriculture. In such circumstances, no matter how careful the use of such minimal local resources of land and labour, there is very little room for manoeuvre in most peasant households. They may always be very close to the margin of survival, so that low prices, taxation, bad health, unexpected weather conditions, civil unrest or any one of many other such hazards may spell disaster. Yet one of the features to emerge most clearly from the study of traditional agricultural systems in different environmental conditions (see Chapter 4) is the use of local resources in such a way as to support the local community by production of foodstuffs both for internal consumption and for sale or exchange, and also to maintain the productive capacity of local resources for future generations. The *primary* aim of production has traditionally been for subsistence, either at the level of the domestic household or at the level of the community as a whole, but most peasant societies must also produce an additional surplus for sale, exchange or the payment of rents and taxes. The balance between production for sale and production for subsistence is, however, changing in most Third World countries. Second, traditional peasant systems relied *primarily* on inputs drawn from the internal resources of the community: energy in the form of human labour or animal power; materials for fuel and construction; and the maintenance of fertility by fallowing, regeneration of the forest, or by application of green or animal manures.

It is also necessary to stress that rural communities and peasant households are not simply agricultural units. They are also concerned with many other activities which in the advanced industrial world have become specialized industries. To a greater or lesser extent, depending on the degree to which local economies have become commercialized, peasant households may be involved in food processing, and the domestic production of textiles, clothing, and household and agricultural implements. They may carry out much of the work in house construction, using predominantly local materials and perhaps with the help of specialized artisans such as carpenters or metal workers. Agriculture is an essential part of rural life, but it is generally a part and not the whole. What in the Western world are separate activities and specializations are often, in the Third World, the concern of communities involved in other activities besides agriculture, entailing an intricate use of time and labour throughout the year, but always closely geared to the insistent demands of agriculture and climatic seasonality.

The domestic household is the basic unit around which these multiple activities usually revolve, though the composition of the household may vary from place to place. The division of labour between the sexes has increasingly engaged attention in the last twenty years or so, particularly in connection with the status of women, but studies have thrown considerable light on the complexities of domestic arrangements in different parts of the Third World. In Africa, women are generally responsible for the care of small children, 'housework', cooking and fetching water or firewood where this is the normal fuel (Box 5.1). The gathering of firewood often takes longer when surrounding areas are denuded of their trees to make way for farmland. In many areas they may also be responsible for many food-processing operations which are normally done in factories in the developed world. The grinding of wheat or maize into flour by hand or the pounding of cassava are laborious and tedious tasks, require time as well as physical effort and this task too becomes more arduous when families of young children as well as crop production grow in size.

In other respects it is difficult to make generalizations about the division of labour between the sexes. Where plough agriculture is practised, it is usually men who cultivate the land, and especially so when mechanization takes over. Where hoe cultivation is normal, in West and Central Africa for example, women usually do a great deal of the agricultural work, including planting, weeding and harvesting, which they may share with the men, who are in turn responsible for clearing the bush in preparation for cultivation. In the intensive irrigated agriculture of China and South-East Asia, men and women share the agricultural tasks. Marketing shows great variations in the role of the sexes. Women may frequently market surplus subsistence crops for cash or barter in local markets, but men may take the income from marketing valuable cash crops or larger items – cocoa in Ghana, barley for brewing in Bolivia, alpaca wool in southern Peru, and cattle throughout Latin America.

There is no clear correlation between the status of women and the economic functions they perform. In East Africa, for example, Maasai women may look after the cattle which are the household's main assets, but only the menfolk have the right to dispose of them. In general, the participation ratio of women in agriculture in the Third World is high, except in Latin America, and in parts of North Africa, the Middle East and South Asia where religious culture may require that women should be confined largely to the home or to 'invisible activities'.

The productive functions of women are changing as peasant farming becomes more commercialized and as standards of living rise. Men take a greater part in marketing and often, therefore, take charge of the cash income that accrues, though not always to the benefit of their families. Commercialization implies the use of manufactured goods bought in the market – processed foods such as flour and pasta, clothing and household utensils such as simple kerosene stoves. Some of the traditional responsibilities of women are thus removed. On the other hand, child care, hygiene and health care become more time-consuming than formerly, and education may reveal new horizons and expectations which may revolutionize women's attitudes, as they have in the region of Lake Titicaca (Benton, 1987, 95).

In most of the Third World the rural community has an identity and a cohesion to a much greater degree than in the Western world. It is usually a unit for religious celebration and for the organization of festivals and ceremonial rites; it may have customary procedures for the appointment of village chiefs or communal officials. Other functions serve also to integrate the community (Plate 5.2). Even where individual peasant holdings are carefully and jealously demarcated, land may be held in common and communally controlled for the grazing of animals, the extraction of wood for fuel and house construction. Where land is irrigated, the allocation of water and the maintenance of canals and ditches is often a collective responsibility. So also may be the construction and repair of local roads or communal structures such as a meeting place, a school, washing places or a football field.

Communal needs of this kind may sometimes

BOX 5.1 WOMEN IN AFRICAN AGRICULTURE

The variation in the type and amount of work done by men and women is of crucial importance to understanding the nature of agricultural production. The common view of traditional divisions of labour between the sexes is that domestic food production and rearing of small livestock, together with the fetching and carrying of water, forage and bedding for animals, and kindling for fires is firmly in the hands of women. Clearly a range of responsibilities associated with child rearing and household management has limited women to those activities which can be integrated with these functions. Equally it has been assumed that men have to carry out the heavier tasks of clearing woodland and farming with animal ploughs, herding, and the cultivation and marketing of cash crops. These social and symbolic relationships involved in the sexual division of labour have not always been clearly understood by planners when new crops or cropping systems have been introduced and social stresses have ensued.

It is now clear that these older patterns have been breaking down for some time and there is a clearer appreciation of the part played by women in farm labour as a whole, even if earlier estimates of between 60 and 70 per cent of all agricultural production have given way to a recognition that there is considerable local variety. In some areas women still concentrate on crops which can be grown close to the dwelling, but especially where male labour migration has been substantial, as in western Kenya or Zambia, women may undertake almost all but the most arduous tasks, cultivating a wide range of crops (see Figure 3.7). Clearly, however, the introduction of new crops can create a complete change in social and economic behaviour as people adjust to new farming cycles and to the relative significance of the new patterns in the cash economy. This is particularly important when a woman undertakes work for pay, becomes involved in cash-crop farming or takes up marketing. In some societies, like Ghana and

Plate 5.1 Women recording their savings at a rural credit scheme in Kenya To counter female illiteracy, a system of symbols is used to record their money

BOX 5.1 ***continued***

southern Nigeria, women have taken over certain types of market trading completely, creating for themselves a level of economic independence unknown in other parts of the Third World. In all these circumstances there is a social impact. Social relations between men and women take on new forms, as when emphasis is placed on encouraging women to take initiatives by forming farming cooperatives. For the first time they may be controlling their own cash flows and this creates options for other independent behaviour (Plate 5.1). Divorce rates may increase and women adopt different strategies of fertility for themselves. Patterns of marriage and child rearing are altered. More farm work devoted to cash crops means less time for food preparation and child rearing. So the steps by the women involved in rural development work to encourage their rural sisters towards greater levels of 'empowerment' have to take sensitive account of the changes in behaviour which such policies bring and to make sure that the infrastructure is provided to take account of such changes.

Plate 5.2 Village scene, Ethiopia Villages are an important focus for meetings, commerce and local administration in rural areas

be met simply by forms of local taxation, but in the central Andes, for example, they have traditionally been met by a levy of labour on the peasant membership of the community. All contribute their labour on appointed days to lay out a new plaza or to mend irrigation ditches, and the occasion is made festive by the provision of food and drink. Peasant households may be able to call in the labour of their remoter kin or neighbours in the community for work beyond the scope of the household itself, in house construction, forest clearance or harvesting, for example. But the provision of help sets up an obligation for reciprocal assistance at some later date, and these obligations are carefully calculated according to such factors as the degree of kinship and social distances of the parties concerned.

The cohesion of rural communities varies enormously in the Third World, but in general they are the units that give a sense of participation, stability and permanence, much more so than loyalty to the national state and its institutions. Agricultural operations are characteristically embedded deeply in the whole fabric of rural society, and it is important, therefore, to recognize that agricultural changes adopted in the process of modernization frequently imply consequential social change in rural society as a whole.

Adaptation to change

Peasant communities are still a dominant feature of the Third World. Yet while they have shown a remarkable talent for survival, they have also shown themselves able to take the initiative in many ways to adapt to the changing pressures of the modern world. Yet the speed of innovation in such communities has often been seen as needlessly slow by 'outsiders' concerned to increase peasant production and welfare. Indeed, the whole field of 'rural development' can be represented as the attempt by merchants, entrepreneurs, agents of government or of international organizations to make rural areas more productive.

Attempts at modifying peasant agriculture may be seen in quite different ways by those directly or indirectly involved. Many peasants see such 'rural development' as the efforts of 'outsiders' to devise sophisticated methods for subverting, circumventing, transforming or accommodating their farming structures in the interests of enlarging landholdings, improving yields or introducing new crops. Conversely, some external commentators see such interventions as the only basis for modernizing rural societies. Others stress how the efforts of business interests and governments have, unwittingly or not, transformed the peasant into a wage labourer in an urban-oriented production system, and have concentrated wealth into fewer hands, increasing social and economic inequalities between rich and poor.

Among the 'outsiders' there are, on the one hand, the forces of free enterprise: the landlord, the merchant, the money-lender and more recently the representatives of multinationals and agribusiness. On the other hand there are the forces of government and bureaucracy: the tax collectors, district administrators, politicians and international aid advisers. All these agents have an interest in increasing the surplus from peasant landholdings or in releasing land and labour for more productive use. Whether this is seen as marshalling the productive energies of the system or merely its exploitation is very much a question of interpretation and of local circumstances, but for the peasant the result is much the same.

The idea that peasant farmers of the Third World are inherently conservative and resistant to change is a myth. Even with limited technology, peasant societies have adapted to the growth of population, the introduction of new crops and the pressures of international commerce and urbanization, though there are limits to the traditional dependence on locally available resources and serious consequences for the environment in many areas. Shifting cultivation has given way to bush fallow in West Africa, producing a greater output from a given area of land even though yields per hectare of *cultivated* land have fallen. In the Far East, complex systems of terracing and water control have enabled larger populations to be supported. In Indonesia the replication of the tiered structure of the rainforest on a small scale by the cultivation of crops at different levels (sometimes known as agricultural involution) makes the maximum use of small plots of land. Over the centuries peasant farmers have readily adopted new crops, not only for cash income but also for subsistence. Though they are not blindly resistant to change, it cannot be said that they are profit-maximizing entrepreneurs. Their strategies have sometimes been seen as 'satisficing' in the sense that they aim at an adequate return for effort. Peasants behaving as '**satisficers**' may be prepared to trade off potential income for extra effort against the leisure that must thereby be forgone. But it is also important to recognize two other factors that affect peasants' attitudes to change. One is that the peasant household is not only concerned with a strategy for the use of its land, but also with a strategy to make the best use of the labour available. Such use of labour from the point of view of the peasant household may not coincide with the perspective of external observers, particularly where seasonal migration demands a careful scheduling of agricultural operations. Second, peasant strategies towards agricultural change can often best be understood in the context of risk avoidance. Farmers in the Third World face natural hazards of drought, pests and plant diseases, unseasonable rainfall, hail or frost; they have to deal in markets which are often unstable and unpredictable; and with governments and government policies which are also unstable and unpredictable. Risk avoidance is the only sensible strategy under these circumstances, and change can only be risked if there is a virtual certainty of success. In short, peasant attitudes often reflect a careful scheduling of time and labour in relation to the resources available to the whole household rather than the individual. Value systems may differ from those of Western Europe or North America, and it has been difficult for scientists to avoid imposing Western concepts in their analysis of peasant behaviour.

Rural societies in the Third World have adapted to change in the past, but what is different today is the rate and scale of change, and the strength of the pressures towards change. These pressures are complex and often subtle and indirect, but two major categories may be identified: the consequences of rapid population growth, and the need to move from subsistence to commercialized farming.

Responses to the rapid growth of population

Rates of population growth in much of the Third World have accelerated rapidly since 1950. Rates of 2.0 to 3.5 per cent per year have outstripped the capacity of many rural communities to support themselves by traditional methods of intensification relying on local resources. In many areas the limits of land cultivation by traditional methods have long ago been reached and further extension requires heavy investment by governments or international agencies in large-scale projects of irrigation, land reclamation and road building, in land reforms that may make land available to peasant farmers, or in off-farm economic activities in the areas of land shortage or beyond.

Meanwhile, in the absence of such help, peasant communities take up whatever spare land they can, frequently by cultivating marginal and submarginal land or by grazing too many stock on natural pastures. The results are all too evident in many areas: soil exhaustion, accelerated soil erosion, degraded secondary vegetation and over-grazed and exhausted pastures. In Kenya, for example,

> the harmful effects of population pressure on Kenya's farmland is currently taking many forms. . . . Part of the marginal (low, dry) land is becoming arid because overpopulation is leading to improper cultivation. Squatter migrants on marginal land often harmfully exploit natural resources, both by destroying forests and by using poor farming techniques. (World Bank, 1980, 55)

In the densely populated provinces of Kenya bordering on Lake Victoria, more than half of the small farm households are estimated as being below the national poverty line and land hunger is rife. In the 1960s and 1970s land hunger was alleviated by planned resettlement movement into the Highlands to occupy fertile and well-watered land vacated by European farmers, and sold or allocated, often but not always in much smaller lots, to family groups from the overcrowded areas. By the 1980s, however, most available land in the Highlands had been settled, and new settlement was going, largely in spontaneous rather than planned movements, to the lower and drier margins of the Highlands, areas more susceptible to environmental hazard.

In intensively cultivated areas, the most obvious consequence of the growth of population has been the fragmentation of holdings as a result of the division of land among the children of successively larger generations. Even allowing for the occasional existence of communal pasture or woodland with which individual farmers may supplement their resources, the scale of the problem for small farmers is acute. On such tiny holdings there are rarely savings enough to invest in improved seed, expensive fertilizers, pesticides, etc. Small-scale cultivators are often too numerous and too scattered to be reached by agencies that offer credit or technical assistance even if the political will exists to reach them.

In the absence of cash crops, or with too little land on which to grow them, cash income must be secured by the sale of labour, usually by migration to large estates, plantations or, above all, to the towns (Chapter 3). Seasonal migration has long been customary where there is a slack period in the agricultural year. The demand for seasonal labour may not be so conveniently arranged, however, and the carefully evolved scheduling of labour in traditional societies may be disrupted by the timing of seasonal employment. Normal cultivation practices may be abandoned, including such vital tasks as weeding and hoeing.

The migration of labour on a temporary or seasonal basis brings income back to the villages; more permanent migration to the towns may yield remittances from the migrant which may sustain tiny and uneconomic holdings in the absence of active youth to work them. There are other consequences. Migration to the towns, offering some possibility of release from the grinding poverty of rural life, becomes normal, so that ambitions for the young focus on educating them for urban life rather than an agricultural future, and the family plot is seen as no more than a subsistence base. Surplus income, often from urban remittances, will tend to be invested in education and school fees rather than on the land, for returns from an urban job for an educated migrant are perceived to be likely to be higher than from a small family farm. The family holding is, in short, seen as an insurance against disaster and a guarantee of survival, but not as a primary source of income. This role should not be forgotten in countries where health and social services, social security and unemployment benefits do not reach the countryside or most of the urban poor. The insurance element is a factor to be taken seriously (though it rarely is) when considering the arguments for and against the consolidation of plots into viable holdings in the interests of agricultural efficiency.

Subsistence to commercial production

As outlined above, subsistence agriculture implies that farmers produce a substantial part of their own food requirements. This does not preclude the production of a surplus for sale or barter, the proceeds of which may yield a cash income to pay rents and taxes, or for other purposes. The difference between subsistence and commercial farmers may thus be seen simply as a matter of the degree to which they participate in the market.

Many traditional farming structures are based on the exploitation of a variety of ecological niches or seasonal changes that allow for the production of a varied and complementary pattern of goods, either at the level of the individual household or of the rural community. The exchange of goods takes place by negotiated barter or conventional money valuations at organized rural markets. Concentration of production on commercially viable crops frequently distorts or destroys this complementarity. For example, maize may be sold in the national market at prices higher than those traditionally offered to producers of, say, barley and potatoes, who may find it difficult to obtain the maize they want; locally produced wheat and barley may be replaced by cheaper imports.

The consequences of such changes may be to set in train agricultural changes towards greater maize production in the first case and the replacement of wheat and barley by something else in the second. Total cash income to the community may be increased, but it is not often equally distributed. The innovators in the community or those fortunate enough to hold or to acquire the appropriate land (irrigated land for maize cultivation in the example mentioned above) may receive substantial rewards for their efforts, but those with land suitable for only wheat and barley (to continue the analogy) will be impoverished by a shift to commercial production. Others may be impoverished by the loss of non-agricultural occupations or by the fragmentation of holdings following the growth of population. One consequence of modern trends has therefore been the creation of greater economic inequalities in rural communities, and these are often accentuated as the fortunate gain access to credit facilities, improved methods of cultivation and even aspire to middleman functions. To be in a position to acquire a lorry and become a trader is commonly an ambition of the aspiring farmer throughout the Third World.

Many studies of rural communities have discovered that modern changes are leading to a greater degree of inequality, and that greater inequality has in turn strained the social ties which integrated the community. Inequality in wealth or land ownership threatens reciprocal labour arrangements and traditional controls over the use of communal agricultural resources, since the relatively well-to-do are often able to pre-empt a major share of communal grazing, or to appropriate for their own use land formerly allocated equally to members of the community for temporary cultivation. At the same time, returned migrants, the better educated and the young innovators are inclined to challenge the traditional order of seniority.

In the transition from subsistence agriculture to commercialized farming and a totally monetized economy there is a threshold to be crossed at the point where the peasant strategy of decision-making shifts from the need to assure the household's food supply to a desire to maximize cash income from the sale of farm products. The peasant must know that it is possible to sell produce for a satisfactory price and also that there is a *reliable* supply of essential goods to buy. It is a step which presumes a fairly advanced commercialization of the national economy. But the crossing of this threshold also carries with it other implications. It implies the replacement of a varied, risk-spreading range of crops by a high degree of specialization on those products which yield the highest cash income. Specialization means a greater vulnerability to climatic hazards and price fluctuations on the national market. In short, specialization may bring with it a greater degree of efficiency, higher total output and greater incomes, but at the price of greater risk.

There is another sense, too, in which an important threshold is crossed at the point where the decision is taken to concentrate almost exclusively on production for the market. In predominantly subsistence societies agricultural inputs of stock, implements, seed and fertilizer are drawn from locally derived sources: stock from natural increase of local animals; implements from simple and often home-produced raw materials with a little imported metal; seed from the previous harvest; and fertilizer from animal manure, domestic wastes, green crops and rotations involving legumes. Crossing the threshold to commercial agriculture makes it necessary to widen the field from which agricultural inputs are drawn if yields and quality are to be raised to satisfactory levels. Seed and manufactured fertilizer must be purchased; agricultural implements and machinery bought and improved breeds of stock introduced (Plates 5.3 and 5.4). To buy these things, credit is usually needed and this, too, increases the farmer's vulnerability – to the weather, and to banks, as well as to the unpredictable operation of market forces. To participate wholeheartedly in the national, commercial economy is to enter a world of new risks associated with price fluctuations, credit, rates of interest, politics and changes in government policy; and in many cases, the necessary conditions of political and economic stability at the national level do not exist.

In spite of the evidence of an ability to make adjustments to the pressure for change, many peasant farmers have been squeezed out by the loss of their tenancies, the fragmentation of their holdings or their inability to adapt to the demands of change, and have joined the ever-increasing ranks of landless rural labour or the stream of permanent migrants to the cities.

Systems of rural exchange

'Traditional' farming systems provided subsistence and were self-sufficient, with little need of exchange. However, the vagaries of climate or the health and fortunes of the peasant family mean that there are always times of famine and times of plenty. Deficits and surpluses occur from year to year, but may be resolved by the

Agricultural modernization

Plate 5.3 Farm mechanization, São Gotardo, Brazil In an area into which the 'agricultural frontier' has recently advanced, in the *cerrado* (savannas) of Brazil, large-scale agriculture has been introduced, using combine harvesters and other machinery for the cropping of maize

Plate 5.4 Improved pastoralism, São Gotardo In the same area, there has also been the introduction of modern pastoralism. Friesian cattle and calves are fed fodder crops. The grassland is cultivated, and in the background there is a silage tower to provide dry season fodder. This contrasts with the natural pasture and unimproved cattle typical of traditional Brazilian cattle ranching

exchange of gifts, though as with the exchange of labour discussed above, the offer of gifts sets up a reciprocal obligation. Further extensions of this principle are the gift-making ceremonies that reinforce ties at all levels, cementing marriage bonds with dowries, placating enemies, winning political support, gaining prestige or impressing a rival. The gift is part of the political, social and economic system of many Third World countries.

Barter is the next stage in the establishment of rural exchange systems. A family with a surplus of beans and a deficit of rice looks for someone with more rice than is needed, and preferably a shortage of beans. They may be exchanged by a conventional, but essentially local, system of equivalence which may have little to do with prices ruling in the national market, or price may be determined by negotiation (or 'haggling'), which is an essential part of many rural transactions in the Third World and has a social, as well as an economic function. Where many items are to be exchanged there are clear advantages in meeting at one place where comparison is possible. The rural market place is a central agreed point for collective bargaining, barter and the fixing of values. It is also a place for the exchange of information, gossip and ideas, for organizing social events, settling quarrels and making political deals (Plates 5.5 and 5.6).

The full development of rural markets has been related to a particular stage in the growth of a local and regional economy when an urban system is emerging and the range of commodities circulating in the rural areas is expanding. Rural markets become part of wider trade relationships that bring brokers from the towns, both to buy local produce for shipment to wholesale markets in the towns and also to sell manufactured products and other goods. The rules that govern the spacing and timing of rural markets operate according to local systems of social, economic and political organization. The distances between markets are governed by the time taken to travel on foot or by horse or mule to reach the scheduled market place. Spacing and size are also conditioned by the ways in which traders circulate. Some larger markets are held each day, but most are held periodically in weekly cycles or periods of 3, 4, 6 or 10 days. Whether a market is held periodically or daily depends on the range of goods exchanged and the distances involved – factors related to the density and wealth of the population. The rural market is deeply embedded, of course, in the history of the Old World, but not all regions are blessed with a rich network of rural markets. Small and scattered populations, a low range of produce or an unfavourable social structure may inhibit their development. In the region of Lake Titicaca, for example, land reform and the break-up of large estates, whose owners formerly channelled produce directly to the towns, was followed by the spontaneous growth of markets as peasants assembled to exchange or sell surplus produce from their new lands.

The importance of rural markets has been somewhat diminished by the development process. New cash crops or larger and more varied harvests for sale may by-pass the local markets altogether in consequence of road building and the activities of local or town-based truckers. The establishment of government-controlled marketing agencies, where prices are determined by national or international norms, diminishes the role of local valuations and places of trade. A pessimistic view is that as more and more traders and producers rationalize sales and purchases in this way, rural markets and their rich web of social functions and connections may become a relict feature of the rural landscape. Modern systems of organization are too well orchestrated to be able to absorb the lively, self-regulating anarchy of the rural market place.

CHANGING AGRICULTURE

From a local to a national economy

In the 1960s and 1970s, former colonial territories in Africa and Asia were striving towards nation-building policies, and in Latin America, economies formerly dominated by export sectors and the elites associated with them followed policies of internal development. These shifts implied strategies aimed at industrialization, energy production, the creation of a

Marketing

Plate 5.5 Fish market, Ghana The gender division of labour has the men at sea catching the fish, and the women selling it on the beach to local farmers and urban-based traders

Plate 5.6 Urban market, Sousse, Tunisia Rural produce, mainly vegetables, is sold to the urban population. In this Muslim society, the men are the traders, and the veiled women the buyers

national network of communications in place of export-oriented links, investment in education, the widening of the political franchise, subsidized credit and rapid urbanization. Rural communities which were formerly no more than marginally affected by the need to produce cash crops are increasingly involved in a wider, national society. Urban expansion has created growing internal markets for agricultural produce; industrialization has created new markets for raw materials; rural education and improved communications have stimulated a wider consciousness of economic opportunities and a 'revolution of rising expectations' in terms of material welfare; subsidized rural credit has tempted farmers to produce primarily for the market rather than for subsistence. In isolated rural communities the stimulus towards commercial production and social change may come by way of a new road and the appearance of motor transport, the return of migrants bringing new skills and experiences, the presence of an influential teacher in the village, the mobilization of the community against the depredations of a neighbouring landowner, and also, of course, through the direct action of government agencies.

The consequences of such innovation may, however, be ambiguous. Road building has rightly been seen as a prerequisite for the extension of the market. There are many examples of the rapid expansion of production for the market following the building of ports, roads and railways. Unused resources may be called into play for the production of foodstuffs or raw materials for urban or international markets. Lower costs of transport for the marketing of surplus products should, in principle, stimulate the concentration of effort on those products for which the community may have a relative advantage, and thus lead to greater income and welfare. This equation also has a negative side. It becomes possible, and cheaper, to *import* goods into rural communities which were formerly self-sufficient. In historical terms, cheap imported manufactured goods (textiles, clothing, plastic goods) have in many places replaced indigenous, traditionally produced goods of homespun fibres, wood or pottery. In one sense, these imports represent a net gain in terms of standards of material welfare, but they also remove many of the non-agricultural functions of the rural population and help to create rural underemployment. Similarly, improved roads also bring in imported foodstuffs which may have a greater prestige value than the traditional diet. One of the effects of road building in some Andean communities has been to bring in rice and imported wheat products at the expense of locally produced potatoes, barley and maize. In the forested eastern slopes of the Andes, one remote and old-settled community virtually abandoned its own domestic production of rice for imported products following the construction of a new road – and promptly turned to the destruction of forest to extract its marketable timber.

Plantations and agribusiness

Large estates, plantations and **agribusiness** make a substantial contribution to the export trade of many Third World countries. Their origin and structure were precisely geared to that end, and such large units of production have commonly been seen as the most obvious locales for innovation and modernization. Plantation agriculture has been defined as the 'large-scale production of tropical crops by a uniform system of cultivation under central management' (Courtenay, 1965, 7; the 1980 edition of this book contains an extensive discussion of the meaning of the term and its evolution over time [Courtenay, 1980, 1–18]). The stress is on the management of the plantation as a business enterprise rather than on the prestige of owning a large estate, and the implication is that such modern and capitalistic structures would be most amenable to new developments.

Although historically rooted in the colonial period, and associated with sugar, cotton and slavery in the Americas, plantation agriculture became of greater significance in the nineteenth century, characterized by foreign ownership, technical innovation and large-scale monocultural production for export. The reasons for their development are bound up with the changing relationship of Europe and North America with the tropical world, which involved:

- an increase in the volume of trade from about 1880, associated with rising standards of living, growth of population, and the increasing complexity of the industrial world. The demand for imports of tropical agricultural produce grew rapidly and became more varied – rubber, coffee, tea, palm oil, for example, as well as cotton and sugar;
- technological change in transport, with railways and improvements in shipping (from sail to steam, the screw propellers, and refrigeration, for example);
- direct foreign investment in colonial or ex-colonial territories had been limited mainly to railways, port facilities and public utilities, but from 1880 there was an increasing tendency for foreign capital to be invested directly in the production of tropical foodstuffs and raw materials;
- the establishment of 'safe' environments for investment through the vast extension of colonial empires in Africa and South-East Asia.

Plantations were not the only means of increasing exports of tropical 'plantation crops' to the industrial world, since some output came from the initiative of 'traditional' farmers. The success of cocoa production in Ghana and palm oil production in Nigeria, for example, depended on the initiative of peasant farmers, though the major export trade links were dominated by Europeans. In Java, peasants were required to grow and deliver cash crops during the Dutch administration, and in parts of Asia and Latin America some production of tea and coffee has always been associated with peasant production rather than with large estates.

As Third World countries have achieved political independence, traditional plantations under foreign management have been undermined as a result of decolonization and subordination to new national authorities. Changes in the terms of trade have also challenged their profitability. It has become increasingly difficult for Third World countries to earn sufficient foreign exchange by the sale of commodities such as bananas, coffee, tea, sugar, cotton and tobacco, for which demand is elastic (i.e. it varies significantly with changes in price). Despite increases of more than a third in the tonnage of such cash crops since the early 1960s, the real income earned has scarcely increased at all.

Governments struggling to improve their foreign currency earnings or to raise GNP seek cost-effective methods to increase the productivity of their export crops. One way to do this has been to facilitate the entry of large-scale multinational commercial enterprises using modern technologies and business organization, which have come to be known as agribusinesses. Large international consortia, often with only a secondary interest in agriculture, increasingly finance, control and organize the production of most tropical crops. Large firms with a base in international finance even come close to controlling the economies of smaller countries, manipulating political conditions to serve their need for economic stability. There has, in fact, been a world-wide growth of agribusiness, which essentially involves the application of modern methods of organization and technology to the creation of vertically integrated enterprises handling agricultural products from the farm to the consumer. Multinational agribusiness is concerned not only with the production, processing, packing and transport of agricultural commodities at source, but also with their processing, marketing and distribution, usually in the developed world (Box 5.2).

It was notoriously in the banana plantations of Central America that American enterprises wielded economic and political power over governments and local populations alike. The United Fruit Company was, in particular, associated with the banana industry. As well as transporting the fruit to northern markets in specialized ships, it came to control the sources of production. Large estates were acquired and devoted exclusively to the cultivation of banana. Exports grew and foreign currency became available for the finance of imports, and to the benefit of local elites. National interests were often, unsurprisingly, subordinated to those of foreign companies. Economic policies were followed which discouraged local industry and favoured the exporters. Land was alienated to the companies and a landless, wage-paid labour force was created, vulnerable to unemployment in slack seasons and at times of recession.

In recent years, at least in Latin America, the operations of agribusiness have begun to recede from land ownership and the direct operation of

BOX 5.2 AGRIBUSINESS IN AFRICA: LONRHO

The multinational corporation, Lonrho, is involved in mining, manufacturing, tourism, services and agriculture in more than 60 countries in the First and Third Worlds. It claims to be the largest commercial food producer in Africa, farming 490,000 hectares of land and 120,000 cattle. It has operations in nine countries of Central, East and Southern Africa (Figure 5.1) in which it has sugar, tea and cotton plantations, and also produces soya beans, maize and wattle. However, even such a major and diverse company demonstrates the variability and vulnerability of Third World agriculture. In 1992 its sugar operations in Malawi, South Africa and Swaziland increased output and made record profits, and its arable yields in Kenya were at record levels. Conversely, low world prices for tea, cotton and coffee affected profits in Malawi and Mozambique, and drought in Zambia, Zimbabwe and Mozambique had negative effects on arable and ranching activities. In addition to its agricultural activities in Africa, the company has industrial linkages in the ginning and production of cotton textiles in Zambia, Zimbabwe and Malawi. There is also a diverse portfolio of mining, manufacturing and commercial activities in these countries.

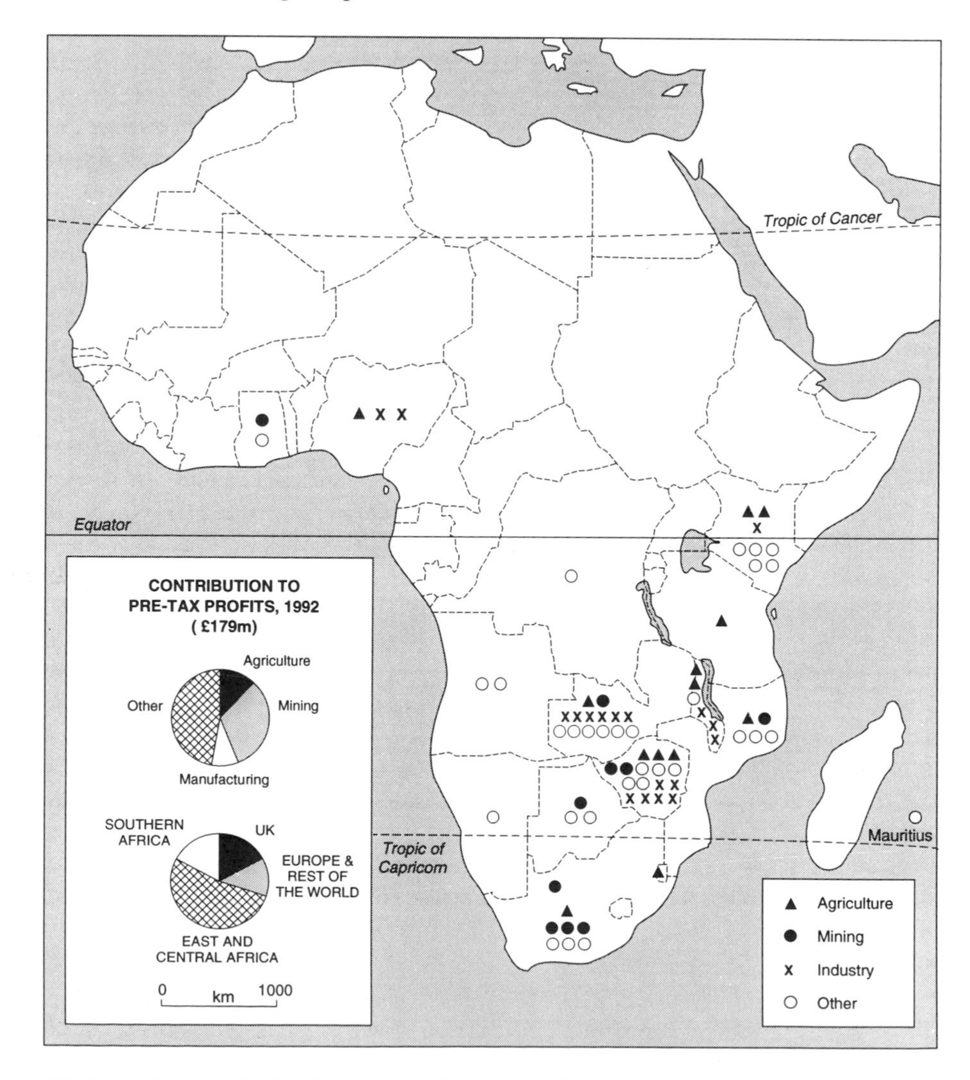

Figure 5.1 Agricultural and related activities of Lonrho in Africa, 1992
Source of data: Lonrho, 1992

large estates, though they continue to control many aspects of production, processing and marketing, and have extended operations to new areas of activity. United Brands, successor to United Fruit, for example, abandoned some of its large landholdings, partly because of the threat of agrarian reform, partly because of labour problems, but above all, because new methods of organization reduce the inherent element of risk in agricultural production and increase profitability.

An essential feature of this new tendency, particularly well-developed in Mexico, is a contractual agreement with small and medium farmers to produce a required crop. The means of production – seeds, fertilizers, pesticides and details of the method of farming, including such matters as the date of harvesting or planting – are determined by the company, but much of the risk falls on the shoulders of the small farmers, whose crop may still be rejected if it fails to reach the minimum required quality. The spread of agribusiness of this type, however, has made possible a greater volume and variety of production by small and medium farmers lacking capital, expertise or marketing facilities. New products have been added to the traditional repertoire of export crops. Vegetables and fruits such as tomatoes, pineapples, passion fruit, oranges or strawberries are produced and processed in Mexico for export to the USA and elsewhere – fresh, chilled, deep frozen, canned or as pulp or juice.

Agribusiness in the Third World is no more than the application of techniques learnt and used in the developed world, tending to increase the scale of operation by mechanization and high technology, but leading eventually to a reduction in the involvement of local people either as a labour force or as managers and independent farmers or traders. Even governments take a secondary role in these operations. Faced by sophisticated technology and high finance, governments are often content merely to take a share of the profits. Indeed, it is ironic that so much foreign exchange acquired in this way is used to buy foodstuffs for the richer elements of an urban population which have grown used to the products of temperate agriculture, such as wheat, dairy products, packaged foods and alcoholic beverages, the consumption of which is a mark of status and prestige. The insertion of agribusiness may ease balance of payments problems of a relatively new country, but it creates a level of economic dependency which may not be in the long-term interests of national agriculture unless steps are taken to transfer technology and know-how.

The state and agricultural change

Intervention by the state has taken a number of forms, including revolutionary transformation. Some of the actions of government are positive in the sense that they are directly and consciously intended to raise the level of agricultural production or to improve the welfare of the rural population. Depending to a large extent on the level of national income and involvement in agricultural exports, governments have sought to establish rural extension services devoted to basic agricultural education, the propagation of ideas and techniques which are thought to assist the participation of small and medium farmers in production for the market (Plate 5.7). Model farms or specialized plots may offer demonstrations of new or improved crops, the effects of fertilizers, pesticides, etc. There is generally no shortage of advice, though it may not penetrate widely or deeply enough to change current practices, and for many peasants it may simply be irrelevant. Subsidized credit may be on offer, usually tied to the adoption of recommended schemes. Government agencies of this kind, the establishment of agrarian universities and the training of agronomists have certainly helped in the development and propagation of high yielding varieties. Often enough, the main beneficiaries have not been among the peasant farmers, but among the comfortably off or the well-to-do. And poor countries simply lack the resources to establish agencies of this kind.

A second kind of direct intervention has been by way of the provision of physical and social infrastructure, with the building of roads, the provision of electricity, clean water supply, irrigation, medical extension services, health posts and rural schools. Since the 1960s considerable progress has been made by most

Plate 5.7 Agricultural education, western Kenya Village polytechnics, designed to provide skills to rural school leavers and discourage urban migration, have plots for training in agriculture for both boys and girls in a region where agriculture has traditionally been a mainly female activity. In this case a seed bed is being prepared

middle-income countries of the Third World in at least some of these areas, though again, low-income countries may be held back by their lack of resources.

On the other hand, there are also negative factors which have hindered rural development in various ways. It is axiomatic that in a country totally reliant on its agriculture, capital accumulation must rely on either sources external to that country or on surpluses extracted from its agricultural production. At a more specific level, Latin American landowners have notoriously extracted surpluses from their estates in order to finance expenditures (or investment) overseas. Direct and indirect taxation may similarly bear heavily on peasant populations. Poll taxes as a regressive tax bearing more heavily on the poor than the rich have their precedents in British colonial practice as well as in medieval times. More importantly, perhaps, the drive to protected industrialization by way of import substitution was achieved by imposing tariffs, quotas or other more drastic means which invariably had the effect of raising the costs of manufactured or imported goods to rural populations whose agricultural produce enjoyed no such protection. Terms of trade, already unfavourable to Third World countries as against the industrial world, were made even worse for rural populations within Third World countries. 'Cheap food' has ever been the cry of politicians dependent on urban voters or vulnerable to metropolitan opinion. Food prices have commonly been controlled for the benefit of the towns, to maintain low wages for industrial sectors, or to placate a restive urban population. Yet it has frequently been shown that the most effective way of improving agricultural output is simply by raising food prices.

Actions of government in two areas of rural development will be given more detailed treatment in order to clarify some of the issues involved: in the organized planning of agricultural colonization and settlement; and agrarian reform as a deliberate attempt to remould rural society on a wider front.

Colonization and settlement

Colonization schemes vary enormously in the degree of planning and the scale of investment. At a minimum level they may involve little more than road building to a new area, land survey and the allocation of plots, or they may involve heavy investment in the provision of irrigation facilities before settlement can take place. More elaborate schemes include the integrated planning of rural settlement patterns with new villages and urban nuclei. Settlers may be provided with land already cleared and prepared, and resources to sustain them until the first crops yield an income, as well as housing, health services and schools. Schemes vary in the care with which potential settlers are selected as suitable for their task. Some colonization schemes have proved excessively costly; most have been less successful than the planners had hoped.

The stated aims of colonization schemes have often been mixed or vague and, indeed, incompatible. They may be briefly summarized:

- to increase national agricultural output either for export or for the domestic market. In Malaysia the Federal Land Development Authority in the state of Pahang successfully organized settlement for the production of rice to replace imports from abroad;
- to consolidate a political frontier zone by the building of roads and setting up settlement colonies, often of veteran soldiers, as in parts of eastern Bolivia and Peru. This military-strategic motivation, based on the principle that occupation of territory will validate political frontiers in sparsely populated territory, seems to underlie a good deal of thinking about colonization in Latin America, notably in Bolivia, and in northern Peru, where the Northern Transandean Highway was constructed by the army in the proximity of the Ecuadorean frontier;
- to solve problems of population pressure in older settled regions. This has been a major motivation for massive resettlement programmes from Java to Sumatra and the Outer Islands in Indonesia, and for colonization schemes in Amazonia to relieve pressure in north-eastern Brazil. In the very densely populated highlands of south-west Uganda, Burundi and Rwanda resettlement schemes started by colonial authorities in the 1950s, with improved drainage in previously uncultivated swampy valley bottoms, have been continued after independence, and have had major effects on the range of crops grown. In particular the growing of vegetable crops (often temperate vegetables – cabbages, potatoes, carrots – at high altitudes) brought nutritional benefits in these areas of chronic under-nutrition, and also cash benefits where the vegetables could reach the urban markets. Schemes on the Rift Valley floor in these regions involved permanent migration from high altitudes to formally organized *paysannats* in Burundi and in the Belgian Congo before 1960, and these too continued into the 1970s and 1980s when they were associated with a rapid expansion of cotton cultivation as a major cash crop for the local textile industry. Cotton could not be grown on the higher and very densely populated shoulders of the Rift Valley, but the extension of irrigation to the rift valley floor, more accessible to urban textile factories, brought cash income to the resettled farmers;
- to divert pressure for land reform of large estates by settling peasant farmers on publicly owned land. Colonization of new land offers a strategy which may relieve the pressure of an increasingly vociferous peasantry for more land, yet at the same time leave the large estates intact. This appears to have been a device followed in Colombia and Venezuela in the 1960s and 1970s. In part, this was also the strategy in Kenya in the 1960s and in Zimbabwe in the 1980s with resettlement on former European-owned estates after their owners had left or been bought out by the state after national independence.

Peasant resettlement on new land is an expensive process, especially if it is fully subsidized by the provision of transport, tools, housing, preliminary demarcation and clearing of the site, and even irrigation schemes and a new road network. Preliminary surveys of resources, feasibility studies and administrative costs to support the inevitable bureaucracy are also costly. In cost–benefit terms, investment on such a scale is rarely justified by a subsequent growth of agricultural production, but planned settlements may later become the nuclei of spontaneous settlement on a much larger scale.

Inconsistencies arise where the aims of colonization are, in fact, incompatible. Military-strategic aims tend to stress road building and settlement colonies in remote frontier areas, but that very remoteness may give colonies little chance of commercial success in view of their isolation and consequent high costs of transport to markets. Expansion of agricultural output may be most easily and cheaply achieved by granting land in large units, which is thus likely to conflict with social policies aimed at resettling small or medium farmers. In eastern Bolivia and Brazil colonization has involved the reproduction, in a new setting, of the social inequalities of old settled regions. Even where projects are organized to settle beneficiaries on small or medium-sized holdings, there can be a conflict of policy. The success of a scheme ultimately depends on the settlers, which implies that they should be carefully selected as to age, skills, ability and motivation. This process may conflict with the idea that colonization schemes should reduce population pressures in older-settled areas, where would-be migrants may lack skills and experience appropriate to their new lands, or be used for political purposes, such as rewarding former guerrilla fighters in the war of independence in Zimbabwe.

As in so many other aspects of government intervention in rural development, success, failure or inaction depends essentially on political pressures, interest groups and the effectiveness of the state. Agricultural colonization of new land is still an important component contributing to increased food production. More has probably been achieved by spontaneous settlement and clearing of new land than by government planning, even in Latin America, where such schemes have played a significant role. It is, however, important (many would say essential) that the new lands should be occupied with careful consideration both of the ecological consequences, and of the kinds of social relations that are to be built in the new zones of settlement (Box 5.3).

BOX 5.3 THE TRANSAMAZÔNICA PROJECT

Brazil's Transamazônica highway project exemplifies many of these objectives (Figure 5.2). Originally the scheme was a rapid response to the impact of a severe drought in north-east Brazil in 1970. The basic idea was to move people from the hazardous conditions of the drought-prone north-eastern *sertão* to the empty lands of Amazonia. This would ease the problem of overpopulation in the *sertão* and help to fill the demographic vacuum of Amazonia. Migrants were initially to help with the construction of the 5000 km highway and later form the labour force to settle along it in organized colonies. The scheme was, on paper at least, neatly planned, with a careful layout of landholdings and a regular hierarchy of settlement nuclei which would provide a range of educational, medical, agricultural and other services. It was also a confirmation of Brazil's possession of the region against possible encroachment by other countries.

Originally the scheme was to absorb 70,000 families by 1974, but it did not approach this number. The reasons for the lack of success are varied. It was hastily planned, without precise knowledge of the conditions of the rainforest environment. One of the consequences of Brazil's activities in Amazonia in the 1970s was a much better knowledge of the region, but initially basic information about environment conditions was inadequate, and there were insufficient technicians, such as pedologists and agronomists, to provide assistance and advice. Because of this imperfect knowledge, settlement projects planned on the drawing board faced various obstacles when they came to be implemented on the ground. The migrants themselves were perhaps not entirely suitable. They were mainly farm labourers from a semi-arid environment who suddenly became smallholders in the rainforest. Success of the schemes was prejudiced by changing priorities on the part of the government and opposition on the grounds of the environmental consequences of deforestation and soil destruction. In addition, the post-1973 oil crisis cast doubt on the logic of developing a major highway network when Brazil was heavily dependent on expensive imported oil. The envisaged scale of the scheme has not, therefore, been realized, and the pattern of land use has been modified to one of much larger holdings used for extensive stock-raising and often owned by companies from São Paulo or abroad.

Criticism has been levelled at this scheme because of its impact on the rainforest environment, but it is probable that major damage may be done, not by organized colonists and large-

BOX 5.3 *continued*

scale pastoralists, but by thousands of uncontrolled migrants who have moved in along the highways, becoming squatters and cutting clearings in the forest. In the territory of Rondonia, opened up by the Cuiabá–Porto Velho road during the 1970s, population increased from 113,000 to 1.09 million between 1970 and 1990. In some areas, the population doubled every two years. Between 1970 and 1985 the number of farms increased from 7000 to 81,500, and the farm area from 1.6 million hectares to 6 million. At least half of these holdings were 'occupied', without formal ownership or legal tenure, and these census figures are almost certainly an underestimate as land occupation was rapid and uncontrolled on this dynamic frontier.

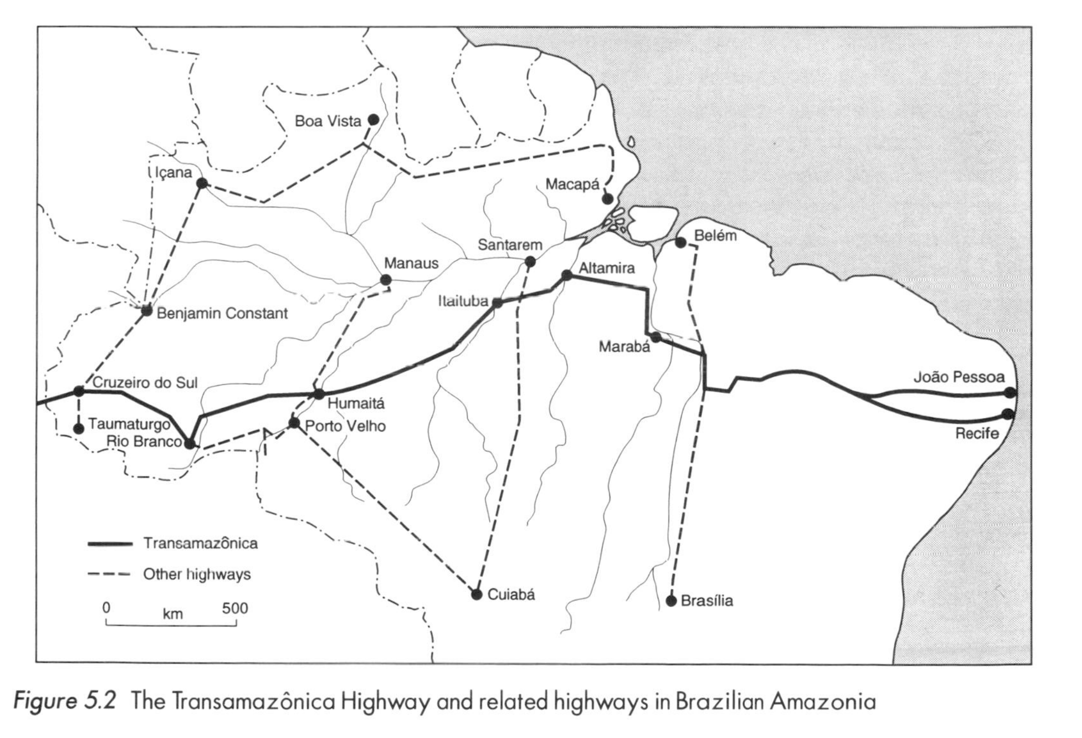

Figure 5.2 The Transamazônica Highway and related highways in Brazilian Amazonia

Land reform and agrarian reform

Agrarian reform is another major area of state intervention in rural affairs. The distinction between land reform and agrarian reform is not always crystal clear. Land reform is understood to mean the reorganization of landholding and tenurial structures, which may involve: (a) the expropriation of large estates, with or without compensation, and their reorganization into peasant farms, cooperatives or collective farms so as to benefit peasants and landless labour; or (b) the consolidation of very small or fragmented holdings into holdings of an adequate size. Agrarian reform is a wider term implying a measure of land reform and in addition the agricultural extension services, rural credit, etc., to make land reform more effective. This distinction between land reform proper and agrarian reform is one which has had important implications for the way in which rural change has been conducted.

Social justice and economic efficiency are the two basic considerations which underlie the political approach to agrarian reform, about which there has often been more rhetoric to secure support than serious intention. For this and other reasons it is useful to distinguish between land reform as part of a radical transformation of society towards public or collective ownership, and more or less 'reformist' development within a capitalistic frame of reference. Most agrarian reforms have been of this latter type.

Pressures for land reform have stemmed from various sources. Since the 1950s there has been a rising groundswell of protest from peasant populations because of the increasing fragmentation of tiny holdings, encroachment by landowners on common lands or their exploitation of tenants and labourers. Protest has taken many forms, as rural populations have become literate and better educated, more vocal, more politically conscious and more cohesive in trade unions or political parties. Rural guerrilla movements, perceived communist threats, and more moderate peasant pressures led some ruling groups to adopt a conciliatory approach by giving support to agrarian reform. On the other hand, the political power of traditional or colonial landowning oligarchies has increasingly been challenged by a new urban middle class representing commerce, industry, the professions and the bureaucracy who saw land reform of a moderate kind as a way of reducing oligarchic power and also of increasing the internal market for domestically produced goods by raising the standard of living of rural populations.

In Latin America, the economic case for 'reformist' measures rested on two major propositions. First, the purchasing power of the vast majority of the rural population was minimal, but if land reform put more income into their pockets, demand would increase for the products of local industry such as clothing, shoes or metal goods. Second, the combination of large estates and small peasant holdings often meant an inefficient use of resources: the underutilization of land and labour on the large estates, and diminishing returns to peasant labour expended on tiny holdings. Where landownership was a matter of prestige and status or even a hedge against inflation rather than the major source of income, estates were inefficiently run by agents of absentee owners. In some areas impoverished tenancies and abundant, cheap, landless labour positively discouraged the introduction of new technologies.

Land reforms involve a number of steps, at each of which there is scope for alternative approaches, for special pleading, pressure by interested parties and, possibly, corruption. Whose land is to be expropriated? How much, if any, compensation is to be paid? Is expropriated land to be divided into small and medium-sized holdings or to be organized as cooperatives, collectives or state farms? What residual rights does the state continue to exercise over the reformed sector? What measures are to be taken by the state to improve levels of agricultural output? All of these questions are important and are answered in different ways according to the social and political aims of those in power – and those in power may change and the nature of reform may change too. In Latin America there have been many examples of reforms which got no further than the statute book, or were diverted to become little more than programmes of supervised credit, the validation of squatters' rights, or programmes for the settlement of public land.

In Africa the problems associated with land reforms have been confined to those countries which had settler populations in the colonial era. Algerian land tenure has been subject to a socialist transformation that has had a major impact on the landscape. In Kenya, on the other hand, the decolonization of the former White Highlands, centred around the Million Acre Resettlement Scheme of the 1960s, served the political aim of providing peasant farmers and landless labour in the former reserves with the opportunity to acquire the freehold of former white land, often subdivided into small family plots. However, about 80 per cent of the land that changed hands was bought by Kenyan individuals or companies, and farmed in much the same extensive and commercial fashion as before independence. Output has not fallen greatly, but great political advantage has been derived from the fairly limited redistribution. More fundamental reforms have taken place in the former reserves themselves since the 1950s, particularly in the Kikuyu areas of Central Province, north of Nairobi, where there was considerable land consolidation and granting of formal title to the customary land, and substantial improvement in the income and productivity of some of the large beneficiaries of the reform; but, for the first time, a significant landless population was created. Central Province now has many of the richest farmers in Kenya, with a strong expansion of coffee and tea production for export, and proximity to the large urban market of Nairobi, and to an international market through air-freighting, for vegetables. However, in 1993 more than 65 per cent of households had cash incomes below the national mean, compared with 63 per cent below that mean for Kenya as a whole.

From the early 1960s great hopes were entertained of land reform programmes in Latin America. These hopes have not been fulfilled. One commentator sums up: 'Reform programs to date in the region have been too small, too late, too underfunded, too dictated from above, too hierarchically organized, and too infrequently responsive to pressure from the grass roots' (Thiesenhusen, 1989, 488). Most beneficiaries from the reforms have improved their standards of living, but nowhere has land reform reached more than a quarter of the rural population eligible for land. Even when a substantial proportion of the large estates in Peru were expropriated in the early 1970s, there was still nowhere near enough land to provide existing peasant families with adequate land to support themselves and their families. Cooperative structures founded after reform have tended to disintegrate in favour of individual holdings.

Land reform of itself cannot produce an expansion of production overnight; it is, after all, no more than a redistribution of property rights. Much more is needed: credit, technical advice, markets and, above all, realistic pricing policies for agricultural produce.

A major consequence of land reform, however, has been generally to increase the power and participation of the state in the agricultural sector. This is particularly true where cooperative structures have been led by managers effectively appointed by government to carry out official policy and practice. Where cooperatives survive, the state can often dictate how land is to be used. Even where land has been divided into peasant holdings, the state often retains residual rights over the future disposition of beneficiaries' holdings (for example, by retaining legal title to the land or requiring that it should not be subdivided on inheritance).

In many rural societies pressures on peasant farmers have periodically exploded into rebellion. For centuries peasant revolts have foundered for lack of cohesion in the face of organized opposition. The most distinctively rural revolutionary movements have taken place in the poorer countries since 1945. The inspiration came from the Chinese rather than the Russian revolution, and has been used as a model for similar revolutions in Korea, Vietnam, Cuba, Libya, Algeria, Angola, Mozambique and Ethiopia, and more peacefully in Tanzania. Land reform followed, with a long and difficult period of transition to new forms of rural organization. The state has played a major role in organizing and controlling the means of production, including land, and there has been a great reluctance to reallocate land to individuals except as a short-term measure. Yet the problem of reconciling the need for increased productivity with state control and personal satisfaction and initiative is one with which no Marxist government has come to terms. The case of China is particularly instructive (Box 5.4) but most attempts to collectivize agriculture have foundered because of the

BOX 5.4 AGRARIAN REFORM IN CHINA

During and immediately after the revolution which resulted in the creation of a communist government in 1949 land was redistributed by expropriating from the landlord class in favour of the landless and poor peasants. The effect of this policy was to deprive the marginally more successful, and to allow the previously landless labourers just enough land for subsistence needs. In this process the average peasant landholding, although improved, did not increase enough to provide a regular marketable surplus. In the politically and socially disturbed post-revolutionary period it gradually became clear to the new communist

BOX 5.4 *continued*

regime that only a programme of major works and social reorganization could cope with the 'inefficiencies' of Chinese agricultural systems. The philosophy of socialist accumulation through central planning required a massive and organized transfer from the agricultural to the urban industrial sector. Consequently, the limited attempts to create mutual aid teams of ten families and to foster group cooperative initiatives in the villages during the 1950s were swept aside by the major social and economic reorganization of The Great Leap Forward of 1958–9. In the first place this was a grassroots movement. Twenty-seven collectives in the province of Hunan, south of Beijing, grouped themselves together into the first large-scale rural collective, put all land into public ownership and set about an impressive programme of public works including land drainage, canal and reservoir construction. The well-established Chinese propaganda machine gave the new large-scale collective a good deal of publicity. An approving visit from Chairman Mao, who renamed the collective a 'People's Commune', was followed by ecstatic adoption of the initiative and within six months the whole country was organized into similar 'people's communes', each with an average population of well in excess of 20,000 and an administrative hierarchy that ran from the village community (now redesigned as a production team) through a grouping of such village production teams as production brigades who were responsible to the central organization of the commune.

The commune remained the main functional and organizational unit, implementing an impressive set of agricultural improvements, reducing the risk of natural disasters by major improvements in water control, extending the area of cultivation into previously unused or degraded land and initiating rural social services and industrial developments. During this period of command communism attempts were made to force the Chinese peasant into a mould that removed all individual initiatives and replaced private desire by collective will. Through it all, and despite the Cultural Revolution of the early 1970s, the peasant held tenaciously to the need for a 'private plot' on which to grow domestic produce and (at times of more benevolent central government) to raise farm animals and cultivate crops for market in the small private sector. After Mao died in 1976 new approaches became possible and his successor Deng Xiaoping inaugurated dramatic changes of attitude and organization. Gradually, over the next five years, the rural collectives (the People's Communes) were supplanted by a contract production system centred on the individual household. The end of this period of China's rural history was signalled when the People's Communes were formally abandoned in 1984.

More recently, this freedom to operate independently has increased as policies have become more liberal. Within the overall structure of management there is now much more freedom to respond to market forces and to follow local preferences in decision-making. One result of these changes is that they have both exposed and helped to create a growing problem of surplus rural labour. Despite considerable success in the off-farm sector in the 1980s when the total employed in rural enterprises, manufacturing and services increased from about 30 million to almost 100 million, it is now estimated that at least a third of the 500 million people who work in rural areas will be forced out of agriculture. This is the product partly of increasing population but also of changes in agricultural practice as production becomes less labour intensive.

At the same time, China moves rapidly towards being an urbanized consumer society. About 60 cities already have core populations of over half a million and they are physically incapable of absorbing a great many more migrants. So one of the most important objectives for policy-makers is to create yet more opportunities for alternative employment in rural areas, especially in the small market towns, so that fewer people are enticed to chase scarce urban employment opportunities. Often this form of initiative is neither economically efficient nor environmentally friendly – producing further severe loss of land and polluting the air and the waterways. It is a considerable understatement to observe that successful navigation along the difficult road towards full modernization of the economy and society is going to demand all the skill and energy of this populous but resourceful nation.

individual's desire to retain a private plot and its produce. In Tanzania, the socialist transformation worked best in the poorer south and west but met great opposition in regions such as Kilimanjaro where cash-crop farming was well established and private initiative had flourished since the 1950s, if not before. Many cooperatives in Peru sank under the weight of peasant insistence on individual holdings. China is finding its way towards a compromise. The basic structures of socialist organization of the means of production are retained but, increasingly, a free rein is allowed to small-scale individual enterprise for horticultural crops and animal husbandry, in response to the influence of an increasingly open market.

THE ADOPTION OF FIRST WORLD MODELS

It has been argued above that the basis for a good deal of what passes for rural development in the Third World has been the extension of commercial enterprise and the creation of larger and more efficient forms of business or bureaucratic organization; that such developments involve the peasant as wage labourer, sharecropping tenant or as a token-receiving 'comrade worker'. Such systems remain exploitative, demeaning and unsatisfying. At the same time, the increasing size and political significance of urban populations in Third World countries makes it imperative that the productivity of land should be improved and that export crops be produced. The world scale of this problem has generated a response from the UN and from relief agencies. Faced with the prospect of large-scale famines, international agencies have disbursed increasingly large sums of aid in the direction of improving agricultural productivity in the Third World. Although we are often reminded that this represents only a small fraction of the wealth generated and spent in the North, these are substantial programmes of assistance. Working through independent governments, rural development aid has been of various kinds:

- large-scale schemes designed to serve national budgeting strategies for import substitution: similar in style and conception to plantation farming, they can immediately transform the peasant into a wage labourer;
- the provision of demonstration farming units designed to form the focus of an innovation process;
- the 'model farm' approach;
- the provision of credit for acceptably innovative farmers;
- the creation of marketing facilities for specific cash crops designed to by-pass middlemen and give 'better' prices.

A number of variants of all these approaches exist to incorporate some regional differences in political and social systems. In some areas the tenant farmer has been the focus of attention, in others settlement schemes attempt to replicate the 'community spirit' of a tribal village. Some experiments have attempted to copy exactly the style of the European idea of a village settlement. One thing they have in common: very few survive more than a few years. The Third World is littered with examples of failed schemes and can show very few that have any hope of a self-sustaining future. Short-term local successes have meant increased reliance on the inputs that dominate European farm-management systems. In this case, however, fuel, pesticides and fertilizers, tractors, harvesters and plant-breeding technology are all imported at high cost and with no security of supply. Every developing country has its piles of rusting machinery, the relics of ambitious schemes to mechanize peasant agriculture.

The sheer scale involved in reaching a largely illiterate population in *all* rural areas makes nonsense of this attempt to transfer modes of production without taking full account of the social formation. The most favourable result from this sort of enterprise is to raise the living standards of a few families and in this way exacerbate rural class differences, which are already beginning to appear in many rural areas.

Some of the interchanges between the rich and poor regions of the world have been overtly or covertly exploitative and can be criticized as such, but many more have been the products of ignorance or conservatism. For example, at each scale of investment of advice and aid, from local water supplies to regional planning policy, far too much foreign expertise has been employed in circumstances in which the experience and wisdom was inappropriate. Far too few have been flexible enough to see why the assistance they bring is ineffective, and few bear any ultimate responsibility for mistakes that are made. It was perhaps inevitable that foreign advisers, from the economist at one end of the scale to the agronomist at the other, should bring with them the knowledge and experience based on the way in which food surpluses are created in the northern hemisphere, and many inexperienced Third World governments, whose administrators were trained in the ways of thinking of the North, and often in the universities of the North, have not always had the strength or insight to resist the blandishments of those who see this process of development as replicable.

Food surpluses in developed countries were created at a particular stage in the development of the world economy and the history of capitalism. Industrialization absorbed labour from the rural areas, and the efficiency of the remaining rural population was improved by the rationalization of landholding, which allowed farmers to increase the scale of their external inputs. This they did through mechanization, by the application of new agronomic research, the development of efficient advisory and marketing services, and the extended use of artificial fertilizers and pesticides. More recently, the increasing efficiency of farming in the developed world has been maintained against potential Third World competition by the sophisticated deployment of central government subsidy and support mechanisms that are the products of national and international trading policies. It is expertise from a highly complex system, evolved over the last 200 years in Europe and North America, that is now on offer as a 'package' for the Third World. So it is not surprising that many schemes for improvement have been based on premises or assumptions that may have some validity in Iowa or East Anglia but bear little relation to circumstances in Bangladesh or Sierra Leone. Despite a continued catalogue of failure, the approach of international aid agencies to rural development persistently follows the assumptions of the mature phases of capitalist development in the North, rather than engaging with the uncertainties of an evolving peasant agriculture in the South. At last it would seem that the Third World is reaching some kind of watershed; a crisis of confidence has been reached in which many strategies for development have been tested and found wanting, while the prospect of greater inequality and resultant social economic and political instability increases.

It might be suggested that the purely *economic* approach to rural development that permeates most rural development initiatives in the Third World has been overly myopic. Real rural development involves more than agricultural productivity. It implies a commitment to the effective provision of a wide range of social services: housing, schools, health centres, electricity, sanitation, water supply. It is the failure to bring effective improvements in this area

This cartoon, from the *New Internationalist*, implies that outside 'experts' from the developed world may not always be properly equipped to deal with problems elsewhere

which is the most important aspect of the crisis now affecting rural areas of the Third World. Rural development may be more broadly defined as the effort to increase social justice and to improve the whole quality of rural life in such a way as not to threaten the ecological basis for subsistence. Within such a definition it is important to focus attention not on the rich and poor *in general*, that is, at the macro scale, for it is just this which in the past has led planners to focus on national rather than local problems. Looking more carefully at the growing gap between the haves and have-nots within rural communities, it is clear that increasing levels of discrepancy between the income earners and the destitute have not been eased by those whose main concern is economic efficiency and the creation of surplus; nor by those who are willing to identify 'success' or 'failure' within such narrow terms. Increasing inequality was identified by Michael Lipton (1977) as an urban–rural problem. He argued that although poor countries have managed to increase their output of wealth considerably since 1945, the poorest rural people have grown poorer. Comparisons between Europe and Japan in their early stages of development and Third World countries today, he suggests, are inappropriate because modern programmes of public investment, aid policies, education and pricing systems all push rural and urban incomes further apart. This imbalance is difficult to change because the urban sector contains most of the 'articulate organized power'. While there is clearly some truth in these generalizations, Lipton presents too simplistic a view of Third World economic landscapes, which involve rich and poor rural areas as well as rich towns and a poor countryside. It also takes no real account of the migrants who move from one milieu to another and send cash to rural members of their families. The true position is obviously more complex and varies from one region to another, but perhaps these broad generalizations serve the purpose of focusing attention on underprivileged and weakened sectors.

The generally poor record of rural development agencies in attempts to reduce disparities between richer and poorer can be attributed also to the circumstances endemic in all Third World countries. The morale of those who do not join the migration streams but live most of their lives in rural areas is increasingly low. If we examine the situation of a poor peasant cultivator, whether he or she be in Ecuador or Somalia, Sumatra or India, certain conditions do not vary. Against a depleted or depleting stock of environmental resources the peasant is generally expected to raise sufficient livestock and grow enough crops to cater for the subsistence needs of his family. Failure to do so can mean starvation. At the same time there is pressure to produce a surplus of cheap staple cereals to feed the increasing numbers of indigent poor who now live in the cities of the Third World. If a peasant is particularly privileged he may be a member of an agricultural scheme or cooperative, able to grow an exotic cash crop that has national export potential, and with an effective marketing system through which to channel produce and receive a reasonable price for its production. The chief beneficiaries have mainly been the richer rural families who can corner most of the production and those urban dwellers who benefit from any spillover effects from its sale in terms of better services. Few of the returns from cash-crop farming reach the poorest sector of the rural population either directly in the form of cash or indirectly, and more importantly, in the form of better social services. It is still only the wealthiest peasants, with access to land and labour, who can accumulate any reserves of capital to pay for education and the chance to migrate to a better social and economic environment.

TRANSFORMING PEASANT AGRICULTURE AND SOCIETY FROM WITHIN

In recent years there have been increasing reservations about the effectiveness of large-scale, top-down interventions, whether motivated by socialism or by capitalist free enterprise. Improvements in production envisaged by state and para-statal schemes carry with them problems of environmental misuse. Tropical soils are often fragile and ill-adapted to large-scale schemes. Nor do they always yield sufficient returns in foreign exchange to justify the investment. Moreover, as the statistics quoted in Chapter 4 have shown, the rise in production brought about by new technologies and improved plant strains may bring foreign exchange to governments and raise the living standards of richer peasants, but they are not reaching the poorest fractions of rural populations. Third World governments are also severely hampered by a rising level of foreign debt and by the harder-nosed attitude on the part of the World Bank and governments of the developed world, favouring only investment in activities most likely to raise productive efficiency and to create exports. In these circumstances the moral initiative for assisting the changes necessary to improve the quality of life for the poorest rural majority has shifted towards more appropriate responses involving local initiative within rural development policy. During the last decade or so a number of writers have drawn inspiration, on the one hand from the disillusion felt among social scientists and businessmen alike for the long-term effectiveness of the large-scale enterprise, and on the other from the recent interest in low-technology solutions to world energy problems, to look afresh at the inter-relationships between man and his environment. It has become increasingly apparent that many rural areas enjoying the benefits of modern economic farming have also had to absorb some kind of environmental deterioration. This experience opens the way to identifying a set of potential solutions to the problem of rural decline in the Third World.

Alternative technologies

This new low-key approach has several characteristics. It identifies alternative technological solutions that rely as far as possible on local expertise and local raw materials; it concentrates on small-scale enterprises; it represents a very low-level approach to advice, which is primarily based on the local wisdom and the indigenous society and economy; and it encourages self-respect and self-reliance. The approach now has the weight of an important international organization designed to promote it – the Intermediate Technology Group – and has

Plate 5.8 Alternative technology, Uttar Pradesh, India Simple machinery, which is low cost and requires little skill, is used to crush sugar cane

been reinforced by a trend in recent thinking and research towards re-evaluation of self-sufficient styles of life.

The main drive towards finding what is now called 'appropriate technology' has come from India, where Schumacher (1974) first developed his ideas of an alternative to high technology. Appropriate technology programmes have sprung up in a number of different parts of India (Bombay, Lucknow, Bangalore, Ranchi) initiating work on the use of solar energy and bio-gas for heat and power generation, the use of local building materials instead of imported cement and steel, the introduction of hand pumps, and the establishment of small-scale rural industries based on local raw materials: soap from fats and oils, sodium silicate from rice husks, plastics from castor oil and cellulose fibre from groundnut shells (Plate 5.8). In Uttar Pradesh in north India improved hand or treadle spinning and sewing-machines have been introduced, as a labour-intensive alternative to the factory system. Similar developments are to be found in a number of other countries: Pakistan, Tanzania, Kenya, Ghana and Colombia. Most projects aim to introduce low-cost labour-intensive small industries, with workshops of no more than 15 or 20 people, to areas where labour is cheap and plentiful. Though initially regarded with suspicion as a 'second best', alternative technology, which is low cost and dependent on local initiative, is now enjoying the support of international aid agencies.

Aid via NGOs

The aid charities, now known collectively as Non-Governmental Organizations (NGOs), have their origins in the work of missions and other religious organizations which go back well into the colonial period. Their new role has been born from a realization of the desperate poverty and hunger in countries tragically affected by drought and political strife. The images of starving African children on the television screens of the developed world during the 1980s increased public awareness. Media events (such as fund-raising concerts with audiences of millions) have highlighted these tragedies and touched the public conscience, raising funds on a scale never before envisaged. Money has flooded into charities and the capital generated by investment during a period of high interest rates has allowed agencies to develop strategies beyond the immediate relief of suffering. The most important feature of these initiatives is that they are directed towards the people least touched by previous rural development policies and are more often designed to improve the quality of life in rural areas, rather than just to raise production. They aim to reach the poor by means of small-scale projects which involve the people themselves at all levels of organization, offering training, work skills, the use of credit facilities, primary health care, social development and awareness. The recipients of aid are given responsibility to identify needs and carry out improvements. People are encouraged to formulate their own policies. This change of attitude now infects all agencies, even the older organizations previously dedicated exclusively to raising productivity and debt servicing.

The effectiveness of individual schemes is difficult to evaluate as there are no generally accepted criteria for success or failure. The limited evidence which does exist suggests that NGO programmes are more effective than any other alternative in reaching the poorest sectors of the rural community and in raising their social awareness. They are more likely to encourage social responsibility, particularly in areas where other approaches have failed, for example the introduction of proper procedures for monitoring credit, and reducing the possibilities for corruption by money-holders (a factor bedevilling many rural credit schemes throughout the Third World). The most successful schemes have been those involving basic health and nutritional information, and those which have improved training in relevant skills. Specific income-generation schemes are less successful and share many of the characteristics of earlier government initiatives which have failed because of low levels of managerial and technical ability.

It is important not to exaggerate the claims made for the effectiveness of NGOs for, in spite their proliferation, they remain of only marginal significance against the scale of the problem of

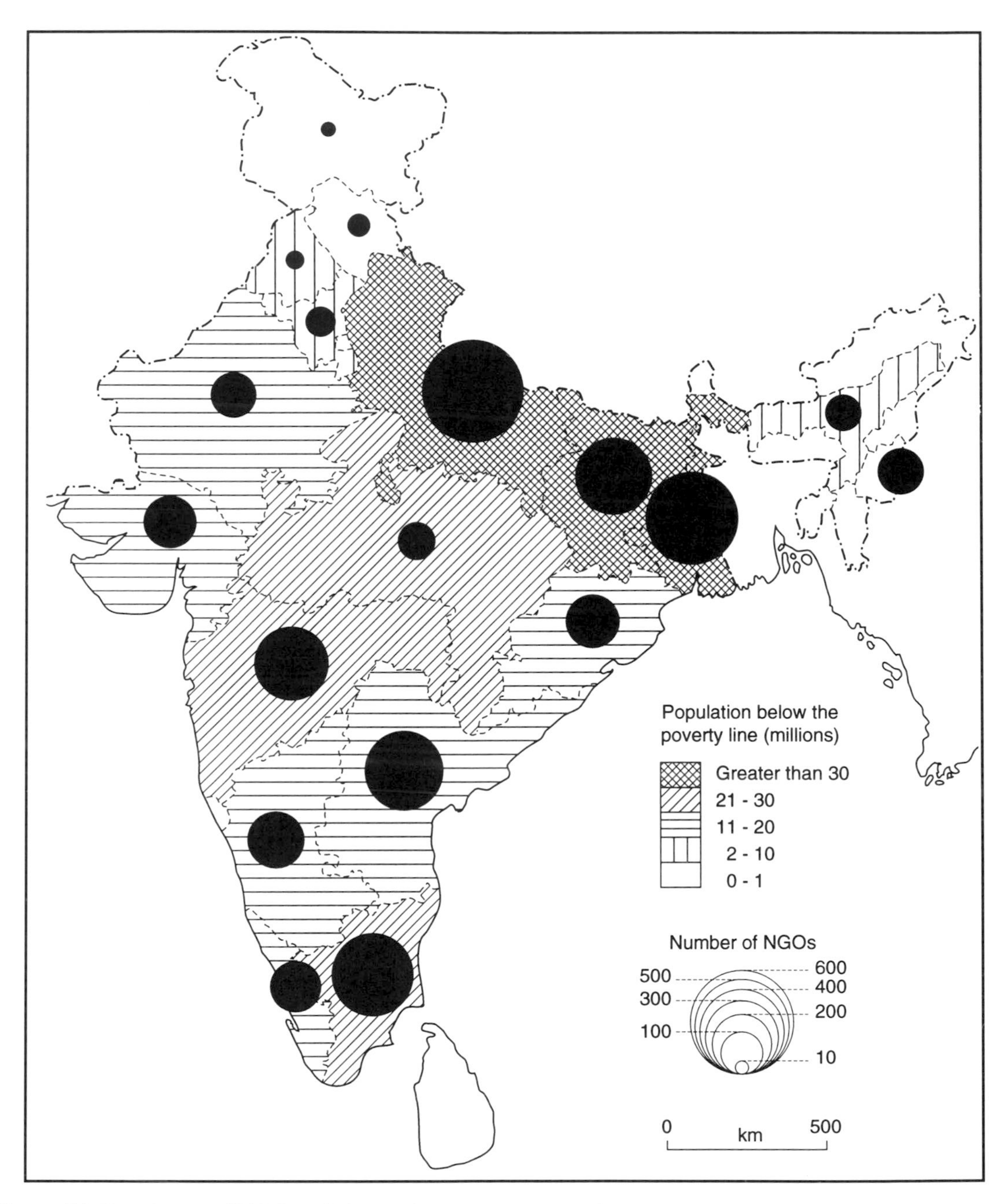

Figure 5.3 Distribution of NGOs in India in relation to proportion living in poverty
Sources of data: CAPART, n.d.; Robinson, 1991

rural poverty. In India, for example, with a long tradition of small-scale low-technology initiatives, there are an estimated 15,000 NGO schemes. This is certainly the largest and most diverse NGO community in the Third World and foreign funding for these initiatives now comprises about US $500 million, a quarter of all the official development assistance (Robinson, 1991). Yet their impact is very uneven and localized. As Figure 5.3 shows the main concentration of NGO activity is in the south, where Christian missionary presence was the strongest. Yet even in these areas there is no more than one NGO for every 12 million or so people living in poverty. So although entirely praiseworthy in intent, in general it seems that one must remain sceptical about the impact of NGOs. Unless particular care is taken to focus on specifically appropriate kinds of initiative, the very poorest still fail to benefit from these interventions. The return on the investment of time and energy is still not high, nor is the diffusion of such innovations very marked. Perhaps the most significant impact of NGOs has been in changing attitudes among donors: the ways in which the relationship between poor and rich is defined; a recognition that much of what passed for rural development in previous decades was both prejudiced and patronizing. Along with this change of attitude there is an awareness that the very term 'Third World' carries with it connotations of moral superiority by those who are privileged enough to live in richer economies in the North. There is also an increased sense of awareness that even in the poorest states, misdirected aid frequently favours an affluent minority rather than reaching those most in need.

If NGOs have not provided the answers in addressing the problems of rural poverty they have raised questions and they represent a most significant change of direction in policy. What is now needed is closer links with agencies of government to spread the influence of NGO thinking more widely through the apparatus of the state.

CONCLUSION

The inhabitants of the rural Third World are increasingly aware of the possibilities for improvement in their condition. In fact they are encouraged to increase their output against a background of media information and evidence concerning the quality of urban life. Moreover, the wage-employed migrant, temporarily returning to the village, brings back not only information but also the tangible symbols of relative affluence – in the form of a watch or transistor radio, or better clothes. It is the 'pull' factor of affluence as well as the 'push' of poverty that impels more and more of the young and more enterprising to leave the countryside.

As a consequence, rural economy and society begin to disintegrate. Weeding and bird-scaring are neglected, rotation practices are abandoned, reciprocal systems of labour exchange fall into disuse and the downward spiral of environmental deterioration and social decay is completed when marginal settlements are left to the old, the infirm and the indifferent. It is this process that development planners have generally done little to halt, and it is a sad fact that many of the communities on the margins of production are sustained indirectly only by the remittances of urban workers. Their only hope for advancement is that this can continue for long enough so that the increasing number of urban dwellers who retain their rural identity can make sufficient investment to lead to some kind of general local improvement in conditions of housing or sanitation, and that more will be tempted to stay. There are those who may see this circular process as eventually bringing some kind of revival to rural areas. Many urban dwellers still retain their rural social and economic links and it is at least a positive and hopeful feature of difference between the developed and 'underdeveloped' world that this is still so even for third or fourth generation migrants.

Clearly, in order to approach the problem of increasing the quality of life in rural areas, it is necessary to approach the rural community *as a whole* – not as a separate entity but as part of a system that now almost inevitably includes migrant wage labourers in varying proportions – and to ask very different questions from those that relate only to the increase of agricultural production. How can the well-being of a whole community be best improved? With so much rural–urban interaction, what *is* a rural community? How can the linkages with urban areas be better orchestrated to improve the conditions of those who stay in or return to rural areas? Can state intervention be anything but ham-fisted, ill-directed and inept? Are there political structures that sustain local self-help and initiative rather than destroy or subvert them? Is there a realistic role for international aid agencies in the development of rural areas, or is this work inevitably either cosmetic (to satisfy consciences or retain influence) or short-term and remedial following a major disaster? In other words, how far can the many mistakes made be translated into lessons learnt so that the approach to the problems of rural improvement in the next decade can benefit from the dismal experience of three ineffective decades of rural development planning in the Third World?

SUMMARY

This chapter has sought to show how and why agrarian structures throughout the Third World sought to change from the 'traditional' structures (that were the subject of Chapter 4) by the processes of rural development. Attempts to promote rural development in government-led initiatives have tended to be faced with severe problems arising out of social, economic and political, as well as environmental constraints. Peasant farming systems, in which the principal objective of farming is not necessarily to maximize production, predominate in the Third World, but the policies and strategies that have been formulated by governments and external agencies have generally assumed the economic superiority of the higher productivity and incomes of plantation and capitalist agriculture and the environmental superiority of 'scientific' knowledge and technology over that of users of the environments. Criteria for success in rural development are levels of output and income, and since these may be inconsistent with the needs and values of rural communities, taken as a whole, they have not achieved the intended results. In particular, they are shown to be associated with widening intra-household inequalities in rural areas, and a widening income gap between town and country. New approaches that can be based on indigenous institutions and values and using appropriate technologies, with a less prescriptive role for the state in designing and implementing rural development policies, are being sought in rural communities throughout the Third World countries. These are the initiatives that suggest a way forward towards more socially sensitive policies for rural development.

6

Mining, Energy and Manufacturing

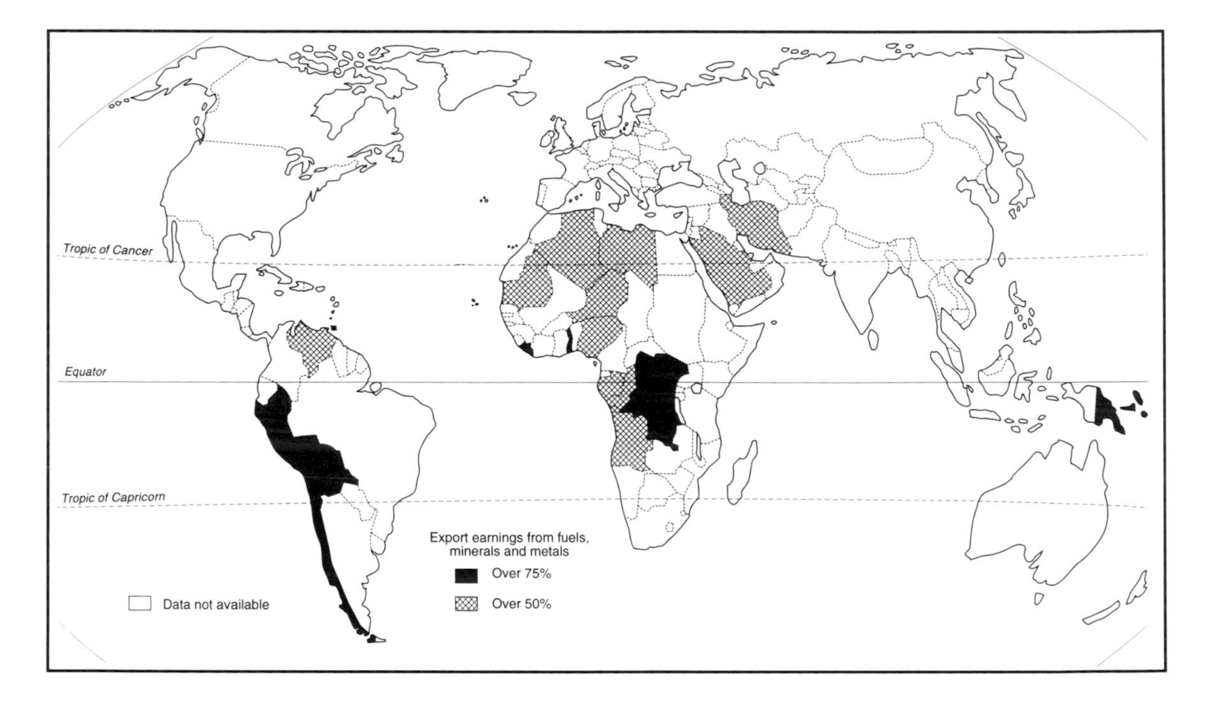

Figure 6.1 Export earnings from minerals, 1990
Source of data: World Bank, 1992

Primary production is a major element in the economies of many Third World countries. Agriculture remains important as the source of rural subsistence, to sustain urban populations, to provide raw materials for industry, and to earn export revenue. The production of minerals is also a crucial sector of primary activity and, as Figure 6.1 indicates, they are a major source of export earnings for some Third World countries, particularly in Africa, Latin America and the Middle East. However, a key element in development strategies since 1945 has been to reduce such dependence on the primary sector, by trying to establish a **secondary sector** dimension to the economy, through the development of manufacturing industry. To sustain this, and to foster a broader social and economic modernization, improvements in infrastructure have also been given priority. A key element in infrastructure for industry and for urban and rural improvement has been in the provision of energy.

◀ Miner, Zimbabwe

Table 6.1 Third World shares of mineral production, 1990

Mineral	*Country*	*Percentage of world production, 1990*	*Rank in world*
Bauxite	Guinea	16.5	2
	Jamaica	8.7	3
	Brazil	8.4	4
Iron	Brazil	18.5	2
Phosphate	Morocco	11.6	3
Silver	Mexico	16.0	1
	Peru	11.7	3

Source of data: United Nations, 1993

MINING

Minerals have been a significant element in shaping the relationship between the developed and less developed countries. The desire to find gold, silver and gemstones was a significant factor in the voyages of the Age of Discovery. In colonial Latin America the finding of mineral wealth helped to shape the pattern and process of development in Mexico, Brazil and the Andes. In the nineteenth century similar considerations contributed to the scramble for Africa, influencing both the economic and political geography of the continent. Minerals continue to provide an important potential resource for the Third World, both as a base for industrialization and as a source of export earnings. For the developed world, minerals imported from the Third World are an important source of raw materials. Table 6.1 indicates the importance of some Third World countries in global mineral production. The region is also an important source of copper, tin, nickel, cobalt and fluorspar.

Despite the wealth and diversity of minerals known to be in the Third World, their full potential may not yet have been realized. There is still a basic lack of knowledge of the geology of these countries, and of the detailed distribution and scale of their mineral deposits. While geological survey in the developed world has been extensive, so that resources are well known, the Third World offers the possibility of new discoveries – of base and precious metals, and of fuel sources. These circumstances offer economic opportunity for otherwise poor countries, and make them attractive to investment by First World mining companies. However, mineral deposits are not distributed equitably; some parts of the Third World appear to be better endowed with minerals than others.

As a basis for development the mining industry has some distinctive and awkward characteristics when compared to other economic activities. Mineral deposits are 'fixed' in nature – they have specific locations, they are finite in size, and the minerals have fixed physical and chemical properties. They are non-renewable – once the mineral has been exploited it cannot be re-used (except for a small degree of recycling). Some minerals are widely available, or 'ubiquitous', such as sand, gravel and brick clay, and because of this they have low values per unit and are generally worked close to their markets, and are of little significance in international terms. Others are more localized in occurrence, and their relative scarcity increases their price and means that they can bear the costs of transport over longer distances. In addition, some minerals occur in remote locations, so that their exploitation requires substantial investment and is a high-risk activity.

The exploitation of minerals has been seen as a potentially important contribution to the development of Third World countries. Besides the contribution made to export earnings, mineral development may attract foreign capital, provide infrastructure, create jobs, provide new skills for the labour force, and generate demands for local goods and services. Yet mines have finite lives; the returns may be short-lived. For many of the more important minerals, production is on a large scale and is heavily mechanized, so that the industry is capital-intensive rather than labour-intensive. Considerable use may also be made of skilled expatriate labour rather than local sources.

Much of the world mining industry is controlled by a few companies, and they have had a major role in the exploitation of many Third World mineral sites (Plate 6.1). In the 1960s eight firms accounted for 50 per cent of the world's refined lead output, and six firms dominated aluminium production. Within some individual countries control was even more pronounced. In Chile two US companies controlled 90 per cent of copper production, in Peru three such companies dominated the industry and in Zambia two foreign companies controlled output. Such companies acquired prominence through their ability to raise the necessary capital to cover the high-cost risks involved in the exploration and preliminary development of minerals, and for the provision of infrastructure and advanced technology. Critics argue that much of the profit from mining goes to the parent company in foreign countries, rather than remaining in the source country. Returns from mineral extraction may not therefore have been maximized by the producing country, but it is possible that had the minerals not been exploited by foreign companies they would not have been exploited at all, since indigenous capital was unavailable to secure their development.

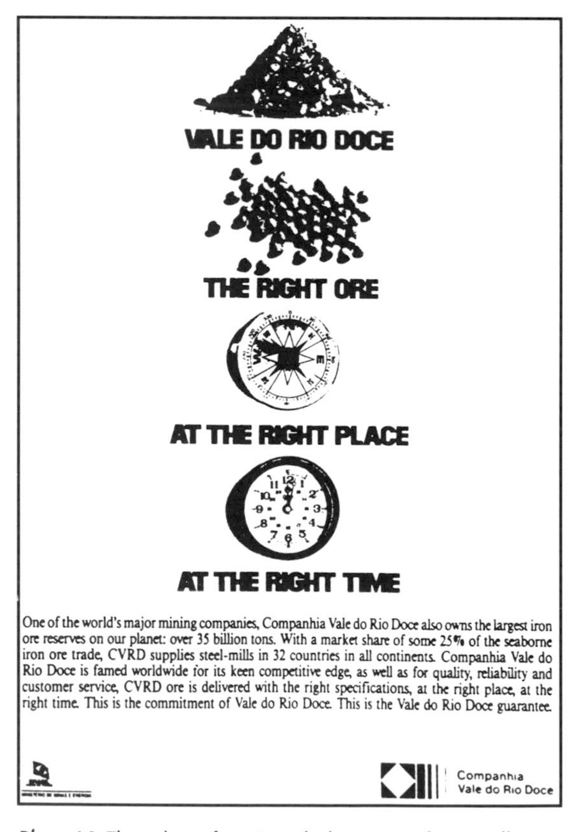

Plate 6.1 This advert, from Brazil's largest, and originally state-owned, mining company indicates the importance of Third World ore production to the world market

Third World countries have been sensitive to the activities of such companies, particularly in the petroleum sector, which was nationalized in Mexico in 1938, in Iran in 1951 and in Brazil in 1953. During the 1970s a number of countries nationalized foreign mining concerns, or required them to participate in joint ventures with government or private local capital. The objective was to increase domestic control over the exploitation of mineral resources – for political and economic reasons, to reduce environmental damage, or to improve the working conditions of miners. By 1977, for example, majority-owned government concerns had taken control of copper production in Chile, Zaire and Zambia, and in the Jamaican bauxite industry joint ventures between government and foreign companies had been created. However, multinational companies may have other mineral sources to exploit, so that nationalizing a resource may not ensure its development or determine the scale at which it is utilized. During the late 1980s there was some easing of this nationalistic approach, as mineral-rich Third World countries sought to attract foreign capital by relaxing their mining laws, even, in some cases, extending this to the petroleum sector.

In the exploitation of minerals in the Third World there are often more specific problems. Most minerals are commodities of high bulk and low value, so that transport costs and difficulties of access have confined exploitation for export to areas reasonably close to the sea or navigable waterways. In consequence much of the mineral production takes place within 300–500 km of the coast. Only with higher

Plate 6.2 Alloys factory, Gweru, Zimbabwe

prices or cost reductions due to transport improvements will more remote resources be exploited. Such factors have, for example, contributed to the exploitation of oil in the Amazon areas of Colombia and Ecuador, and copper in Zaire and Zambia. The remote location of many minerals has, however, given rise to development of a limited nature at isolated sites, with the creation of mining camps, with advanced technologies and often foreign staff, in otherwise undeveloped or backward areas (Box 6.1; Plate 6.2).

The infrastructure which has been developed in the form of railways, ports, power stations and education facilities has been primarily in the interests of foreign companies and only incidentally in those of the host country. Although there has been some 'spin-off' in the creation of demands for local food and equipment, much has been imported. It is also the case that earnings from mineral exports have not been wisely used, with concentrations of investment around the already more-developed core areas of the country, or in conspicuous consumption, rather than in securing development more widely.

A major concern for the producing countries is that they do not derive full benefit from the 'linkage' effects to be obtained from their minerals. In addition to the purchase of foreign mining equipment, much of the smelting and processing of ores is carried on abroad, so that considerable income and foreign exchange is 'lost'. There are various considerations in the location of smelting facilities, and a number of major mineral exports from the Third World, such as tin, bauxite and copper, pose particular smelting problems, which may be more easily tackled in more advanced countries. Considerable raw-material inputs, in excess of the production of a single mine or country, may be required to sustain an efficient smelter; while small units of production may also make for high costs of production. Considerable fuel

BOX 6.1 MINERAL DEVELOPMENT IN PAPUA NEW GUINEA

Papua New Guinea has been dependent on mineral exports for over half its national export revenues since national independence in 1975. In 1992 66 per cent of export revenues were derived from minerals, 55 per cent from copper ore and 11 per cent from gold, contributing to a per capita GNP of almost $1000 that places the country in the Middle Income group of the World Bank, even though it is a country with great extremes of modern and traditional societies, and the majority of its 4 million inhabitants are subsistence farmers. Minerals generate substantial revenues to government for provision of infrastructure and economic and social services, but the exploitation of national mineral resources has also created seriously destabilizing political, geographical and environmental problems.

The political problems are most evident in Bougainville, North Solomons Province, the location of the oldest copper mine in the country, a major support of the national economy since independence (Figure 6.2). The Bougainville mine, owned by Rio Tinto Zinc, has met resistance from the local population, initially over pollution of fishing grounds, but more fundamentally over the distribution of benefits between the company as profits, the national government and the provincial government as revenues. Conflicts over ownership of the mineral rights have led to a strong secessionist movement in North Solomons, leading to a formal declaration of independence in 1990 which was forcibly and firmly resisted by the national government, anxious not to threaten the attractiveness of the country and its resource base to foreign mining companies, as well as to secure its substantial revenues from this resource.

The geographical problems of isolation are most apparent in developments from the 1980s in the highlands of the mainland, at Ok Tedi (mostly copper with some gold) from the early 1980s and at Porgera (gold from the late 1980s) with the promise of making Papua New Guinea the largest gold producer after South Africa by 2000. Both these sites are rich in mineral reserves, but are very isolated, in rugged and largely uninhabited mountain areas with no access by road, to the extent that the mines are supplied entirely by air. These are effectively island sites, with no spin-off to the local

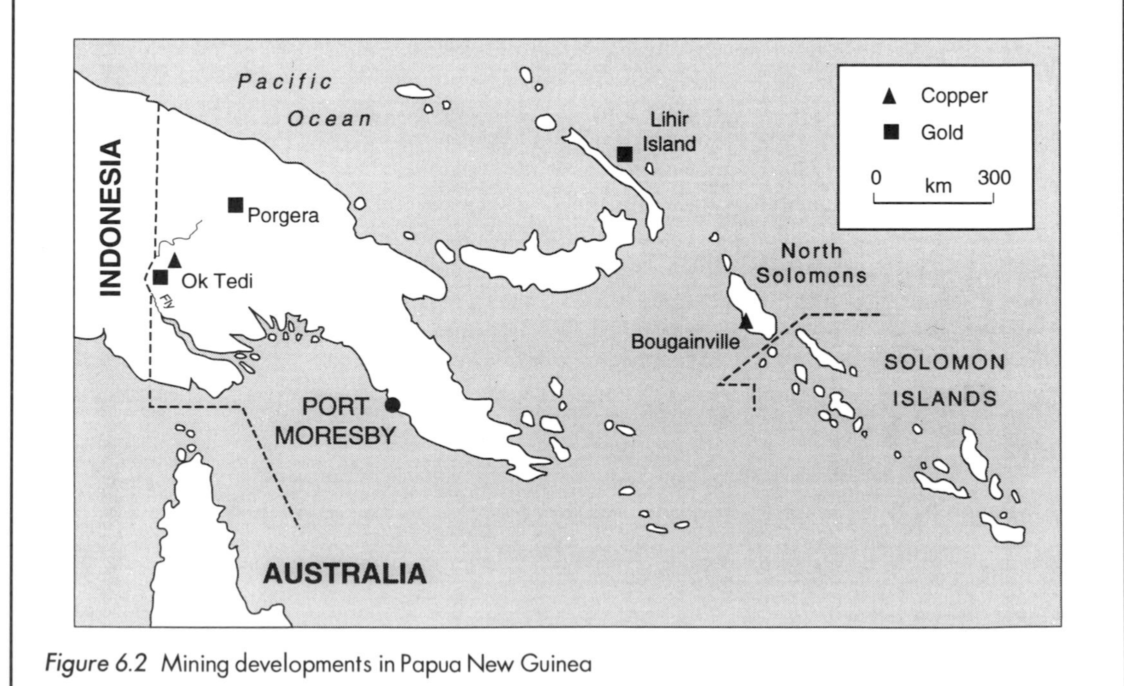

Figure 6.2 Mining developments in Papua New Guinea

BOX 6.1 ***continued***

economy, either in jobs or in services. The principal direct beneficiaries are in the capital, Port Moresby, where the foreign MNC concession companies have their bases, and where they hire local Papuan labour. In the Porgera area, however, there are also local small-scale miners, attracted to the highlands in a mini-gold rush in the late 1980s, but even then local interests derive little benefit and there are conflicts between local communities and individual miners, as with the large companies, over mineral rights.

Although there have been legislative attempts to impose environmental safeguards on these mining developments in the highlands, these have been difficult to enforce, and there is evidence of substantial environmental degradation. Both Ok Tedi and the Porgera mine discharge toxic waste into the Fly River, a very large catchment of the highland region, with high copper pollution levels extending to the sea. The mining companies claim that having dams to trap the mine waste in their mountain-top sites would prove technically difficult and financially impractical, so that pollution remains a problem to be tackled by government. New gold developments, potentially even larger than at Porgera, are planned for Lihir Island, again in one of the outer provinces, raising the possibility of further political and environmental costs as a consequence of exploiting abundant and internationally attractive mineral resources.

demands may also be significant, particularly in coal-deficient countries. In addition, it is in the more developed countries that the principal markets for the finished metals and metal goods are found. Transport costs and freight policies are such that it may be cheaper to export ore, even with a high proportion of impurities, than to ship metal.

There is evidence that only about one-third of the mineral output of less developed countries is processed locally and this pattern has changed little since the 1950s. Nonetheless some Third World countries have begun to establish processing plants and ship at least part of their ores as concentrates or refined metal. In 1990 Chile, Peru, Zambia and Zaire produced 24 per cent of the world's smelter copper, but their production of refined copper, the more highly processed product, was only 16 per cent. Similarly, there is now more local processing of tin by producers such as Indonesia, Thailand, Malaysia, Brazil and Bolivia. However, a considerable part of iron ore production is still exported as crude ore; in 1989 Brazil produced one-fifth of the world's iron ore, but only 3 per cent of its pig iron.

Aluminium represents a good example of the factors involved, in that its production requires the processing of bauxite to yield alumina, which is further processed to give aluminium. Some 2–3 tons of bauxite yield 1 ton of alumina, and 2 tons of alumina give 1 of aluminium, so that there is considerable weight loss in the production process. Alumina is also worth twice as much per ton as bauxite and its production generates additional demands for labour, equipment and supplies. Before 1939 much of the world's bauxite and all of its aluminium came from the developed world, but increasingly the Third World has become a major source of bauxite, with Guinea, Jamaica and Brazil as important producers. In consequence these producing countries have been able to exert some pressure to secure a degree of processing capacity close to the mines. The high cost of transporting bauxite made this attractive to the aluminium companies.

The further processing of alumina to aluminium requires large power inputs and to date the majority of the world's aluminium smelters have been located at cheap power sites in the developed world. In addition, technical complexities in the smelting process make for economies of scale, such that large units of production are required. However, increased power availability and rising domestic demand, together with the desire to secure a greater degree of processing of the resource within the producing country, has led to some establishment of alumina and aluminium plants within the Third

Table 6.2 World bauxite and aluminium production, 1990

	% of world output	
	Bauxite	*Aluminium*
Third World		
Latin America	23	7
Africa	18	2
Asia	9	7
Developed world		
Australasia	37	6
Japan	–	5
Eastern Europe	9	15
EU	3	16
Rest of Europe	–	6
North America	–	32

Source of data: United Nations, 1993

World. Nonetheless, as Table 6.2 indicates, there is a marked imbalance between the Third World's role as a source of bauxite, and as a producer of aluminium. Except for Australia, the developed world mines little bauxite, but the bulk of aluminium output comes from the USA, former Soviet Union, Canada, Germany and Japan.

To date, exploitation of the minerals of the developing world has been primarily in the interests of the consuming nations of the developed world, which do not necessarily accord with the best interests of the producing nations. Decisions to exploit particular deposits, the pattern of investment, the nature of the extraction and beneficiation of minerals, and the apportionment of earnings have not always been consistent with the development needs and objectives of the Third World.

ENERGY

Advances in the use of energy sources such as water power, steam, and coal played a vital role in the Industrial Revolution in Europe; and it has come to be accepted that provision of energy is an important factor in the development process. There is certainly a good correlation between levels of energy consumption and economic output, with the industrialized countries having levels of consumption greatly in excess of those of the developing world. Figure 6.3 shows some startling contrasts in the provision of energy between First and Third Worlds, with much of Sub-Saharan Africa with levels of commercial energy per capita below 100 kg and Europe and North America with levels above 3000 kg. In broad terms it can also be shown that there is still a heavy dependence on primitive energy sources in the Third World, and that conventional energy sources are unevenly distributed. In the developed world energy is derived mainly from coal, oil, gas and electricity from fossil fuel, water and nuclear sources; in the Third World such sources are significant in supplying major urban areas, and in providing fuel and power for industry and modern transport, but manual and animal 'power' remain important in the countryside, with much use still made of more traditional fuels such as wood, charcoal and dung for heating and cooking (Box 6.2). At a broad scale Table 6.3 might suggest that the Third World is rich in energy sources, since Asia is the largest producer of coal and the Middle East of petroleum. However, as in the case of mineral reserves, the distribution of these resources is uneven.

When Third World countries began to promote economic development after 1945, energy prices were relatively stable and crude oil was abundant and inexpensive (at US $2 per barrel). In consequence, they developed energy strategies largely dependent on cheap oil, and paid only limited attention to other sources of fuel and power. For those countries without oil deposits, or with small and economically non-viable oil reserves, the bulk of their energy needs was met mainly by imported oil and traditional fuels. However, the quadrupling of world oil prices in 1973–4 by the OPEC countries had a major impact on those countries without oil resources of their own. They were faced with substantial increases in their energy bills and the need to borrow abroad to pay for the oil necessary to sustain the rate of economic growth, or else accept a slow-down in that growth rate. More significantly, they were stimulated to search for their own sources of oil or alternate forms of energy.

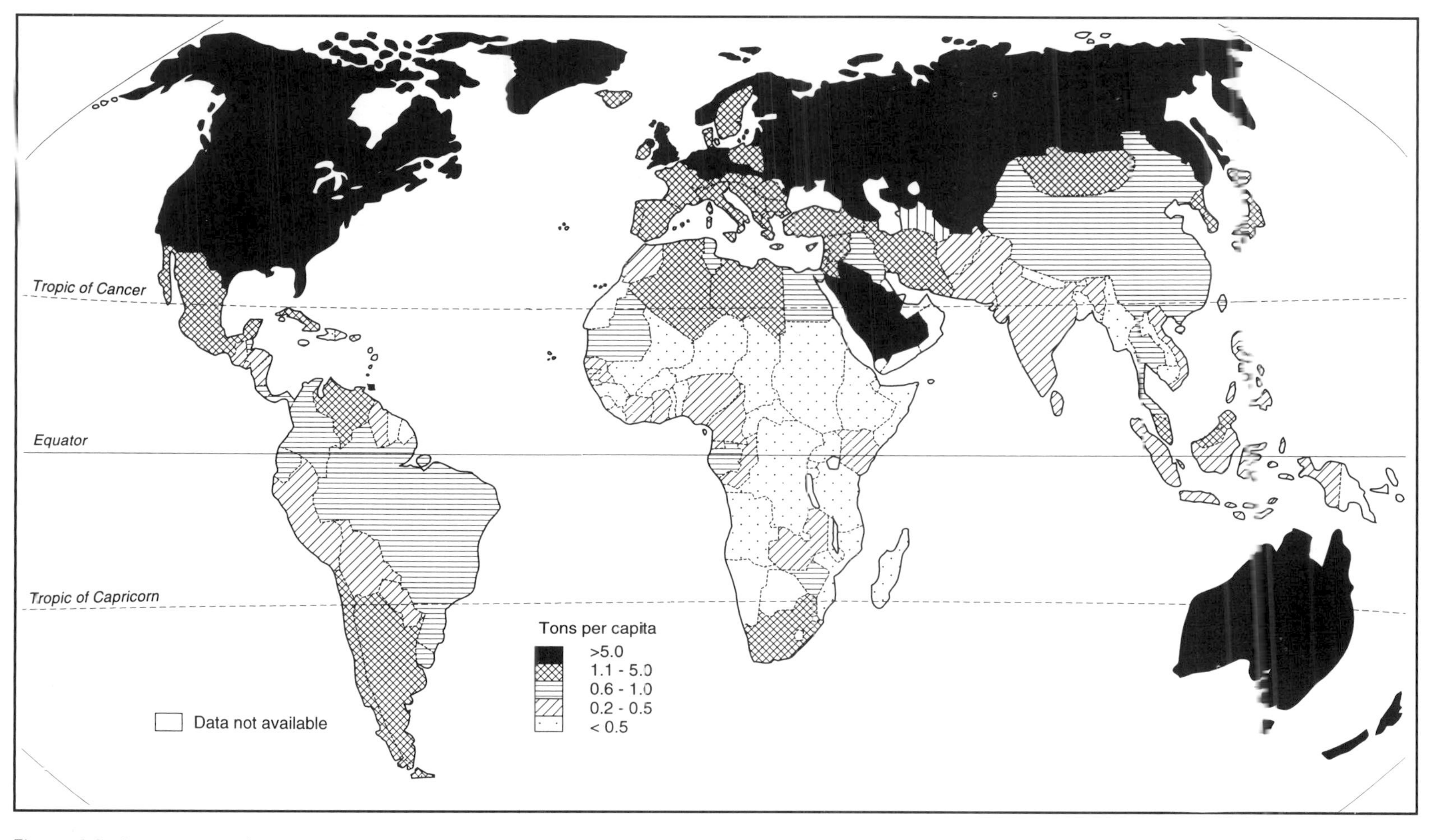

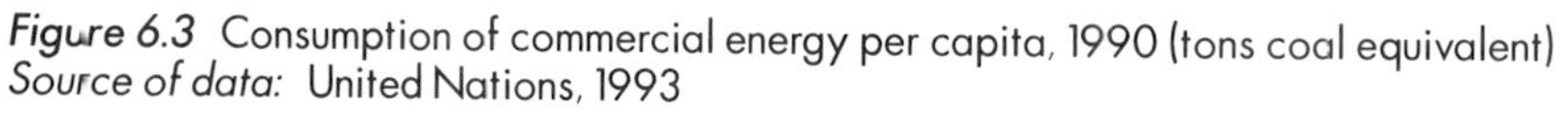

Figure 6.3 Consumption of commercial energy per capita, 1990 (tons coal equivalent)
Source of data: United Nations, 1993

BOX 6.2 NON-COMMERCIAL AND ALTERNATIVE ENERGY SOURCES

It is easy to overlook the fact that in the less developed world considerable inputs of energy still come from the muscle power of men and animals, and that seemingly crude energy sources such as wood, charcoal, vegetable waste and dung are important. Since these are simple and do not enter into commerce it is likely that their use is underestimated, but it is probable that such materials provide one-third of China's energy, half that of India, and in some countries, such as Chad, Upper Volta, Nepal and Haiti, over 90 per cent. Such fuels are inefficient energy sources, their collection is often time-consuming as they are gathered from field and forest, and their use may be detrimental to the environment. Clearance of woodland for firewood encourages soil deterioration and erosion, and the use of dung deprives the land of scarce fertilizer.

Though transmission grids and rural electrification schemes are beginning to push into the countryside, there is still scope for more efficient use of traditional energy sources or the development of alternate forms of energy. Improvements in wood-burning stoves, or the use of charcoal rather than timber, might make more efficient use of diminishing forest reserves. It is possible that intermediate-technology forms of water, wind or solar power might be utilized. The use of dung or vegetable wastes to produce bio-gas appears to offer possibilities as a rural energy source, and in India a substantial programme of this kind has been developed. In the medium term the diffusion of conventional power, via transmission lines, may bring energy to the countryside, but it is possible that alternative small-scale technologies may provide output appropriate to the needs and level of development of rural communities. In the remote mountain communities of the Karakorams in northern Pakistan, for example, the provision of small-scale hydroelectricity schemes as part of irrigation developments that channel the meltwater of glaciers to the fields in the valley floors is an integral part of village-based rural development initiatives.

Table 6.3 Production of selected energy commodities, 1990 (million metric tons of Coal Equivalent)

	Coal	*Lignite*	*Wood*	*Oil*	*Electricity*
Africa	139.3	–	152.4	421.1	38.7
Latin America	33.9	–	76.5	509.3	73.8
Asia	1010.1	32.4	262.6	401.6	184.9
Middle East	–	–	–	1195.1	25.5
Japan	7.2	–	–	–	105.3
Australasia	130.2	15.0	2.9	38.9	23.2
Ex-Soviet bloc	453.8	80.0	22.5	807.8	212.0
Western Europe	311.8	246.5	20.9	297.0	345.1
North America	773.7	67.8	30.9	637.2	431.5

Source of data: United Nations, 1993, 752–78

Petroleum

The growing dependence on petroleum by both the developed and developing countries after the Second World War was a consequence of the discovery of abundant oil resources in the Middle East, greatly in excess of demand, with resulting low prices, coupled with the utility of oil as a high-quality energy source which is easily transported and stored, and which can be processed easily to yield a variety of fuels and industrial raw materials. From the early 1970s, with rising demand, OPEC policy, and warfare in the Middle East, there was an escalation in price. This considerably increased the wealth of those countries which possessed oil, so that some now form a distinctive group of countries, the capital surplus oil producers, with very high per capita GNP levels – in the case of Qatar and the United Arab Emirates in excess of $15,000 in 1992. Lesser beneficiaries include Venezuela, Gabon, Iran and Brunei. Such prosperity sustained considerable conspicuous consumption, financed major development projects, and provided funds for investment in the developed world and aid to less fortunate Third World countries. Although the former USSR and the USA are respectively the two largest producers, the other main oil producers are within the Third World.

The converse consequence of the oil crisis, however, for those countries lacking oil was a closer search for domestic oilfields, or for alternative sources of energy. A number of countries, including Benin, Cameroon, Brazil, Guatemala and Vietnam found oil in commercial quantities sufficient to diminish their dependence on imports, which benefited their balance of payments and acted as a stimulus to oil-based industries (Plate 6.3).

Coal

In contrast to the considerable, if unevenly distributed, oil reserves of the Third World, its share of known coal deposits is limited. Most of the world's known coalfields are in temperate latitudes, and the former USSR, USA and Australia have three-quarters of the known reserves. Production within the Third World is principally in those countries where coalfields were developed before the era of cheap oil – in China, India, Nigeria, Zimbabwe and Korea. In 1990 China and India were ranked respectively first and fourth as world coal producers, contributing almost one-third of world output. Few other Third World countries are so well endowed and their coal production is often small and of poor quality; in Africa 95 per cent of output comes from South Africa alone. It is possible however, that the availability of cheap oil has discouraged careful searches for coal deposits and new sources may be found. Recent evidence suggests an increase in the known reserves of Colombia, Brazil, Bangladesh, Indonesia, Nigeria, Zimbabwe and Botswana.

Hydroelectricity

Hydraulic power offers better possibilities for energy supply in the Third World. Water power potential is considerable but utilization to date is limited and there are also variations in the distribution of reserves, in response to climatic conditions and the pattern of river flow. However, the fact that water power is a renewable resource, and that there is large untapped potential in the Third World, makes it likely that there will be a substantial increase in energy supply from this source. In some cases a single dam site, such as those at Aswan (Egypt), Akosombo (Ghana), Kariba Falls (Zambia and Zimbabwe) and Owen Falls (Uganda), make a considerable contribution to national total power capacity, with Owen Falls making a contribution to the Kenyan supply. The largest source of South Africa's electricity is the Cabora Bossa dam on the Zambezi in Mozambique. The Third World has also developed some of the world's largest hydroelectric power projects, as at Itaipú on the Brazil–Paraguay border, with an installed capacity of 12,600 MW. The construction of power dams can also bring other benefits in the form of flood control, improved navigation and the provision of irrigation water.

In development terms, however, it is significant that whatever the broad significance of primary fuel production such as oil and coal in the Third World, generation of electricity as an energy source for industry and urban needs is of lesser significance. Electricity contributes less than 10 per cent of the total energy sources listed in Table 6.3 in Africa, Asia and Latin America, but over one-quarter in Western Europe and North America. The number of countries in the Third World with more than 5 million kilowatts of installed electricity generating capacity is small (Figure 6.4).

Other energy sources

Largely as a result of the oil crisis there has been a search for other sources of energy, both conventional and unconventional. The price of oil encouraged an interest in oil shales, of which Brazil and Zaire are known to have substantial deposits; however, technology for their exploitation remains experimental. There are a few examples of the transfer of developed world technology to the Third World in the form of

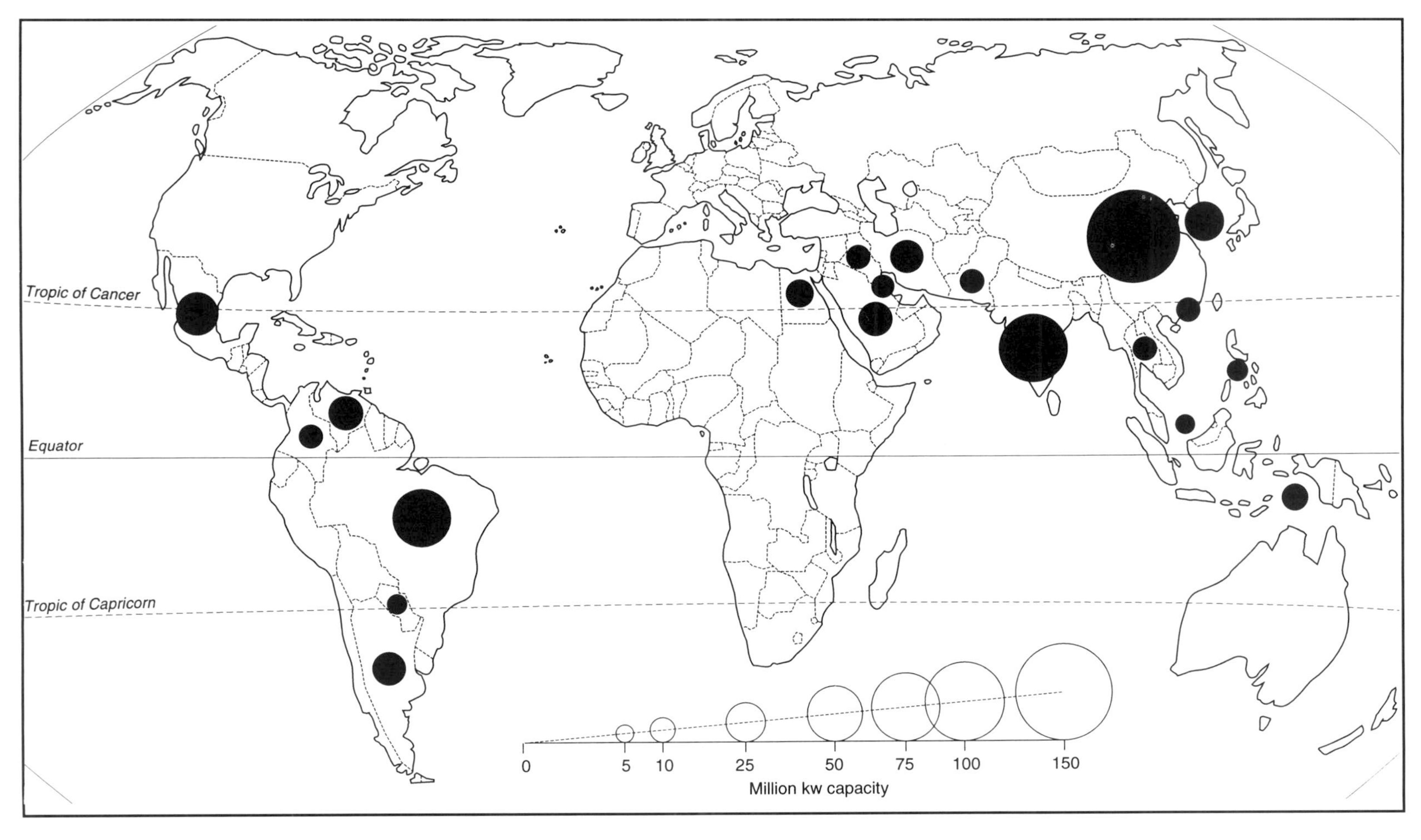

Figure 6.4 Third World major electricity producers, 1990
Source of data: World Bank, 1994a

Plate 6.3 Oil rig, Rio de Janeiro, Brazil Brazil was severely affected by the oil crises of the 1970s, as its known on-shore oil resources were small. However, the discovery of off-shore oil near Rio in 1974 has considerably increased domestic contribution to the country's oil consumption. The search for oil, its production and processing, and most of the distribution of petrol is in the hands of a state-owned company

nuclear power stations, but their economic viability is uncertain, and they pose questions of safety and security. As of the late 1980s Brazil, Argentina, Mexico, India, Taiwan, Philippines and South Korea had established nuclear power stations.

Exploitation of wind and solar energy, despite the seeming promise of the latter in tropical latitudes, is still at a preliminary stage and they, and other sources such as tidal power, probably require considerably more development in the advanced nations before they can be used in the Third World. Successful use of geothermal energy in countries such as Iceland and New Zealand suggests that this may be a useful energy source for those areas located in volcanically active parts of the Third World, such as the circum-Pacific 'ring of fire' of western South America, the Pacific islands and the Far East.

Increasing attention is also being paid to biomass energy sources, from the forests, biological wastes and energy crops. More efficient use might be made of the tropical forest or of planted woodlands. Crop residues and urban waste could also be put to effective use. Brazil, where cane trash had long been used as a local fuel in sugar mills, began to use alcohol obtained from sugar cane as motor fuel to ease the need for the increasingly expensive oil imports.

Energy sources and the location of economic activities

During the Industrial Revolution in the developed world coal was a significant energy source, and a powerful influence on industrial location; many of the traditional industrial areas of Europe and the United States were coalfield-based. Such ties have diminished with transport improvements and the development of other energy sources, and particularly the availability of electricity, which can be transmitted over great distances, and thus diminishes the 'pull' of energy sources. Some ties remain, where there are significant coal deposits and thermal power stations and coal-using industries such as metallurgy have been established. Similarly, oilfields have provided the foci for refining and petrochemical industries. However, oil is a mobile energy source, capable of relatively low-cost transfer by road, ship or pipeline to the point of consumption. Thus, although Brazil has refining capacity adjacent to its north-eastern oilfields, the largest refineries are now to be found at the principal markets such as São Paulo, Rio de Janeiro and Belo Horizonte.

The development of electric power has afforded the greatest freedom. Early power stations were located close to the existing markets, that is, the largest cities, with coal or oil-fired power stations, or hydroelectricity brought short distances from small dams. However, the improving efficiency in transmission and increasing size of power stations, particularly those using water power, have extended the availability of electricity, though the largest cities tend to be best supplied, and some parts of the rural Third World remain without access to electric light and power (Box 6.3).

INDUSTRIALIZATION

Why industrialize?

Agriculture and mining were important in the colonial and post-colonial economies of Third World countries, and in many cases the primary sector remains important in terms of production, export and employment. However, particularly since 1945 industrialization has been a fundamental objective in the development strategies of the Third World. Manufacturing has been seen as essential to the achievement of high rates of economic growth, to diversify the economy, to create jobs, to meet the basic needs of the population, and as a symbol of independence from the developed world.

The latter provided a seemingly obvious role model; the Third World depended on the primary sector for employment and output, whereas the developed countries had strong industrial sectors. Even as late as 1965 manufacturing provided 30 per cent of GDP in the high-income countries, as opposed to 20 per cent in the low- and middle-income group. More tellingly, in the First World only 5 per cent of GDP at that date came from agriculture, as against 19 per cent in middle-income and 41 per cent in low-income countries. Given these patterns, it is not surprising that the Third World countries tended to equate *industrialization* with *development*.

A crucial problem for these countries was the creation of jobs. Many of them are faced with rapidly increasing populations (see Chapter 3), and there is substantial unemployment and **underemployment** in rural areas, while agricultural modernization serves to shed labour (Chapters 4 and 5). In consequence, therefore, there is a search for additional employment, both in a broad sense by planners and, more specifically, by young people coming on to the labour market and by rural dwellers displaced or unable to find land or work. A seemingly

BOX 6.3 ELECTRIC POWER IN MINAS GERAIS, BRAZIL

In 1950, the state of Minas Gerais in south-east Brazil was estimated to have one-third of the country's hydroelectric power potential, but this resource was little exploited. Existing capacity was mainly in the hands of small private companies, which had concessions to supply individual settlements. There were 359 such companies, with over 400 power stations and few transmission lines, and a total installed capacity of 206 MW. This small and unreliable power system was a disincentive to the industrialization the state government was trying to foster, and the government therefore set up its own power company, Centrais Elétricas de Minas Gerais (CEMIG), in 1952 to generate and distribute hydroelectricity. It was recognized that it was impossible to create immediately a single-grid system to serve the entire state, so a number of regional systems supplied by moderate size dams were established, which could be interlinked at a later date. The main concentration of early growth was on the state capital, Belo Horizonte, which was regarded as the core area for any development programme,

BOX 6.3 *continued*

as it had a substantial population, local mineral resources, and an established industrial structure. By 1957 CEMIG's dams had a capacity of over 100 MW, with a power net focusing on Belo Horizonte, together with a number of small regional systems (Figure 6.5). Such a pattern, of early, reliable energy supply, gave these areas an initial advantage in the industrialization process.

The system was then amplified, with a number of larger dams, such as Três Marias (390 MW), and by 1968 installed capacity was over 583 MW. The regional systems were interconnected, and the distribution system began to extend across the state. In 1961 the federal government had set up a national power company, Eletrobrás, to produce and distribute power on a national scale, and in 1967 the grid integrating the electricity system of south-east Brazil was completed. With the development of a number of large power stations on the Rio Grande, on its western border, Minas Gerais's power production and distribution became part of a regional system. Figure 6.5 shows the way in which the state has been integrated into what has become a national power grid, and that the intra-state system has become both more dense and more extensive. As well as receiving electricity through the Eletrobrás grid from major plants such as Itaipú, CEMIG owns over 40 power stations, with an installed capacity of 4464 MW. By 1991 it was supplying over 3 million consumers, in 652 municipalities. It now has a

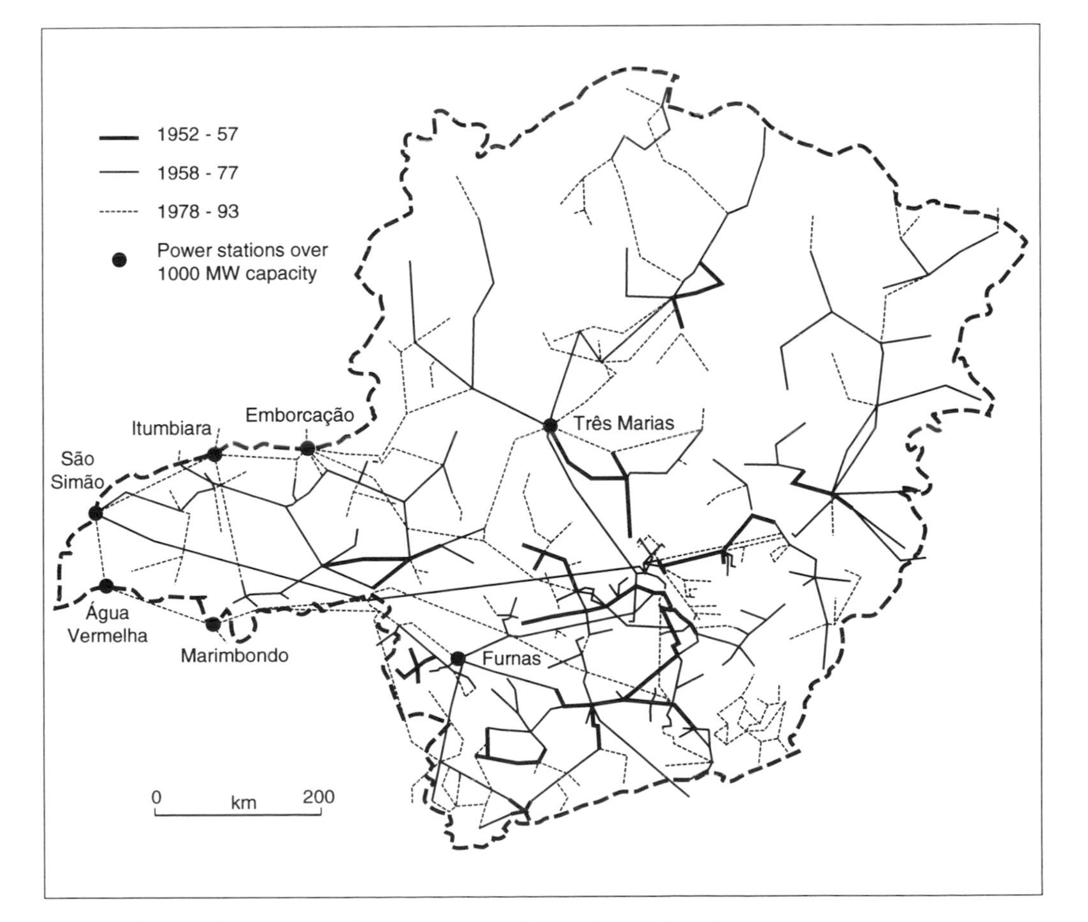

Figure 6.5 The evolution of the electricity grid in Minas Gerais, Brazil
Source of data: CEMIG annual reports, 1952–93

BOX 6.3 ***continued***

transmission system of over 20,000 km, one of the most extensive in Latin America, and supplies Brazil's second largest consumer market. Its target is to provide 100 per cent coverage for the state's urban areas, and since 1962 it has operated a subsidiary company to bring electricity to villages and isolated farms through a rural grid system, which has grown from 44,000 to 185,000 km between 1981 and 1991.

The CEMIG system shows the way in which a power infrastructure evolves, providing an essential resource to a territory and its population. Its necessarily evolutionary nature, however, gave initial impetus to the central part of the state, compounding its economic advantage. The significance of the system for the economic progress of Minas Gerais can be seen by the fact that in 1992 manufacturing consumed 68 per cent of production, and that despite fluctuations in the national economy, energy consumption from CEMIG grew at an average of over 4 per cent per year between 1972 and 1990.

obvious solution therefore has been, and remains, to create *new* employment opportunity in industry and factories.

It was also argued that industry would assist agriculture, not only by removing surplus labour, but by providing new inputs such as machinery and fertilizer; while the industrial workforce of the towns would create larger markets and better prices for farm output of food and raw materials. It was also suggested that industry would be a more secure base for economic advance, as it is less vulnerable to the vagaries of output, to which agriculture was prone because of variations in climate.

Given the dependency on primary production of agriculture and mining noted above, proponents of industrialization stated that it would serve to diversify these economies. Such dependence on primary-product exports made them vulnerable to fluctuations in demand and prices on the world market, over which they had very little influence. Some evidence suggested that export earnings for these primary products had been falling relative to the prices of the manufactured goods they imported. It was also claimed that there would be other economic spinoffs and spread effects. There would be forward and backward linkages to other sectors such as transport, energy, finance and commerce. Less tangibly, it was argued that industry would be a focus of dynamism and change and a source of 'modern' attitudes, in contrast to what were perceived to be the 'traditional' and conservative attitudes of agriculture and the countryside. There were also symbolic considerations, that countries were independent, not only politically but economically, of former colonial powers.

For many countries the Soviet Union provided a role model since, after the Revolution of 1917, it had given high priority to industrialization. It was believed that in order to survive economically and politically, the USSR had to develop a large and self-sufficient industrial sector. In consequence strong emphasis was placed on industry, at the expense of agriculture, in Soviet planning. Between 1918 and 1970 over one-third of all capital investment went to industry, as against 15 per cent in agriculture. There was particular emphasis on heavy industry, such as steel and chemicals, which received over 80 per cent of industrial investment from the First Plan, in 1928, until the early 1950s.

In addition to its basic role of job creation and economic diversification, industrialization might have other effects on the population. The industrialization process has generally been associated with urbanization. The agglomeration of people into towns and cities makes it easier and cheaper to provide them with basic services of electricity, water supply, education and health care than if they remain relatively dispersed in the countryside. The experience of the now-developed world also suggested that urbanization would tend to lead to a fall in birth rates, and thus a slow-down in population increase.

CHECKS TO INDUSTRIALIZATION

Given these seemingly strong arguments in favour of manufacturing, why does a low level of industrialization remain a defining characteristic of the Third World? Why are Third World countries 'under-industrialized'?

A range of constraints is apparent, a significant one being the legacy of colonial rule. The colonizing powers of Europe – Britain, France, Belgium and Germany – had experienced an industrial revolution and were important producers of manufactured goods in the nineteenth century. They saw their colonies as sources of raw materials for industry, and as markets for manufactured goods. It was thus not in their interest to encourage indigenous industries which might compete with such exports.

Also linked to the impact of colonialism are the Third World's infrastructures, which may be inadequate, or inappropriate to the needs of industrialization. Much of the transport system and the provision of energy was created by the colonial powers and foreign investors, in their own interests, primarily to extract raw materials (as in the development of railways; see Chapter 8), and not to provide a base for the industrial development strategies of independent nations. Moreover, the provision of new infrastructure is expensive and involves what is termed 'lumpy' investment; that is, it cannot be provided gradually but must be a total project which, initially at least, may provide capacity in excess of demand – for example a major dam or highway. Provision of such utilities is frequently spatially concentrated, whether as a colonial legacy or from more recent development, so that some regions are more favoured than others.

To provide factories or infrastructure a developing country needs capital, and capital deficiency is again almost a defining characteristic of underdevelopment. Since Third World countries are poor, there is little surplus for saving. Capital has, then, to be sought by savings from limited resources, taxation, foreign loans and aid. Not only is capital for investment limited, but the cost of providing an industrial job is high, and given the advanced technology currently available, is more expensive than during the industrialization of Europe.

Some general characteristics of Third World populations limit their potential as industrial labour forces. High levels of illiteracy and low general levels of education are barriers to the use of modern machinery, and the training of labour in basic skills incurs an extra cost for the industrialization process. Poor health may be a factor in levels of productivity and absenteeism, while there are also problems of adjustment from traditional rural lifestyles to factory routines. There may also be a lack of entrepreneurial skills for large-scale industries. Labour and capital must be combined with management skills to organize, plan and cope with the decision-making and risk-taking involved in industrialization. Not only are such skills in short supply, but the 'brain drain' may see the loss of qualified personnel to the First World.

A crucial problem for many Third World countries is the small size of their domestic market. Despite notions of 'teeming millions', many of the less industrialized countries have small populations; in 1992 48 of the 88 countries classed by the World Bank in the low- or middle-income category had populations of below 10 million. Markets are not only small but poor. In 1992 the GNP per capita of the low-income countries averaged \$390 and the middle-income group \$2490. Such figures were respectively 2 and 11 per cent of the average for the high-income, industrialized countries. In addition, internal income distribution is commonly highly skewed, so that many people have little or no money to spend on manufactured goods. Such limitations prevent firms from taking full advantage of economies of scale associated with manufacturing, and give rise to high costs of production and under-use of capacity.

It is often claimed that the inadequacy of the resource endowment is a major barrier to Third World industrialization. It is perhaps better argued that while individual countries may be deficient in particular resources, in total the Third World is not resource-deficient – such a notion is, after all, hardly compatible with the importance of primary products such as minerals, fuels and agricultural produce in its export structure. Much of the developed world is heavily dependent on the developing world for resource inputs into its economy. In this respect, then, the Third World is resource rich. However, some countries do appear to be poorly endowed. In addition, it should be noted that the resource endowment of most countries is imperfectly known. Comprehensive surveys of geology, hydrology, soils and land-use potential are still incomplete. They are also expensive to undertake and require technical skills which are deficient. However, recent major discoveries of petroleum, iron ore, bauxite and other minerals, and increased identification of hydroelectric potential suggest that the resource base may be an asset for future industrialization.

THE ORIGINS OF MANUFACTURING IN THE THIRD WORLD

Though Third World countries are characterized as underdeveloped or as primary producers, there is a tendency to assume that they have little industry, and that what there is is of recent origin. This is not so, for the processing of crops and minerals prior to export dates back, in some cases, to the early colonial period. As we have seen in the case of bauxite and copper, the trend to local processing has risen.

Furthermore pre-colonial populations had developed significant craft industries. Before the Age of Discovery native craftsmen were producing textiles, metal goods, pottery and similar items. Colonial settlers imported manufactures from Europe, but also relied on local production of food, clothing and building materials, and were involved in processing commodities for export. In the nineteenth century the European Industrial Revolution tended to undermine the artisan industries of the Third World. For example, Indian craft textiles had been an important trade item in the East in the seventeenth and eighteenth centuries, but were severely hit by competition from the machine-made cottons of Lancashire. Even so, a factory-scale textile industry emerged in India in the mid-nineteenth century, based on local raw materials of cotton and jute and on local markets, such that there were over 160,000 workers in cotton mills by 1900.

Elsewhere in the Third World local raw materials and markets sustained a similar emergence of modern industrialization. This tended to be in non-durable consumer goods, such as textiles, clothes and foodstuffs, or in low-value high-bulk goods such as building materials, where even inefficient producers could compete against imports. The mechanization of agricultural processing activities, for example coffee in Brazil and Colombia, and the development of railways, led to the establishment of repair shops and then simple engineering works. However, such innovations were unevenly spread through the Third World, with greater progress in Latin America and Asia than in Africa. Though most of the activity was concerned with consumer goods, there were some early developments in heavy industry; Brazil and India, for example, had established modern steelworks before 1890.

Artisan industries, export processing and consumer industries linked to local resources and markets were the pioneers of manufacturing activity in the Third World, and since 1914 there has been increasing interest and progress in industrialization. The two world wars revealed the dependence on imported manufactures, and the consequent difficulties when supplies were cut off. The inter-war depression, with the slump in trade and prices of primary exports, demonstrated the danger of dependence on such a structure, and encouraged independent countries to try to diversify their economies. Those securing political independence after 1945 similarly sought to establish economic independence. This was further encouraged by the emerging 'development economics' of the late 1940s, which tended to view the 'solution' of the problem of the 'backward' countries as industrialization.

For all of these reasons most Third World countries therefore sought to industrialize. In this process alternative strategies have to be considered. What should be the balance between 'heavy' and 'light' industries? What should be the relative roles of private, state and foreign investment? Should the technology used be capital- or labour-intensive? Should the goal be self-sufficiency or should there be production for export? And so on. Such choices are not necessarily mutually exclusive; they may be appropriate to different sectors or stages of the industrialization process.

Plate 6.4 Making drainage pipes, Sousse, Tunisia Very simple manufacturing uses manual labour, simple moulds and local materials to produce pipes for urban drains

Plate 6.5 Cotton mill, Biribiri, Brazil Cotton manufacture is a common pioneer in Third World industrialization. This mill is at an upland site to use water power. In consequence, besides the mill, a small 'company town' was built for the labour force, consisting of houses, and a school, church and store

THE IMPLEMENTATION OF INDUSTRIALIZATION

Import substitution

Third World countries formerly imported many of their manufactured goods in return for their primary products. Increasingly, they have sought greater self-sufficiency by producing their own manufactures, often behind some form of tariff protection. In the import-substitution process, light industries such as food, drink and tobacco, textiles and clothing tend to be the first to be substituted. These tend to be technologically simple, requiring relatively few engineering skills and low levels of capital investment, both of which are scarce (Plates 6.4 and 6.5). Markets already exist, and as these are consumer goods, demand is direct. Many countries chose this relatively 'easy' strategy, leaving more complex industrialization until later. Such a strategy was important in the early industrialization of Argentina, Brazil and Mexico, and for many Third World countries manufacturing remains limited to import substitution products.

Heavy industry

Following the Soviet model, some countries have given high priority to developing basic industries, post-independence India providing a classic example. Its large population implied a very large demand for consumer goods, but in order to provide these a certain input of capital goods was required. For many Third World countries these would have been imported, but it was argued that Indian demand was sufficiently large to sustain internal production, and that it was also unlikely that the country could expand its exports sufficiently to pay for the large volume of capital goods required. In consequence, the country's second Five-Year Plan, 1956–61, gave very high priority to developing capital-intensive heavy industries such as steel, chemicals, coal and petroleum. Mining and manufacturing received 20 per cent of planned investment during this period, compared with 7.6 per cent under the First Plan. Subsequent plans had similar priorities, though by the mid-1970s the government was giving more scope to market forces. Nonetheless, such a strategy did provide a capital-goods base for industrialization, and by 1990 India ranked as a major world producer of iron and steel, cement, and nitrogenous fertilizers. In this process, the country was aided by good resource endowments of coal and iron ore in the Damodar valley, and by technical assistance from various developed countries. Conversely, this resource-based industrialization tended to bring progress to a limited area, and thus intensify regional inequity.

State participation

In the industrialization process in developing countries, the state has often played a major role. Where resources for development are scarce, central government may be the only agency capable of marshalling the savings and taxation, or securing the foreign capital or aid necessary for industrial investment. The state makes decisions on the allocation of resources (see Chapter 10) between sectors of the economy or within industry. It may assist industrialization through financial aid or tariff protection, or it may involve itself directly in certain industries, which are politically or strategically sensitive, or where private capital is reluctant to invest.

Such involvement was frequently within the framework of development plans, as in the case of Brazil (see Chapter 10) or India. The latter's Second Plan, 1956–61, distinguished between essential industries such as steel, coal, heavy machinery and munitions, which were to be controlled by the state, an **intermediate** sector, including machine tools, fertilizer, synthetics, rubber and ferro-alloys, which was to become progressively state owned, and left the remaining areas of industry to private capital. In other cases a state development agency was active in fostering particular sectors or specific projects. In Chile the Corporación de Fomento de la Producción (CORFO) was set up in 1939 to plan electrification, develop the oil industry, and establish a steelworks, and sugar-beet and paper factories.

Plate 6.6 This advert for the construction of an off-shore oil rig for Brazil's Petrobrás oil company reveals the complexity of multinational industrial linkages. Petrobrás collaborated with a Singapore-based engineering firm to build the rig. It involved finance from Japan, and inputs from British, French, German, Norwegian, American, Japanese and Malaysian companies

Such strategies had considerable success in establishing infrastructure and basic industries, particularly in the relatively more advanced countries of Latin America and Asia. However, by the late 1970s economic circumstances and changing development notions had modified this pattern and aggressive state capitalism has become less common. The drive to independent, integrated industrial economies has diminished, and countries have sought a strategy to meet domestic needs and to seek niches in the world market for manufactures. The element of state control has diminished, with countries such as Brazil, India, Pakistan and Singapore selling off at least part of the state-controlled sector. Such trends have been intensified by the market-driven philosophies of the IMF and the World Bank.

Foreign investment

Investment by colonial powers and foreign capitalists was significant in the early exploitation of the Third World. In the nineteenth century there was a substantial amount of private foreign investment in plantations, mining, urban services and infrastructure. In the present century there has often been hostility to such investment, but it remains significant in certain industrial sectors, through the activities of multinational corporations (MNCs). These large firms, usually with their headquarters in the USA or Europe – but with the recent emergence of Japanese, and even Third World-based companies – are responsible, through their subsidiaries for a wide range of goods, especially in the consumer-durable sector, such as vehicles, chemicals and pharmaceuticals, and electrical goods. Many MNCs have annual sales volumes greater than the GNP of most Third World countries. In the mid-1980s only the largest Third World economies, such as those of China, Brazil, Mexico and India could rival the scale of such global corporations as General Motors, Ford and IBM.

It is estimated that around one fifth of direct foreign investment in manufacturing from

Europe and North America is in the Third World (which means that much of such investment is, in fact, intra-First World). Such investment is important to the industrialization of the developing countries, in that the MNCs provide access to capital, technology, 'know-how' and, given that they have become world-wide economic systems, they provide crucial access to world markets (Plate 6.6). On the other hand, MNCs have been criticized because their capital-intensive technology provides relatively few jobs, and on the grounds that they seek to exploit 'cheap' labour in the Third World. They are also criticized for producing goods aimed at only a segment of the domestic market, for discouraging domestic producers, and for inhibiting development of domestic skills and entrepreneurship. Their dominance in some sectors of industry, and the fact that their strategies may be guided more by corporate goals than the interests of the host nation, has also made them targets of hostility. In some cases they have been accused of avoiding tax and other levies on their profits, and of interfering in domestic politics.

MNCs, however, have provided industrial advances which might not have occurred otherwise. This is particularly the case in the capital- and/or technology-intensive industries such as vehicles, engineering and chemicals, where they are often the sole source of the necessary skills and capital. They have also been important in the increasingly important labour-intensive component manufacture and assembly industries, such as electronics and vehicle components. In such cases the low cost of labour, especially of women, has been attractive to MNCs and a source of job creation, even at low wages, for Third World countries. Such industries have been of particular significance in the NIC countries of Asia, such as Hong Kong, Singapore and Taiwan, as well as in Mexico, Malaysia and the Philippines. In the same way production of simpler labour-intensive light manufactured goods, such as plastics, toys, clothing and sports goods has also been significant in the industrial success of South Korea, Hong Kong and Taiwan.

In some cases MNCs were attracted to Third World countries by the potential of their markets. This was the case, for example, with the Brazilian car industry, where restrictions on imports and the existence of a considerable

BOX 6.4 *MAQUILADORA* INDUSTRIES IN MEXICO

A major example of this type of activity is provided by the *maquiladora* industries which have developed along the Mexico–United States border. In 1965 the Mexican government initiated a Border Industrialization Programme to attract US industries to the region. Its aim was to attract American labour-intensive industries, in search of cheap labour, to locate in Mexico to provide employment and create industrial exports, principally back across the border. Firms were allowed to import equipment and materials duty free. Initially this was only for the production of export goods, but in 1983 this was relaxed to allow 20 per cent of production to be sold within Mexico. Such a relaxation represented a change from the original intention to protect Mexican industry from the competition of American firms. Legislation had also been required to remove protectionist legislation which allowed only Mexican-controlled firms to operate within the country. The number of *maquiladora* firms has grown rapidly, from 70 in 1966 to over 1400 in 1990, and employing over 375,000 workers. Location of these firms was restricted to a zone within 20 km of the border and, in an attempt to influence location within that area, Free Ports and Free Trade Zones were established, particularly in the north-west of the country. These regulations were later relaxed to allow firms to locate in other parts of Mexico; however, over a third of the jobs created were in the two border cities of Cuidad Juárez and Tijuana. The principal industries which have been attracted are concerned with electrical goods and clothing. Although aimed initially at American-based MNCs, and attracting firms such as General Electric, Westinghouse and ITT, access to the American market has more recently attracted Japanese assembly plants owned by Sony, Toshiba and Hitachi, and even, in an extension of one NIC into another, some Korean ventures.

market created by the affluent sector of its large population, attracted major European and American vehicle manufacturers. Elsewhere, particularly in South-East Asia, it was the attraction of low-cost, semi-skilled labour. In such cases the MNCs were interested in the potential of lower-cost exports from such countries onto the world – and First World – markets. By the late 1970s such MNC manufactures accounted for over one-third of the exports of Brazil, Mexico and Korea, and close to 90 per cent in the case of Singapore. Furthermore, whereas MNC factories in Third World countries were often subsidiaries producing finished goods from imported components, they have increasingly begun to use locally made components; more crucially they have become part of the global strategies of such corporations, producing components for plants elsewhere, or specializing in particular products within an overall corporate strategy.

Export-oriented industry

An important dimension in the activities of MNCs in the Third World has been their association with a rise in the export of manufactured goods. The availability of cheap labour has encouraged them to set up factories to assemble imported components or, increasingly, to manufacture complete products within Third World countries. The rapid growth of the electronics and similar industries, which are labour intensive and usually with high-value, low-bulk products, have been particularly important. With the availability of low-cost air-freight, such plants have been able to serve global markets. Such industries, whether controlled by MNCs or by local capital, or as joint-ventures, have been responsible for considerable industrial growth in the Third World. Though dependent on foreign capital and imported technology, they have created employment and export earnings. However, such innovation has tended to be restricted in its location. The most striking examples are South Korea, Taiwan, Singapore, Hong Kong, Brazil and Mexico (Box 6.4). There is indeed a close link between this type of industrialization and their status as Newly Industrializing Countries. Significantly, in 1991 machinery and transport goods made up over one-fifth of the exports of Singapore, Hong Kong, Korea and Brazil. Some differential may be noted within these NICs, in that in 1992 manufactured goods made up just over half of the exports of Brazil and Mexico, 75 per cent for Singapore, and over 90 per cent for Hong Kong and Korea.

Such a process represents an 'internationalization of capital', as MNCs seek to maximize their profits, by producing at points of cheap labour cost. The corporation provides the capital, technology and managerial skills, but the strategy is in essence one aimed at maximizing corporate profits. Theoretically such development offers considerable potential for further industrialization in the Third World, given the large pools of cheap and untapped labour available in many countries. However, the NICs which have emerged do offer other attractions, in terms of infrastructure, location and some previous industrialization, which are not immediately available in, say, the poorest parts of Africa or Asia. Moreover, as MNCs become more 'global' in their strategies, their investment policies are more governed by the desire for corporate profit than the long-term interest of countries in which they invest. It is also the case that this type of industrialization *is* dependent on exports; it is vulnerable to fluctuations in the global economy and, as its principal markets are in the developed world, it has to face the concerns of older-industrialized countries to protect their domestic producers.

Given the potential for generating an industrial sector offered by this type of manufacturing, some countries have established **Export Processing Zones (EPZs)**, which are tracts within a country which offer conditions to attract export-oriented industries. These include incentives for investment, the provision of a modern infrastructure of roads, electricity and low-cost factories, concessions on trade and relaxation of legislation on foreign ownership or pollution. They have tended to attract the kind of modern, footloose industries in which cheap labour is important, whether in electrical and electronic goods, or textiles and footwear. EPZs have become common in the Third World, but

Table 6.4 Structure of production, 1965 and 1990

	Share of GDP (%)							
	Agriculture		Manufacturing		Services		Other	
	1965	1990	1965	1990	1965	1990	1965	1990
Africa	40	32	7	–	39	40	14	28
E Asia	37	21	24	34	30	36	9	9
M East/N Africa	20	–	10	–	40	–	30	–
L America	16	10	23	25	50	54	11	11
Third World	29	17	20	25	40	47	11	11

Source of data: World Bank, 1992

especially in South-East Asia, the Caribbean and Central America. EPZs have been a particularly important element in the industrial growth of the Asian NICs such as Hong Kong, Singapore, Taiwan, Malaysia and South Korea. They have also formed part of China's programme to develop an export-industrial sector, with the creation of 'Special Economic Zones' in which foreign capital was allowed to invest. Four such SEZs were established in south-east China in 1980, at Shenzen, Zhuhai, Shantou and Xiamen, and a fifth at Hainan, in 1988. The aim has been to use cheap labour and a range of incentives to attract foreign capital and technology.

One facet of this type of export-oriented, labour-intensive industry has been that it has created industrial employment for women. In part this has been based on assumptions that textile and electronic industries offer particular opportunities for the alleged manual dexterity of women, but another important factor has been that in the search for cheap labour, females have been even less expensive to employ than men in the Third World. In export-oriented industries such as textiles and clothing, and electronics, women may make up to 90 per cent of the workforce – but their wage rate may be much lower than that of men; in 1985 female wage rates in the NICs of Hong Kong, Singapore and South Korea averaged between 30 and 50 per cent less than those for males.

ACHIEVEMENTS AND PROBLEMS

As a result of their efforts to industrialize, Third World countries have achieved high rates of growth in manufacturing in recent years. Over the period 1965–80, their manufacturing sector grew, though from a low base, at an average rate of 8 per cent a year, while the rate in the high-income (and thus 'industrialized') countries was only 3.2 per cent. However, the rate slowed during the decade of the 1980s, a reflection of the slow-down in the global economy, to 6 per cent a year. In both periods, however, the rate of growth of manufacturing in the Third World was higher than that in either agriculture or services, which indicates some success in shifting their economies towards the non-primary sector. This is borne out by the changing structure of their GDP, from 1965–90, in which the share of agriculture fell from 29 to 17 per cent, and that of manufacturing rose from 20 to 25 per cent.

However, as Table 6.4 indicates, there were some regional variations in these changes. These figures show that the shift to the non-agricultural sector has been most marked in East Asia and Latin America; indeed the shift in East Asia between 1965 and 1990 is particularly striking. The rate of growth and diversification was most rapid there. This in turn reflects the emergence of the Asian NICs, with South Korea and Singapore, for example, achieving growth rates in excess of 10 per cent a year even during the 'slow-down' period of the 1980s.

A further important measure of success has been the ability of Third World countries to reduce their dependence on primary product exports. For the region as a whole, manufacturing's contribution to export earnings rose from 26 per cent to 50 per cent between 1965 and 1990. By the latter date, manufactured goods constituted 54 per cent of export earnings in the

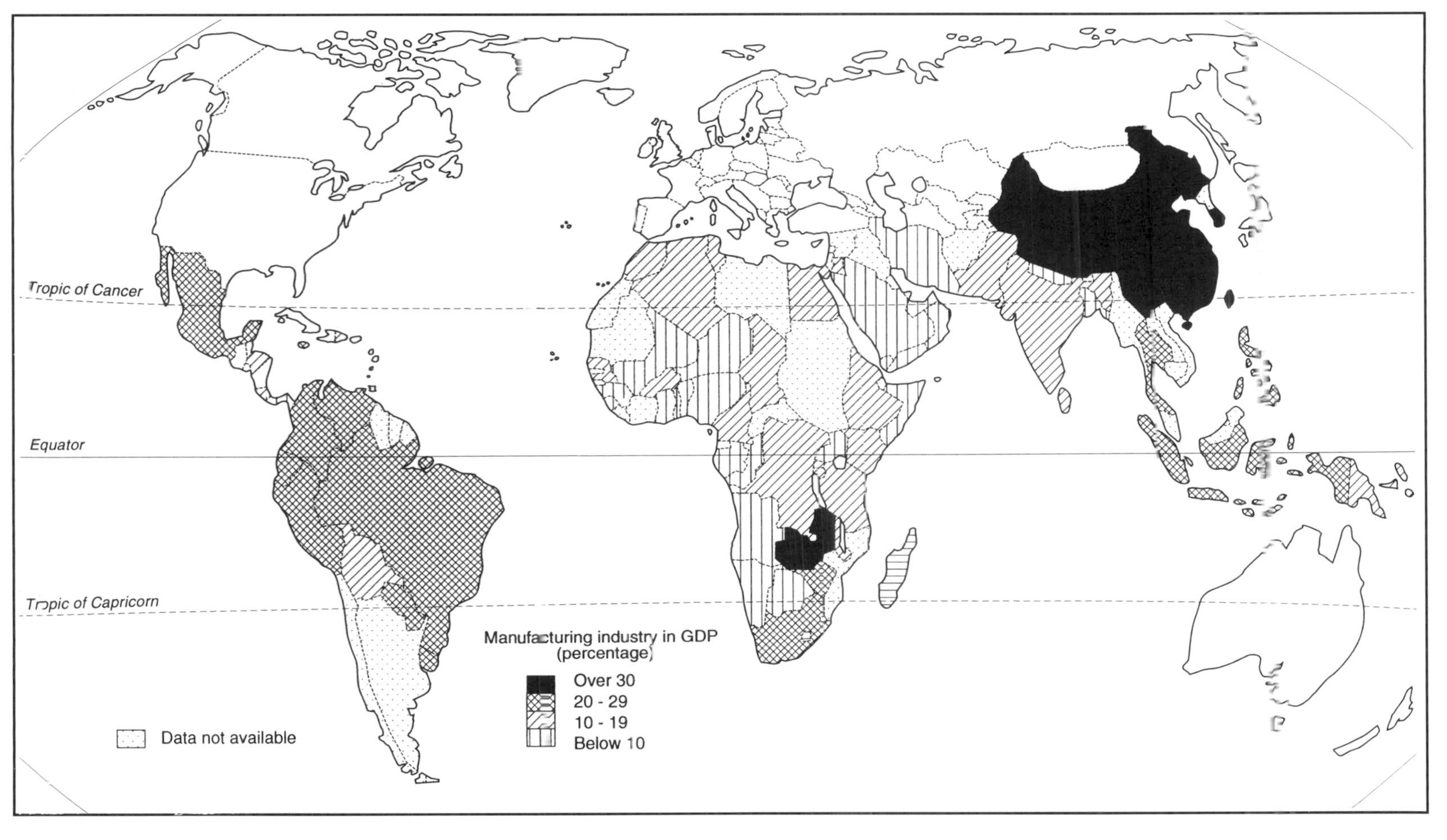

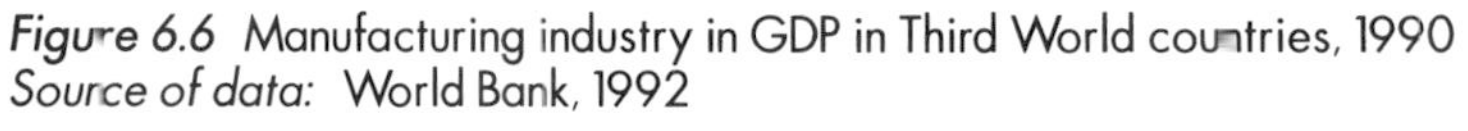

Figure 6.6 Manufacturing industry in GDP in Third World countries, 1990
Source of data: World Bank, 1992

Plate 6.7 Brick-making, near Bulawayo, Zimbabwe

Table 6.5 Value added by manufacturing, 1990 (selected countries)

	Contribution of (%)	
	Food, drink and tobacco, textiles and clothing	*All other industries*
Madagascar	**75**	25
Ethiopia	**67**	33
Tunisia	36	**64**
South Africa	22	**78**
Sri Lanka	**74**	26
Nepal	**60**	40
Malaysia	19	**81**
Singapore	7	**93**
Bolivia	45	55
Ecuador	44	56
Brazil	25	**75**
Venezuela	22	**78**

Source of data: World Bank, 1993

low-income countries, 50 per cent in the lower-middle-income countries and 57 per cent in the upper-middle-income countries. For a number of countries in each category, manufactures have come to lead the export sector – in India, China, Bangladesh and Pakistan among the low-income group, the Philippines, Tunisia and Jamaica in the lower-middle-income group, and Brazil and South Korea in the upper-middle category.

There is, however, still striking variation in the levels of industrialization which have been achieved (Figure 6.6). Sub-Saharan Africa remains the least industrialized: in most countries manufacturing contributed less than 20 per cent to GDP in 1991, the exceptions being Zambia, Zimbabwe and South Africa, where mineral-based industries are significant (Plate 6.7). In Asia some countries, such as Bhutan and Nepal, still lack a significant industrial sector; conversely there are several – China, the Philippines, Thailand, Singapore and South Korea – where industry provides over one-quarter of GDP. Latin America is the most industrialized of the three continents, though only Brazil, Mexico, Uruguay and Argentina derive more than 25 per cent of GDP from manufacturing.

Besides the variations in the level of industrialization, there are also some considerable variations in the structure of industry which has been established. For most countries the import-substitution strategy was the easiest to adopt. Such industries were the easiest and least risky to develop. They required modest levels of capital and generally low levels of technology. Demand already existed and output could be more easily geared to these markets, whereas export-oriented industries required the securing of external markets. Domestic markets could sustain such industries, but were inadequate for the production of intermediate and basic goods. In consequence, for many Third World countries, production is dominated by consumer goods industries supplying everyday necessities, such as food and drink, textiles and clothing, building materials and furniture (Table 6.5). Such examples indicate the limited range of industry established in many cases, but also that some countries have made more progress in the development of consumer durable, intermediate and basic goods industries. In some cases these have been sustained by higher levels of economic development and larger markets; in others by the creation of an industrial sector geared to exports. Some element of these contrasts is evident in Figures 6.7–6.9, which show that while many Third World countries have developed a consumer industry such as cigarette manufacture, essentially for the domestic market, relatively few are involved in the more sophisticated production of TV receivers, or in heavy industry. The technology involved in making cigarettes is simple, and can be adjusted to market size, whereas a steel mill requires large capital investment, complex technology and a substantial market. Because of the investment required and a perceived strategic role for steel, state capital has often been involved in the establishment of Third World iron and steelworks. The pattern evident in Figure 6.9 reveals some interesting elements. The manufacture of television receivers is a modern, consumer durable component-assembly industry, which has involved both national and multinational capital. The map indicates both production for substantial domestic markets, such as those of Brazil, Mexico and Argentina, and the development of such industries in the export-oriented industrialization of Asian NICs such as South Korea, Singapore and Malaysia.

In spite of the progress in industrialization, in terms of high rates of industrial growth, changing economic structure and trends to self-sufficiency, the Third World remains of limited significance on a world scale. In 1977 the developing countries contributed around 9 per cent of world manufacturing output; by the late 1980s this had risen to 14 per cent. This is well short of a target for the year 2000 of 25 per cent, and even were that to be achieved, which seems highly unlikely, it would still need to be set against an estimated 75 per cent share of world population. There was also some progress from the 1970s in the Third World's share of global manufacturing exports, which rose from 5 per cent in 1970 to 13 per cent in 1986. However, as was implicit above, this progress has not been uniform across the Third World. Much of the growth in output and exports has concentrated in the NICs of Asia (Hong Kong, Singapore, South Korea and Taiwan) and Latin America (Brazil and Mexico). In fact there is evidence that some of this expansion, especially in exports, was at the expense of the less industrialized countries. Industrialization has made little progress in Sub-Saharan Africa, where acute poverty, small markets, and deficiencies in skills and capital are major constraints. In 1987 the World Bank indicated that 43 countries accounted for two-thirds of all manufacturing exports from the Third World, and that 60 per cent came from just 15 countries. Of these, 9 were in Asia, 3 in Latin America, and 3 (Yugoslavia, Turkey and Greece) on the European semi-periphery.

Despite the relative success of the NICs, a

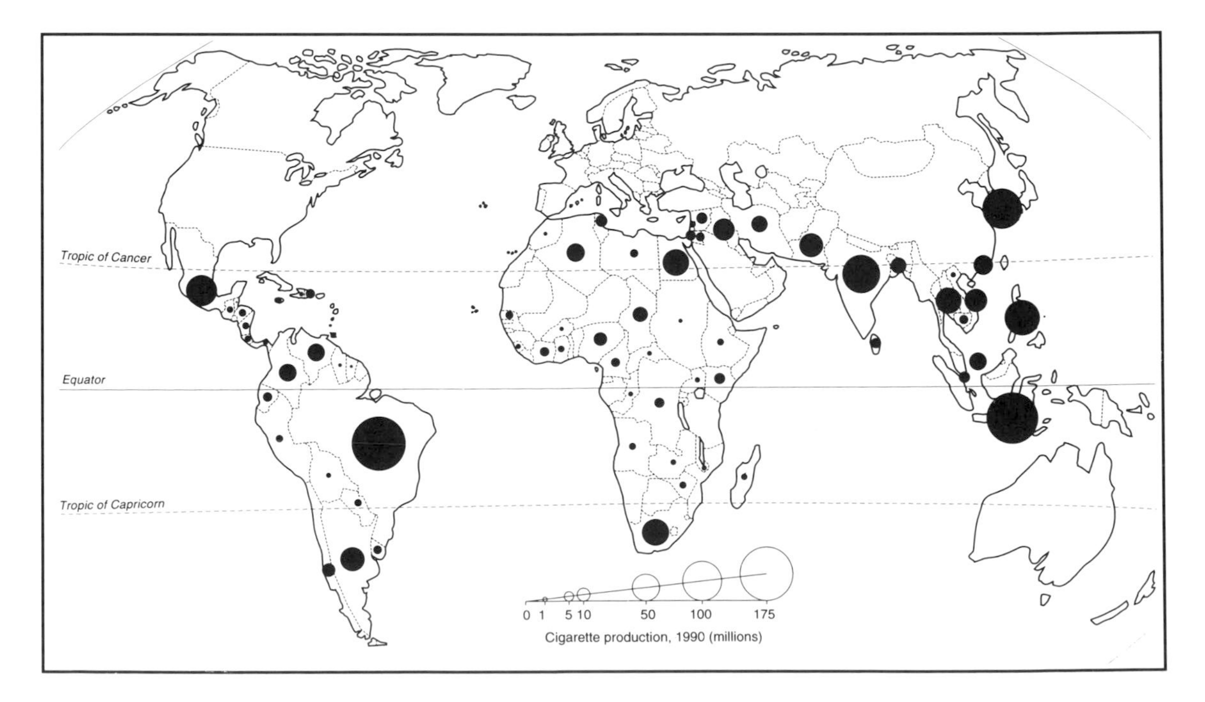

Figure 6.7 Third World cigarette production, 1990
Source of data: United Nations, 1992

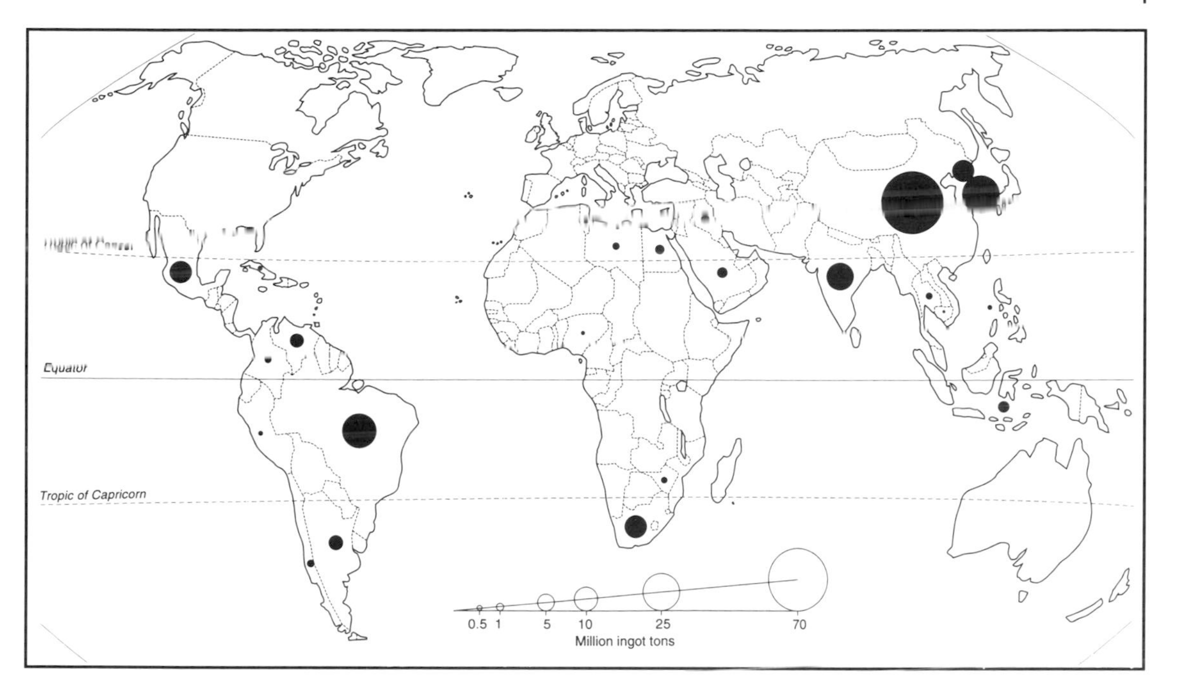

Figure 6.8 Third World steel production, 1990 (million ingot tons)
Source of data: United Nations, 1992

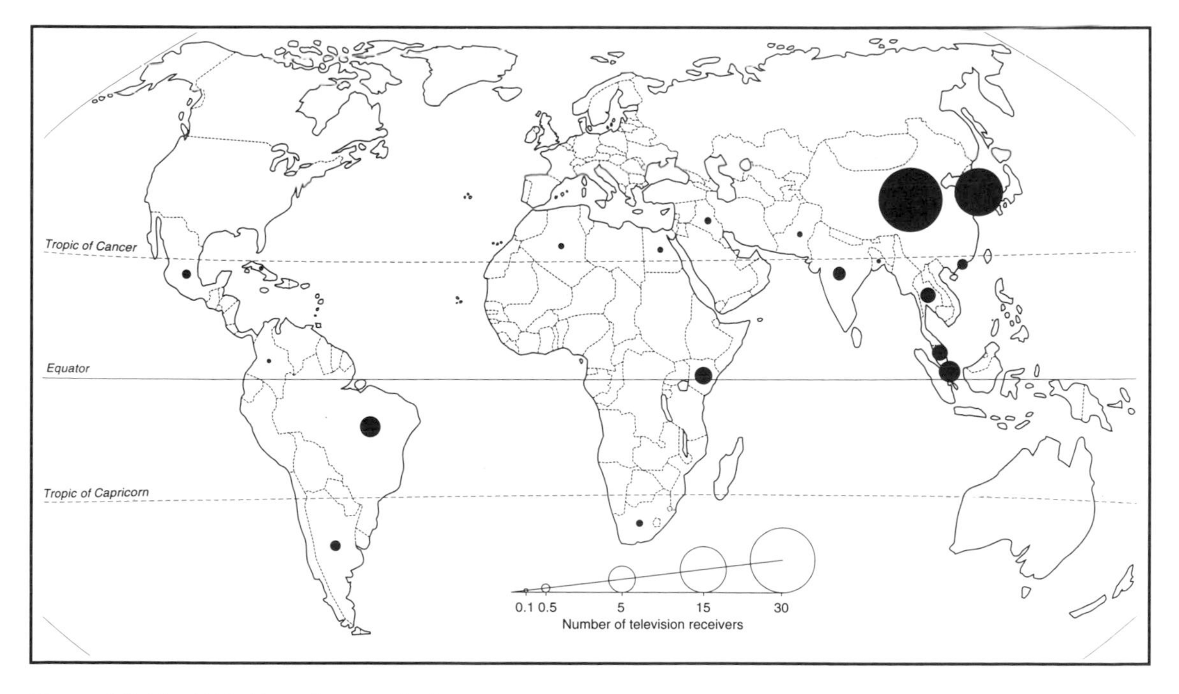

Figure 6.9 Third World production of television receivers, 1989
Source of data: United Nations, 1992

crucial problem for Third World countries trying to develop export-oriented manufacturing is the hostility of developed countries seeking to protect their own industries, especially in times of recession and high unemployment. The imposition of quotas, price controls and subsidies all served to protect First World industries, especially in areas such as textiles and footwear. It is possible that the GATT agreement of 1993 will bring some liberalization to such protectionism (see Chapter 10).

THE PATTERN OF INDUSTRIALIZATION

The spatial pattern of manufacturing industry shows a marked degree of concentration at various levels. Manufacturing makes a varying degree of contribution to the GDP of the countries of the Third World. As has been indicated above, Latin America and the Asian NICs have made most progress in industrialization, and Africa the least, with the other Asian countries occupying an intermediate position.

At a second level, within the main continental areas, spatial concentration is again notable. In Latin America, Brazil, Mexico and Argentina account for three-quarters of the value added by manufacturing, and in Sub-Saharan Africa, South Africa alone accounts for half of the industrial value added.

Spatial concentration is also evident at a third level, within individual countries. In India 60 per cent of the value added by manufacture is generated by the five states of Maharashtra, Tamil Nadu, West Bengal, Gujarat and Uttar Pradesh. These states contain the major industrial centres of Delhi, Bombay, Ahmedabad, Madras and Bangalore. In Latin America Greater São Paulo and Mexico City account for over half of the industrial output and employment of Brazil and Mexico respectively. Concentration is even more pronounced in Argentina, Venezuela and Peru, where Greater Buenos Aires, and the Caracas–Valencia and Lima–Callao axes account for over two-thirds

Plate 6.8 Industrial estate, Pirapora, Brazil To encourage industrialization and to attract factories to particular locations, state and municipal governments provide estates with essential services of energy and transport, and often custom-built factories and tax incentives. This estate was created by the state government to attract industry to the less-developed north of Minas Gerais. The slogan reads 'More industries, more progress, more jobs'

BOX 6.5 INDUSTRY AND POLLUTION

One of the issues of increasing importance arising from the industrialization process in the Third World is the concern for the global environment. During the Industrial Revolution in Europe many rudimentary industries polluted their local environments, but there was little concern or control. Progressively, however, legislation was introduced to restrict pollution of water and the atmosphere. Such legislation was primarily concerned with the local impact of pollution. In recent years there has been growing awareness of the wider implications of pollution through recognition of the impact of acid rain, carbon dioxide and CFCs, as well as the dumping of toxic wastes and the burning of the rainforests. Effects are no longer only local; they are seen to be international.

The countries of the developed world – Europe, the United States, Japan, Russia – are the major sources of industrial pollution, but the prospect of *additional* pollution from industrializing countries, when combined with greater concern for 'green' issues over the past decade or so, has meant that there is increased international concern about new industry in theThird World. China, as a major user of fossil fuels, is a significant contributor to acid rain; Brazil, India and Nigeria are also regarded as significant polluters. A survey in the early 1980s revealed Shenyang, Tehran and Seoul as being amongst the cities of the world most polluted by sulphur dioxide, and Cubatão, in Brazil, is regarded as one of the world's most polluted industrial areas, derived from its complex of steel, chemical and petrochemical industries (Plate 6.9).

The extent to which this constitutes a problem, however, is complex. Some Third World countries see developed world proposals to introduce stricter global controls on pollution as an attempt to inhibit their essential industrialization programmes. The introduction of stricter controls also implies constraints which industrializing countries in the First World were not subject to, and higher costs of production. Conversely it is suggested that as pollution controls become stricter in the developed world, firms may seek to

Plate 6.9 Cement works, Montes Claros, Brazil This factory, also in Minas Gerais, points to the paradox of industrialization. The modern cement works provides an essential raw material for the infrastructural projects, industrialization and rapid urban development of the state. It also creates considerable pollution and uses as its raw material some very spectacular karst limestone deposits

BOX 6.5 *continued*

locate in poorer countries, where legislation is weaker, or less rigidly enforced. Indeed, some countries, such as Nigeria, have indicated that they might welcome such industries, as a route, albeit a potentially dirty one, to industrial development.

In addition to general concern about industrial pollution, the development process also carries with it other hazards. Industrial accidents are not peculiar to the Third World, as Three Mile Island, Chernobyl and Seveso indicate, but there have been major industrial accidents at places like Cubatão and Bhopal, in India. In the latter case, in December 1984 a gas leak from an American-owned pesticide plant caused over 2500 deaths and affected the health of over 200,000 people. The cause of the accident was attributed to design faults in the plant's safety system, compounded by a decline in maintenance standards.

Significantly, in the United States, plants producing the chemicals concerned are required to be located 50 km away from settlements and agricultural activity. Most of Bhopal's victims were slum dwellers, attracted to the city by its rapid industrial growth, and unaware of the risks of living adjacent to a factory producing toxic chemicals.

of industrial activity. In essence, such dominance reflects the role of the primate city (see Chapter 7). These capital cities provide the largest and most affluent markets; they are the best points of assembly for raw materials and for the despatch of finished goods; they have the best provision of economic and social infrastructure; and are the seats of political power and of financial capital. With their better access to global markets via international airports and communications, and their higher level of services, they are particularly attractive to MNCs.

The advantages of concentration in the major cities are thus considerable, but there are also dis-economies because of traffic congestion, pressure on land, and higher prices for labour, services and housing. In consequence there has been some decentralization of industrial activity out of some of these major industrial centres (Plate 6.8). In the case of São Paulo, Brazil's car industry developed from the late 1950s in the peripheral towns of São Bernardo do Campo, São Caetano and Santo André. More recently, newer industrial axes have emerged, linking the industrial towns of the Paraiba valley to the east, and north-westwards towards Campinas and Piracicaba. Similarly the industrial concentration of Mexico City has extended towards the city of Puebla. A significant disadvantage of such industrial agglomerations, however, is that they bring with them concentrations of industrial pollution, especially if control regulations are deficient or less rigidly enforced than in the now more environmentally aware countries of the developed world (Box 6.5).

However, the advantages of the primate cities are such that they seem likely to remain the foci of Third World industrialization. The exceptions are those cases where the pull of raw materials, in the case of heavy industry such as steel-making, is considerable, or where government policy seeks to create alternative patterns. This may be through EPZs located close to ports and airports, industrial estates designed to relocate industry out of the big cities, or programmes of small-town industrialization designed as an alternative to the continuing industrial concentration in the big cities, and as an effort to encourage people not to migrate to such cities but to seek employment in their home districts. Just as attempts to industrialize have been part of *national* development strategies, they have also formed part of regional development plans in some countries; the result has tended to be the same – to foster new industry in the principal regional capitals.

THE PROBLEM OF EMPLOYMENT

It can be argued that attempts to encourage industrialization in the Third World have met with some success. The importance of agriculture has diminished as a contributor to employment and GDP; high rates of growth in the manufacturing sector have been achieved; many countries have established import-substituting industries, and some have developed significant heavy industrial, and, in the case of NICs, export-led industrial sectors. Some countries, such as China, India, Brazil and Mexico, are, in gross terms, major industrial producers. In 1991 the value added by manufacture in Brazil and Mexico was greater than that of Australia, Sweden or the Netherlands; China was the world's third largest pig iron producer, Brazil was the tenth largest car producer, and Malaysia the leading producer of radio receivers. For the NICs at least, manufactured goods had come to dominate their export structures.

Yet this pattern of industrial development has been very uneven. Though there are 'success stories' like Singapore, Malaysia and Brazil, and major industrial conurbations such as São Paulo, Mexico City, Singapore and Hong Kong, there are numerous countries and regions where there is little industrial activity. In many cases industry remains limited in its contribution to the national economy, lacking in diversity, small in scale and spatially concentrated.

Most crucially, whatever success there has been for Third World countries in diversifying their economies, creating some form of indus-

BOX 6.6 WOMEN IN INDUSTRY

The position of women in the workforce of developing countries has varied. Though they were employed in the early industries of textiles and food-processing, when the strategy of import substitution began, it is argued that 'new' industries such as metal-working, shipbuilding, and heavy engineering, which had traditionally been male-dominated in the First World, offered few job opportunities for women. The more recent export-oriented industries have seen an increasing use of women, though on the basis that they were cheaper to employ and likely to be less militant than men. They were absorbed into so-called 'unskilled' assembly industries or those where the perceived stereotypical trait of feminine manual dexterity was considered useful.

The generally low status given to women in the industrial workforce may be attributed to a number of factors, besides the lower value given to their labour. They may (or are perceived to) lack physical strength; they may lack education and training; they are perceived to be suited to repetitive jobs using household skills which can be transferred to the workplace, such as sewing; while their household obligations may mean they are more likely to work part time. Thus, even if jobs are available for women, they often receive lesser returns than are available to men.

Their comparative 'cheapness' has made female employees attractive to multinational companies setting up in the export zones of Third World countries, where low labour costs help to maximize the comparative advantages of such locations. In consequence women have become significant in the workforces of producers of clothing, assembly industries, and the electrical and electronics industries. Their share of employment in such industries in Malaysia, the Philippines, Korea, Mexico and Taiwan has risen to levels of 40–80 per cent.

In broad terms the pattern varies considerably. In South-East Asia women may form 30–40 per cent of the industrial labour force; in Latin America and South Asia between 10 and 30 per cent, but in North Africa and the Middle East less than 10 per cent. The latter figure reflects societal views on the employment of women. In Malaysia Buang (1993) has noted that although the proportion of female workers has risen markedly since the late 1950s, there are negative 'social-moral' views regarding Muslim females working in factories; Buang suggests that foreign multinationals which employ the bulk of migrant Muslim women workers articulate values and rituals which conflict with Islamic ideals.

Thus, although manufacturing has seemingly offered employment opportunities for women, these have often been based on low wages, monotonous jobs and, in some cases, social stigma. Furthermore, even though the cheap labour of women is an attraction, modern capital-intensive plants may require fewer workers, even in 'traditional' female-employing industries such as textiles, food and cigarettes, while the industrialization process per se threatens the older 'cottage' and household production of textiles, clothing and food which were a source of income for women.

trial activity, and generating some export of manufactured goods, their industrialization strategies have had limited success in their anticipated role in creating new jobs, as an alternative to rural labour and to absorb growing populations. The population of the low- and middle-income countries grew at 2.0 and 1.85 per cent a year respectively over the period 1980–92; their urban populations grew at respectively 4.1 and 3.2 per cent over the same period. However, the contribution of manufacturing to employment remains low. In Sub-Saharan Africa in 1986–9 less than 8 per cent of the workforce was employed in industry, with figures below 3 per cent in Guinea, Mali and Burundi. Similar figures can be found in some parts of Asia – of 0.6 per cent in Nepal and 2.8 per cent in Bhutan. Even in Asia's NICs industrial employment was only around 20 per cent in Hong Kong, South Korea and Singapore. The most significant shift has often been not from agriculture to industry, but to the service sector.

The failure of manufacturing to generate employment can be attributed to a number of factors. These include the sheer numbers of people generated by continuing high rates of population growth, and the 'push and pull' factors which expel people from the countryside and draw them into the cities more rapidly than industrial jobs can be created. Industrialization has thus failed to keep pace with overall population growth and with the concentration of people into the conurbations and major cities. This may be compounded by the nature of the industrialization process. One of the alleged benefits for Third World countries seeking to industrialize is that they do not have to 'invent' equipment as did the countries of Europe in the Industrial Revolution; instead, they can 'borrow', or more realistically buy, technology from the First World. The complication is that most of that technology is designed for First World circumstances, where labour is scarce and capital available, whereas the Third World has abundant labour to employ, but scarce capital with which to purchase sophisticated equipment. Even in the case of the export sector industries attracted by cheap labour, it is the cheapness of labour, rather than its abundance (though they are clearly closely related), which is significant in what are often production-line industries. The search for cheap labour has been a significant factor in the generation of jobs for women in Third World manufacturing, though the issue is subject to debate (Box 6.6).

Therefore, whatever the success of industry in diversifying economies and generating exports in some countries at least, its failure to generate employment, compounded by high rates of population growth and migration to the cities, means that the Third World continues to be faced by high levels of unemployment and underemployment. Much expansion of non-agricultural employment has been into the service sector, rather than in manufacturing, and even this has been into marginal jobs rather than in productive and secure employment in the urban **tertiary sector**.

SUMMARY

The Third World is relatively well endowed with mineral resources. These have been developed in order to generate income from export revenues and also internally to create an industrial base and associated employment. However, the potential benefits of mineral exploitation to the national economy have seldom been fully realized, partly as a result of global mineral exploitation and markets being largely controlled by developed country MNCs and partly as a result of the limits to mineral-processing before export. There are similar problems in the development of some energy resources, notably oil, though a few favoured countries have been able to generate substantial revenues from oil, diverted to finance a range of related industrialization projects, such as oil-refining and petrochemicals. Other energy sources, notably coal and hydroelectricity, are of greater significance to individual countries for their direct value as a base for local industrialization and for generating local employment, though these have been associated with important environmental and pollution issues.

Third World countries seek to promote local industry for three principal reasons. First, they see a growing industrial base as being an essential ingredient in the development process overall, with backward linkages to agricultural and mineral development and forward linkages to more sophisticated industries and to the service sector. Industry creates greater national self-sufficiency with less dependence on imports. Second, they see industry as a major source of exports and therefore foreign earnings. Export-processing was important in early industrialization strategies, especially in Latin America, but since exports failed to penetrate the often protected First World markets sufficiently, this aspect had been rather played down in the 1980s. However, recent successes of the NICs have been based on the exports of manufactured products. These have been rather different from the traditional export products in that they have not normally been based on local mineral or energy resources, thus weakening the traditional link between minerals, energy and industry. Third, industry has been seen as necessary for job creation, especially in urban areas. Early industrialization emphasized the creation of jobs for men, especially in heavy industries and traditional engineering, but though manufacturing jobs have more recently been created in the NICs for women, they often involve very low wage levels.

7

Urbanization

Since the Second World War, urbanization – the concentration of people in cities and towns – has taken place in the Third World at a pace and on a scale paralleled only by the experience of Western Europe – and outstripped by that of North America – in the nineteenth century. The crisis of Third World urbanization, however, is expressed not in the speed or short history of city growth (Table 7.1), but in the wide margin by which urban population increase (usually about 3 per cent per annum) has outstripped employment opportunities in the formal sector, most notably in manufacturing industry. Even Third World countries which were deemed to be 'socialist' in the 1980s have been marked by a strong process of urbanization, especially where their colonial experience has been recent (Figure 7.1).

PATTERNS AND PROCESSES OF URBANIZATION

United Nations data for urban areas show that 34 per cent of the people in the Third World were living in towns in 1990, compared to 73 per cent in developed countries; in 1950 the percentages were respectively 17 and 54. Only Latin America of Third World regions closely approaches the level of urbanization in Europe (68 per cent) and North America (74 per cent), with Africa bringing up the rear (30 per cent). It is estimated that by the year 2000, urbanization levels in Latin America will exceed those in both Europe and North America. Latin America is not only the most urbanized part of the Third World, but has the longest record of post-colonialism (since the independence of most republics in the 1820s), and is the most industrialized (see Chapter 6) and 'Westernized'. Not surprisingly, urbanization has been more thoroughly studied in Latin America than elsewhere in the Third World, and most of the theoretical breakthroughs in our understanding of the processes of city growth and organization in the Third World have been achieved by social scientists working on this area.

Third World cities are remarkable for their size. In 1900 there were few cities in the world with more than a million inhabitants; there are now about 400 such cities, more than 250 of them in the Third World. In 1960 less than half of the world's 25 largest urban agglomerations were to be found in the Third World; by the year 2000, 20 will be there (Habitat, 1987). Latin America records 4 of the world's 14 largest cities – Buenos Aires, Rio de Janeiro, São Paulo and Mexico City – which form the cores of massive agglomerations, the two latter approaching the 20 million mark (Table 7.1; Box 7.1).

A major feature of Third World urbanization is the enormous concentration of each country's population in one city, usually the capital. Systems of cities in developed countries are generally graded, following the **rank-size rule**, with the second settlement recording half the population of the largest, and so on. But in Third World countries this rule breaks down, especially near the top of the urban hierarchy,

◀ Urban scene, Singapore

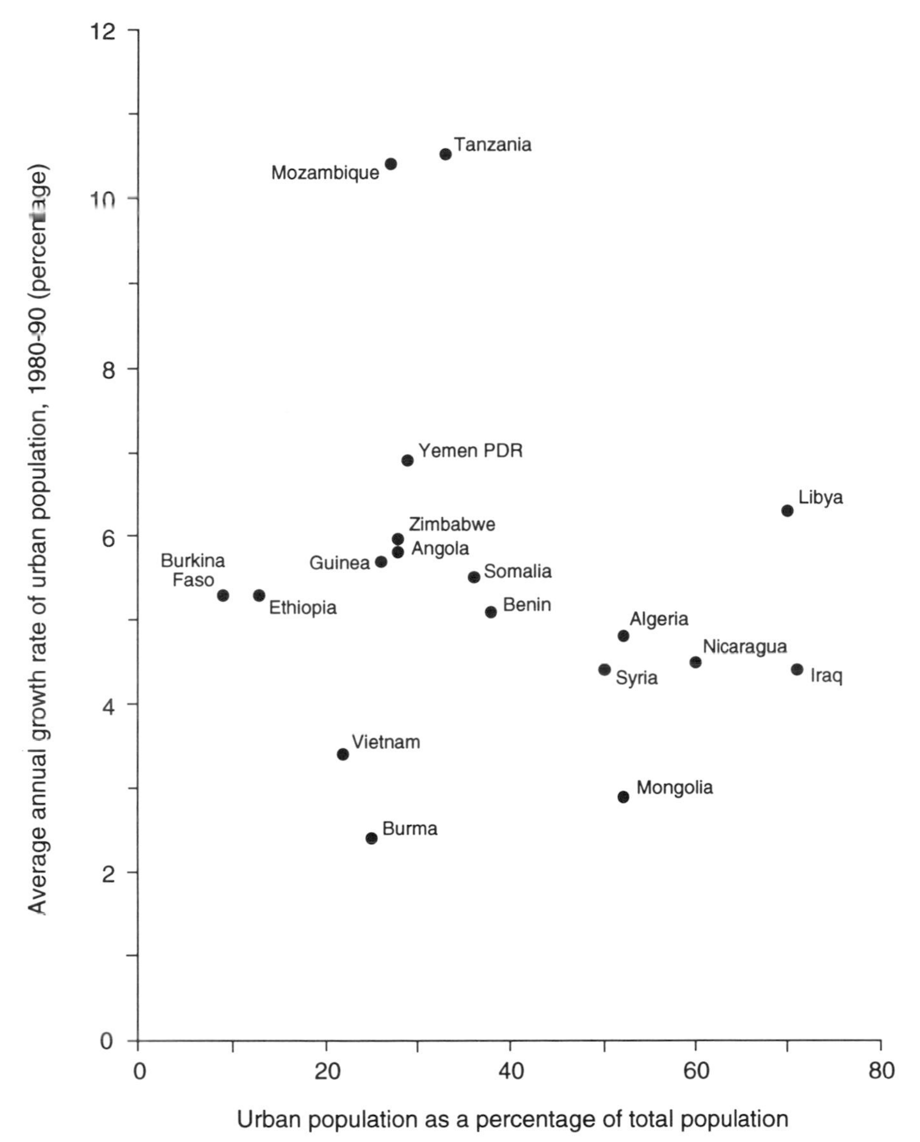

Figure 7.1 Urban growth in socialist developing countries, 1980–90
Source of data: World Bank, 1992

with a high propensity to **'primacy'**, a disproportionately large proportion in the largest city. The largest city in most Third World countries typically contains between 10 and 30 per cent of the entire population, over half the urban-dwelling population, and more than four times the number of inhabitants in the next largest city. Primacy occurs in an extreme form in the city-state of Singapore and city-colony of Hong Kong. Under more normal, yet exaggerated, circumstances Montevideo, the capital of Uruguay, contains well over half the country's population. Exceptions to a high degree of primacy include Brazil, where São Paulo and Rio de Janeiro rival one another, though neither is the capital; China, where Shanghai, Tianjin and Beijing, each with around 10 million, vie for first place; and Turkey, where Ankara, the relatively recent capital, is growing more quickly than long-established Istanbul.

Table 7.1 Ranking of city agglomerations by population (millions), 1960, 1980 and 2000

	1960		*1980*		*2000 (projected)*	
1	New York/NE New Jersey	14.2	Tokyo	17.7	Mexico City	25.8
2	London	10.7	New York/NE New Jersey	15.6	São Paulo	24.0
3	Tokyo/Yokahama	10.7	Mexico City	14.5	Tokyo/Yokahama	20.2
4	Shanghai	10.7	São Paulo	12.8	Calcutta	16.5
5	Rhein/Ruhr	8.7	Shanghai	11.8	Bombay	16.0
6	Beijing	7.3	London	10.3	New York/NE New Jersey	15.8
7	Paris	7.2	Buenos Aires	10.1	Seoul	13.8
8	Buenos Aires	6.9	Calcutta	9.5	Tehran	13.6
9	Los Angeles/Long Beach	6.6	Los Angeles/Long Beach	9.5	Shanghai	13.3
10	Moscow	6.3	Rhein/Ruhr	9.5	Rio de Janeiro	13.3
11	Chicago/NE Indiana	6.0	Rio de Janeiro	9.2	Delhi	13.3
12	Tianjin	6.0	Beijing	9.1	Jakarta	13.3
13	Osaka/Kobe	5.7	Paris	8.7	Buenos Aires	13.2
14	Calcutta	5.6	Osaka/Kobe	8.7	Karachi	12.0
15	Mexico City	5.2	Bombay	8.5	Dhaka	11.2
16	Rio de Janeiro	5.1	Seoul	8.5	Cairo/Giza	11.1
17	São Paulo	4.8	Moscow	8.2	Manila	11.1
18	Milan	4.5	Tianjin	7.7	Los Angeles/Long Beach	11.0
19	Cairo/Giza	4.5	Cairo/Giza	6.9	Bangkok	10.7
20	Bombay	4.2	Chicago/NE Indiana	6.9	Osaka/Kobe	10.5
21	Philadelphia	3.7	Jakarta	6.7	Beijing	10.4
22	Detroit	3.6	Milan	6.7	Moscow	10.4
23	St Petersburg	3.5	Manila	6.0	Tianjin	9.1
24	Naples	3.2	Delhi	5.9	Paris	8.7
25	Jakarta	2.8	Baghdad	3.9	Baghdad	7.4

Source of data: United Nations, 1986

Table 7.2 Third World urban populations, 1992

	Urban pop. as % total	*Capital city as % urban pop.*	*% urban pop. in cities of over 1 million*
Sub-Saharan Africa	29	33	34
E Asia/Pacific	29	12	37
South Asia	25	8	38
Mid East/N Africa	55	26	41
Latin America	73	24	46
World	42	15	38

Source of data: World Bank, 1994a

Where there is a pairing of cities of approximately equal size, commonly one is the capital and the other the major port (e.g. Yaoundé and Douala in Cameroun) or else they each head rival growth regions (e.g. Rio de Janeiro and São Paulo in Brazil). Primacy, on the other hand, is linked to the amalgamation of commercial (usually port) and capital functions in one city, for example, Dakar or Kingston. This condition is particularly acute in small countries

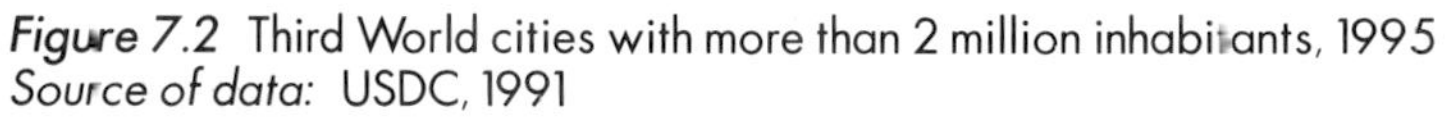

Figure 7.2 Third World cities with more than 2 million inhabitants, 1995
Source of data: USDC, 1991

BOX 7.1 THE '10 MILLION' CITIES

As the twentieth century draws to a close, more than 20 cities in the world have populations near or above the 10 million mark. Most of them, with the exceptions of the conurbations of Tokyo, New York, Los Angeles, Osaka and Moscow, are in the Third World (Figure 7.3). The developed world agglomerations were among the world's largest in the 1940s but are now dwarfed by Third World rivals such as Mexico City and São Paulo, and only New York and Tokyo rank among the world's biggest.

However, much of the global economy is still managed from the cities of the North, particularly New York, Tokyo and London (which is too small to rank in the top 20), whose stock exchanges act as key integrators of the world economy. Third World cities may be growing, but they still lack industrial and financial power. Their claim to significance is almost purely demographic, and reflects the persistence of high birth rates and falling mortality, coupled with city-ward migration. Measured by Third World standards, however, the biggest – all Latin American or Asian in location – are major industrial and financial centres in their own countries.

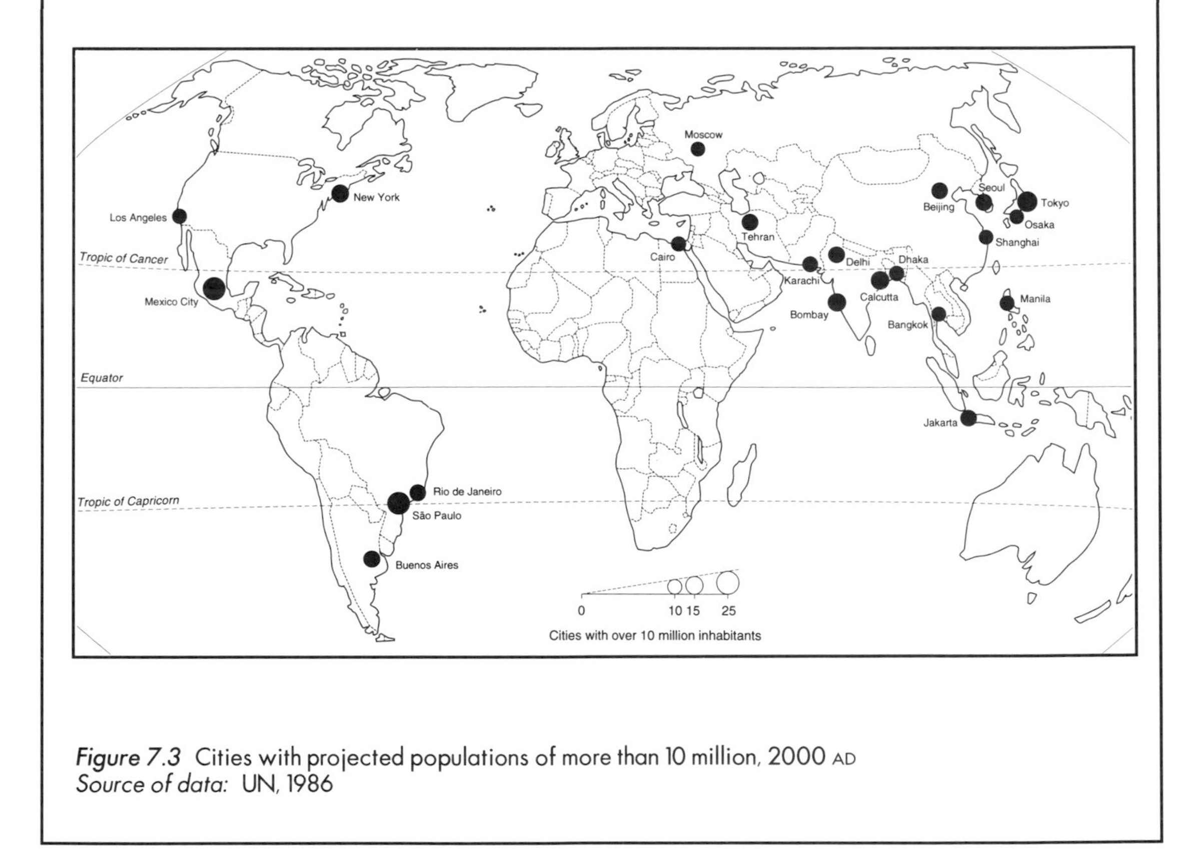

Figure 7.3 Cities with projected populations of more than 10 million, 2000 AD
Source of data: UN, 1986

with undiversified economies, especially – as in Africa and the Caribbean – where colonialism persisted until the 1960s, or, in some instances, until the 1970s. There has been a strong trend since 1970 for more and more of the urban population of Third World countries to live in large cities, of over 1 million inhabitants, the proportion reaching 34 per cent in Africa, which is the least urbanized continent (Table 7.2). This process of increasing growth is reflected in the enhanced importance of the major cities (Figure 7.2), and the rapidity of urbanization is such that the Third World is projected to have a majority of the world's 'mega-cities', of

Plate 7.1 Urban scene, Peshawar, Pakistan Peshawar lies at the mouth of the Khyber Pass, a major routeway between Europe and India, and a city has existed there since Alexander the Great passed through in 328 BC. Many parts of the modern city, in particular the military garrison of the cantonment, were laid out during British colonial rule, but old Peshawar exemplifies many of the features of the morphology, land use and activities of a pre-industrial city

over 10 million inhabitants, by the year 2000 (Box 7.1).

Despite their recent growth, many Third World cities have long histories. This is obviously true of pre-colonial settlements, but, Africa south of the Sahara excepted, many colonial towns established by Europeans also have origins going back to the voyages of the Age of Discovery in the sixteenth and seventeenth centuries. Towns flourished in India, China, the Middle East, and in parts of North and West Africa long before European contact (Plate 7.1); colonial settlements are the norm in Latin America, the Caribbean and most of Africa. Even where there were indigenous traditions of urban-dwelling, it is common to find European colonial settlements added to the original town, as in the case of Ibadan in Nigeria (Box 7.2) and the company town of Abadan in Iran. Delhi and (planned) New Delhi in India exemplify this arrangement on the grand scale.

Third World cities for many decades, and often for centuries, have been agents of European imperialism, facilitating trade penetration and acting as gateways for the export of primary products and the import of manufactured goods. Many Third World cities remain essentially pre-industrial in their economic base, while their coastal locations in South-East Asia, Latin America, Africa and, above all, in the Pacific and Caribbean, indicate their long association with colonial dependency (Figure 7.2).

Neo-Marxist economists extend the relationship between Third World towns and dependency further to argue that not only are they agents of imperialism and external *neo-colonialism*, but they also dominate their hinterlands through relationships of internal colonialism. According to this perspective, urban markets are locations at which unequal exchanges occur and are the means by which mercantile capitalists extract surplus value from the labour of agricultural and handicraft workers. Clearly, the degree of domination depends upon the socio-economic gulf between town

BOX 7.2 THE GROWTH OF IBADAN

The Yoruba city of Ibadan in south-west Nigeria is more recent than the Hausa cities of the north of the country. There were probably two earlier settlements on its site, but the present city began to grow in the early nineteenth century, a time of great rivalry between Yoruba factions, when it served as a war camp. It developed military and commercial functions, expanding outwards from its focus around Oje market, and extending beyond the city wall, which had been completed by 1858 (Figure 7.4). At the end of the century its population was between 150,000 and 200,000. By that time the city was under British colonial rule as part of the Protectorate of southern Nigeria, and was acquiring new districts set aside for the colonial government and foreign commercial activities, with residential areas for expatriates sited away from the indigenous town. The railway reached the city in 1901 and its already extensive trading activities were expanded, stimulated by the increasing cultivation of cocoa as the main export crop of the south-west.

As the indigenous and immigrant Nigerian population approached 500,000 in the 1950s, the old town became more and more congested, with closely packed mud-walled buildings, originally thatched but now with roofs of

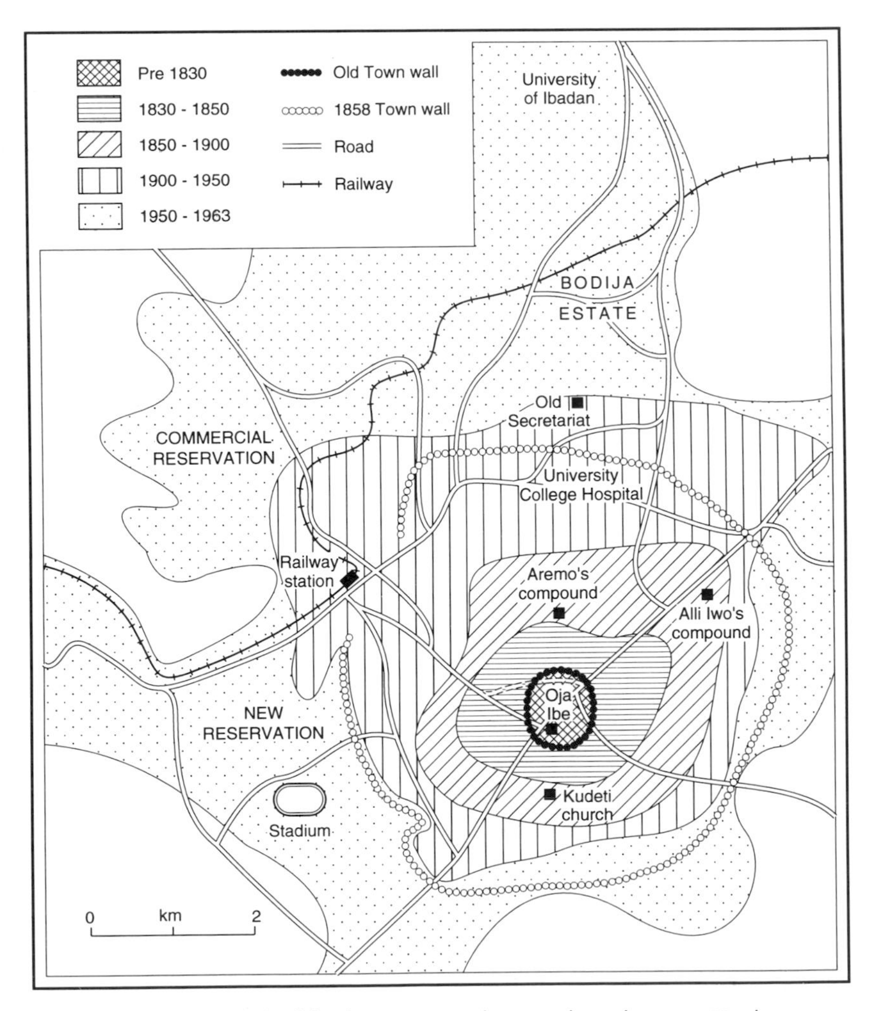

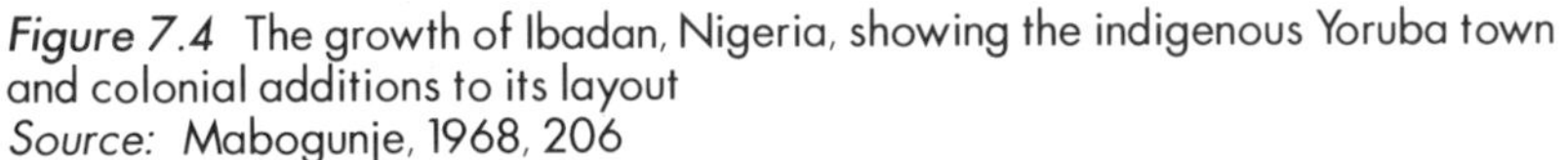
Figure 7.4 The growth of Ibadan, Nigeria, showing the indigenous Yoruba town and colonial additions to its layout
Source: Mabogunje, 1968, 206

BOX 7.2 *continued*

corrugated iron, and deficient in water supply, sanitation and waste disposal (Plates 7.2 and 7.3). The expatriate 'reservations', with low housing densities and better amenities, stood in sharp contrast. This duality persisted after independence, with the expatriate population being replaced by a Nigerian elite. The population has increased to over a million, and the urban area has continued to expand, particularly to the north, associated with the University of Ibadan, and in other directions in response to commercial and industrial development.

Plate 7.2 Urban scene, Ibadan, Nigeria The old part of the town in the 1960s, showing the densely packed houses with corrugated iron roofs, and an absence of paving

Plate 7.3 Houses and alley, Ibadan The ornate carving of the house in the foreground contrasts with the mud-brick dwelling behind. The street is unpaved, and the open drain is crossed by a plank 'bridge'. Some household chores are carried out in the street and water is being carried by a woman as a head load

and country and upon distinctions in culture and language between the groups involved. Domination, the norm in colonial times, is now often mediated by the post-colonial state and class negotiation.

THE IMPACT OF MIGRATION

Whatever the origins of, or links between, Third World cities, their growth has been precipitated by high rates of natural increase and large inflows of migrants (Chapter 3). Young migrants expand the urban population both by their presence and by their reproductive capacity; male migrants are usually in their 20s, but females are on average even younger, often predominantly in their teens. Conditions vary in detail from country to country, from town to town, but migration is undoubtedly a major component of urban population growth throughout the Third World, though its significance has declined with development since the 1950s (Table 7.3). A notable exception are the countries of the communist Third World, such as China and Cuba, where mobility has been constrained by authoritarianism.

In Africa, Asia and the Middle East migration is a male phenomenon, and especially so in Africa and India where movements are often circular rather than permanent: women and children stay on the land, maintain the cultivation of subsistence crops, and provide a secure village base to which men can return (Chant, 1992). Many women migrate as part of the family and do not have separate motives for mobility, but such is the volume of independent female migration in Latin America and the Caribbean, largely due to historical process of male and female proletarianization, that most cities and towns in these regions record an excess of women (Box 7.3).

Cityward migration operates in a number of ways. Those born in rural areas are drawn into the urban system and move either by way of smaller towns up the urban hierarchy towards the largest city – a common pattern in Latin America and the Middle East – or they may settle directly in the capital – a situation typical of West Africa and East and South-East Asia – depending on the distance and attraction of alternative destinations. Another possibility is for migrants to move through the settlement system in a less structured way in response to the availability of work and the location of friends and relatives from back home to assist them.

Return migration from capitals and smaller towns to rural areas is the norm in many African countries; in Latin America, the Middle East and Asia it is generally a smaller-scale process. But even in these latter instances, rural links are maintained by urban migrants, who band together in regional associations for mutual support, remit cash to their families and occasionally raise funds for development projects in their areas of origin. A unique variant on this theme obtains in Mexico, where male cityward migration is often directed, illegally, to the US. Women and children usually remain in the rural source areas, maintaining agriculture and other local activities at a low level, and awaiting remittances and the periodic return of their menfolk (Clarke, 1992).

Size and spacing of towns and the way in which they are interconnected undoubtedly affect the type of migration recorded at various levels of the settlement hierarchy. Information about the processes of urban migration is mainly derived from the largest cities, and evidence for urbanization in general tends to be based on studies of capital cities. In some Third World countries with established urban systems, migration is step-wise, the larger settlements drawing on smaller ones, and so on down the hierarchy. Major cities therefore receive many migrants who are already familiar with urban life, though selectivity of this kind has invariably declined as the movement has continued over time.

There is no reason, of course, why families or individuals should divide their migration into stages, completed over the years or generations, if transport is cheap, there is an important city only a few hours away by bus or train, and relatives are at hand – as they so often are – to help a migrant in difficulties. Many Third World cities experience a mixture of direct and stage migration. Nevertheless, the distance-decay function usually holds good, and migratory

Table 7.3 Urban population growth and the contribution of migration in some African cities, 1960–85

	% annual growth rate		Estimated % growth due to migration
	1960-70	1970-85	1970-80
Abidjan	10.0	6.5	59
Accra	6.3	5.6	44
Addis Ababa	6.3	4.6	56
Dakar	4.6	4.6	47
Dar-es-Salaam	7.3	5.2	45
Freetown	4.9	4.6	65
Harare	8.5	4.3	22
Kampala	5.8	4.6	29
Kinshasa	8.3	4.7	23
Lagos	9.8	5.4	40
Nairobi	10.3	4.6	41

Source of data: Makannah, 1990

BOX 7.3 WOMEN IN THE CITY

In most parts of the Third World men predominate in the urban areas, but in Latin American and Caribbean cities women are more numerous than men, largely because of their withdrawal from agricultural labour and the stereotyping of domestic work as women's work. Whether they are in a majority or not, urban women in the Third World carry a heavy burden, associated with both reproduction and production. However, their roles are perceived principally as domestic, associated with child rearing, food preparation, clothes washing and similar chores. Working outside the home, if it occurs, involves a 'double shift'.

Though men predominate in the urban labour force, and are more numerous than women in the informal sector, women who work tend to be concentrated in that sector. Some writers have commented on the 'tertiarization' of female labour, because of their tendency to work in domestic service, as waitresses and seamstresses, and in market selling. This reflects the tasks they traditionally perform in the home. Indeed, many women work from home, carrying out the hand manufacture of small items, such as clothing, which can be incorporated into their household activities. Structural adjustment policies, more local crises, and basic poverty have forced ever-increasing numbers of women into the urban labour force to supplement men's earnings – or simply to provide single women and single parents with some income.

Women are particularly affected by the poor housing conditions in which the majority of the Third World urban poor live. This is most acute in the squatter settlements, where there are deficiencies of water, electricity, sewerage, transport and other services. Under these circumstances basic chores of child rearing, clothes washing and food preparation become immense tasks, and it is not surprising that women feature prominently in group projects to improve conditions in shanty towns.

It is in the cities, too, that women are increasingly becoming heads of their own household. This trend is well developed in Latin America and the Caribbean, but less common in India. Although women earn lower wages than men, female-headed households often incorporate other adults who earn, or child mind, thus boosting the overall income.

flows generally cover short distances.

Reasons for the rural exodus are essentially economic, though in the Middle East and Asia political problems have generated vast flows of refugees (Chapter 3). Per capita agricultural output is either static or declining in many Third World countries. In parts of Africa, where access to land is often communal and relatively free, there is a lower level of permanent migration and of urbanization (Peil, 1991), but in India and Latin America the bulk of arable land is in the hands of a small elite. Even in Mexico where a land-reform programme operated from the 1920s to the early 1990s, the penetration of the countryside by consumer goods has displaced peasant craft workers and helped to create a surplus population that has responded by out-migration to take its chance in the cities. Wage differentials everywhere explain the move to the towns, even in Jamaica, where unemployment is far higher in Kingston than in the rural areas, though it is rarely perceived to be so by prospective migrants (Clarke, 1975).

An additional stimulus to cityward migration is the educational system. Secondary schools and institutes of higher education – Cuba and China excepted – are located for the most part in large towns. Moreover, the value system in most Third World countries produces adolescents whose aspirations are to urban jobs. Their main objective is to live and work in a city or small town in preference to a rural locality, where roads, electricity, piped-water supply and health facilities may be non-existent or only slowly improving (Figure 1.13 and Table 7.4), the social status of families is common knowledge, and opportunities for non-agricultural work are negligible, unless economic decentralization is taking place. In Kenya, for example, most rural secondary school leavers aspire to an urban job, and many immediately go to towns, though their aspirations are likely to be dashed (Gould, 1985).

EMPLOYMENT AND UNEMPLOYMENT

Concentration of manufacturing activities in large Third World cities is as striking as primacy itself, and has occurred because the major long-term strategy for economic development has been one of industrialization by import substitution (Chapter 6). This policy has concentrated food, tobacco, drink and textile industries in port-commercial centres or in pre-existing urban agglomerations where there has been a ready market for consumer goods. In Latin America, where import substitution was introduced after the First World War, the largest cities, such as São Paulo and Mexico City, now have branch plants of multinational firms manufacturing motor cars, electrical goods and other consumer durables, in many cases stimulated by government incentives and investments in infrastructure.

Planned industrialization, such as port-

Table 7.4 Access to sanitation in some Third World countries (% of population)

	Total		*Urban*		*Rural*	
	1980	*1990*	*1980*	*1990*	*1980*	*1990*
Sierra Leone	12	39	31	55	6	31
Mali	14	24	79	81	0	10
Egypt	26	50	45	80	10	26
Bangladesh	3	12	21	40	1	4
India	7	14	27	44	1	3
Indonesia	23	45	29	79	21	30
Honduras	31	62	40	89	26	42
Bolivia	19	26	37	38	4	14
Argentina	79	89	89	100	32	29

Source of data: World Bank, 1994a

located steel and fish-meal processing plants at Chimbote in Peru or oil refineries in the Middle East as at Abadan in Iran or Dhaharan in Saudi Arabia, have created economically specialized urban settlements. Indigenous raw materials have stimulated inland concentrations of manufacturing industry, exemplified by the towns of the Zambian copper belt or the more complex textile, mining, and iron and steel towns of central Minas Gerais in Brazil.

Excluding towns developed to exploit specific local resources and others like those in Hong Kong, Taiwan, South Korea and Singapore, which produce and export manufactured goods based on cheap labour, it is generally true that the accumulation of population in Third World cities has outstripped by a large margin the capacity of their manufacturing economies to absorb labour. In some countries like Zaire and Haiti, sheer economic stagnation is responsible for the gap between the number of town dwellers and jobs; but in Brazil, India and Mexico – the Third World's principal manufacturing countries by volume of output – labour surplus in the towns has been exacerbated because their industrialization programmes are based on imported technology, which is capital intensive and employs relatively few workers.

Comparison between indices of urban population concentration and industrialization compiled for Latin America about 30 years ago, produced the then-puzzling conclusion that, at the macro scale, disharmony between these two social and economic processes was more acute in the most advanced countries. Now urban labour surplus is diagnosed as inherent in 'dependent' or 'peripheral-capitalist' development and associated with the prominent role played by the multinational corporations, many of whose factories are located in the cities of most advanced Third World countries.

Population growth in Third World cities has greatly inflated the service sector – in lieu of industrial employment – and in the largest settlements tertiary activities may account for more than half the jobs that are available. The ratio of employment in the service sector to that in manufacturing, in general, is higher than in Western Europe: the ratio approaches, and for many countries exceeds, that in the USA, where economic development has permitted a proliferation of services.

Most Third World countries are characterized by burgeoning urban bureaucracies that have expanded to administer newly independent or increasingly complex nations. Ministries, departments and agencies have provided white-collar work for the educated, and unskilled jobs for porters, messengers and cleaners. National and local government are major, sometimes *the* major, employers of urban labour, and even small towns may have large numbers of officials to administer public works and collect taxes. However, IMF-imposed structural adjustment programmes may have led to severe retrenchment in civil service employment since the early 1980s.

The largest Third World cities have sophisticated infrastructure – motorways, metros, bus services, telephone communications, water and sewerage facilities, electricity supplies – and require a substantial labour force to maintain and operate them. Finance houses, building societies and the head offices of national firms provide white-collar jobs in the principal cities, but even small towns linked by dirt roads may have branch banks. In most big cities, universities, polytechnics, schools and hospitals are urban magnets for professionally qualified staff, while the retailing sector, whether in the centre or in suburban malls, requires tens of thousands of shop assistants.

In addition to the existence of occupational activities comparable with those of cities of developed countries, employment in Third World cities is heavily reliant on petty services and petty commodity production on a small scale – what has been called the **'bazaar' economy**. Employment for the rapidly expanding labour force depends upon the absorptive capacity of small industries with high labour requirements but low productivity; shoemaking, traditional and modern handicrafts, metal work and machinery repair; trade, especially market and street selling of food and clothing; services, including domestic service (for women in Latin America and the Caribbean, for men elsewhere), shoe-shining, rickshaw operating, car watching; and casual labour of all kinds

Table 7.5 Income opportunities in a Third World city

Formal income opportunities
(a) Public sector wages
(b) Private sector wages
(c) Transfer payments – pensions, unemployment benefits
Informal income opportunities: legitimate
(a) Primary and secondary activities – farming, market gardening; building work; self-employed artisans, shoemakers, brewing and distilling
(b) Tertiary enterprises with relatively large capital inputs – housing, transport, utilities, commodity speculation, rentier activities
(c) Small-scale distribution – market sales, petty traders, street hawkers, food and drink sales, bar attendants; carriers; commission agents
(d) Other services – musicians, launderers, shoe-shiners, barbers, night-soil removers, photographers, vehicle repairs; brokers and middlemen; ritual services, medicine and magic
(e) Private transfer payments – gifts and similar movements of money between people; borrowing; begging
Informal income opportunities: illegitimate
(a) Services – receiving stolen goods, usury and pawnbroking; drug-pushing, prostitution, smuggling; bribery, political corruption, protection rackets
(b) Transfers – petty theft (e.g. pickpockets), larceny (e.g. burglary and armed robbery); speculation and embezzlement; confidence tricksters; gambling

Source of data: Hart, 1973, 69

(notably in construction work and street cleaning) (see Plates 5.5, 5.6 and 10.2). At the lowest level of living the urban poor have no alternative but to raise and deal in small livestock or to scavenge the streets and garbage dumps for discarded food or for scrap material, such as metal and paper, which can be used to construct a shelter or sold for small sums of money. Even a beggar in Cairo or Manila may do better than a landless labourer in the countryside, though a peasant will, of course, have a subsistence base that may be more secure.

The term 'the tertiary refuge sector' has been coined to describe these heterogeneous activities, which include small-scale workshops, repairing and recycling as well as services. A more appropriate term is the '**informal sector**', which Hart (1973), in his work on Ghana, contrasted with the '**formal sector**' (Table 7.5). The latter implies large-scale activity, permanency of employment, set hours of work and pay, and sometimes the provision of pension and social security rights. The informal sector lacks all these features and is, or appears to be, characterized by family labour, small-scale enterprise and self-employment. It is easier to enter informal than formal occupations, since the former are at best semi-skilled and require little if any capital outlay. Women are essentially a subordinate population *vis-à-vis* men in most aspects of life in the Third World, and in the cities are markedly concentrated in the informal sector. A notable exception is the Caribbean, which has a history of greater female equivalence based on common gender exploitation during the period of plantation slavery. Santos (1979) has divided Third World employment into two circuits, upper and lower (Table 7.6), which are equivalent to the formal and informal sectors outlined above, though he contrasts the circuits more systematically than other authors, drawing attention to differences in technology, organization, capital, labour, wages, prices and credit.

This dichotomy is extended to the global scale by Roberts (1995), who argues that formal sector employment has largely been pre-empted by the developed world, leaving Third World cities – and the towns at the lower levels of the urban hierarchy in particular – essentially dependent on small-scale, marginal employment. This pattern is allegedly being disrupted

Table 7.6 Characteristics of the two circuits of the urban economy in Third World countries

	Upper circuit	*Lower circuit*
Technology	Capital intensive	Labour intensive
Organization	Bureaucratic	Primitive
Capital	Abundant	Limited
Labour	Limited	Abundant
Regular wages	Prevalent	Exceptional
Inventories	Large quantity and/or high quality	Small quantity and poor quality
Prices	Generally fixed	Negotiable (haggling)
Credit	Banks/institutional	Personal/non-institutional
Profit margin	Small per unit, but large turnover and considerable in aggregate	Large per unit; but small turnover
Relations with customers	Impersonal and/or on paper	Direct, personalized
Fixed costs	Substantial	Negligible
Advertising	Necessary	None
Re-use of goods	None (waste)	Frequent
Overhead capital	Essential	Not essential
Government aid	Extensive	None or almost none
Direct dependence on foreign countries	Considerable	Small or none

Source: Santos, 1979, 22

by the New International Division of Labour (NIDL), whereby modern manufacturing is being displaced to cheap-labour locations, using 'just-in-time' methods of assembly. However, this is not a uniform process and is concentrated in cheap labour zones in the Americas close to the US, or in sweat-shop enclaves in South-East Asia (Cohen, 1987).

The basic difference between the two sectors or circuits can be reduced to finance capitalism versus **penny capitalism** (Table 7.6). But how, if at all, do the sectors relate? In the early 1970s it was usual to treat the two sectors as though they were separate, and to focus on the absorptive capacity of the informal sector, whose employees subsisted largely by earning from one another or, on a casual basis, from the formal sector. In this way the urban lower class, and many of the middle class too, were able to make ends meet by stringing together a number of low-paid jobs. Santos has made an important contribution through his perception that growth of small-scale manufacturing and services, which cater mostly to the urban poor, does not block expansion of the formal sector, or upper circuit; rather, the lower circuit responds to changes in urban consumption patterns and general conditions of employment and capital.

Most commentators now reject the dichotomy between formal and informal employment, arguing that both are characterized by the profit motive, and that finance and industrial capital dominate both sectors (Bromley and Gerry, 1979). Moreover, the two sectors inter-relate in complex ways, the lower circuit selling cheap services and transformed goods to the upper circuit, and purchasing from it consumer goods such as electrical items. In some of the more advanced Third World cities, an interstitial set of activities, part formal, part informal has emerged. But since the oil price rise of 1973, World Bank policies of economic restructuring have been imposed on many struggling Third World countries, and contraction of the formal sector has had to be compensated for by the expansion of informal activities, sometimes aided by the support of Non-Governmental Organizations (NGOs). In Oaxaca City, Mexico, for example, formal sector activities accounted for 44 per cent of the male and 63 per cent of the female labour force in 1977, rising ten years later to 66 and 73 per cent, respectively (Murphy and Stepick, 1991).

Work in Colombia has shown that street sellers are rarely self-employed and self-financed, but act as a front for highly capitalized organizations. Moreover, paper pickers on refuse dumps in the Colombian city of Cali are not independent operators but outworkers for the local paper mill: by engaging their labour through middlemen and by keeping them off the official pay-roll, the company appropriates the surplus value created by the scavengers and avoids minimum-wage regulations and paying social security contributions (Birkbeck, 1979). Evidence from Lima, Calcutta, Dakar and Kashan confirms the existence of similar outworking systems in shoemaking, carpentry, tailoring and carpet-making, but it is too early to say firmly how formal and informal employment interlock in the context of street selling and petty services. However, it is clear that domestic service depends almost entirely upon wages channelled directly from formal sector employment.

Whether there is, or is not, linkage between the sectors hinges on the presence or absence of the formal sector and, in the former case, the need to cheapen the production process by tapping cheap labour by outworking. Where – as in some parts of Central America and the Caribbean – the modern manufacturing labour force (mostly of women) works in free-port industrial zones, in which it is denied labour unionization, social benefits and a minimum wage, foreign investors may conclude there is no need to extend operations outside the factory. In Mexico, however, Japanese capital is currently penetrating many provincial towns – notably Nissan in Aguas Calientes – with a view to entering immediately into cheap-labour outworking systems of 'flexible accumulation'. In this **post-Fordist** version of what otherwise looks like a traditional feature, the modern factory is modest in scale, and its connections reach into the local small-scale urban economy.

It has been argued that this path is also being followed by Third World manufacturing concerns hit by economic restructuring, and that the growth of the informal sector is sometimes closely connected with outworking, as the original plant seeks to cut costs. De Soto (1989), a right-wing economist, drawing on his experience of Lima, has argued for the de-regulation of formal economies, hedged around as they

A First World view of what constitutes 'useful' employment

often are by complex governmental restrictions, to force them to use whatever techniques may improve their competitiveness. In a similar vein he has advocated the unleashing of what he envisages as the natural vitality of the informal sector, untrammelled by government or NGO involvement.

If the dichotomy between formal and informal employment is being increasingly revealed as less functionally than morphologically significant, so too is the distinction between employment and unemployment. Measures of unemployment, if they exist, are likely to be inaccurate; where they are high, they are indicative of informal employment, much of it socially stigmatized or illegal. In the Caribbean, for example, where it is common for cities to record rates of unemployment of 20–30 per cent, idleness, in reality, is impossible, since in common with most Third World regions, very few countries provide unemployment benefit, and activities such as begging, stealing, prostitution and marijuana peddling are widespread. These activities illustrate Hart's informal illegitimate sector, and reach their apogee in Kingston, Jamaica, where the slums have been polarized by political affiliation, and heavily armed gangs (many of which have allegiance to one political party or the other) are engaged in internecine struggles to control the drug trade.

Under such circumstances it is impossible to differentiate clearly between employment, underemployment and unemployment, since some of the urban poor change their position within this range week by week as their fortunes shift. Third World cities, then, are characterized by enormous marginal populations. In Marxist terms, they form a **lumpen proletariat** that endures persistent poverty: the **reserve army of labour**, which in advanced capitalist countries is waiting to be drawn into use by new enterprise, is on permanent leave in Third World cities, where it has the overall effect of cheapening labour.

As most urban employment is in manual and service work, migrants reflect the same basic occupational pattern as the urban born. However, two additional factors help to explain this similarity: some migrants possess high levels of education relative to their areas of departure and are able to acquire secure employment; considerable numbers of migrants have been city dwellers for a long time and have achieved promotions in their urban jobs. Even in the large cities of Latin America, where educational credentials are important in controlling entry into formal sector jobs, newcomers rapidly resemble the local population in their employment, since migrants usually obtain work through the advice and influence of friends and relatives who are familiar with the system and can vouch for them. One migrant group in Mexico City, which originated in a particular rural community, has settled the same neighbourhood in the capital, and now works in collaboration as carpet-fitters. In many African cities, residential neighbourhoods have developed a marked tribal character and occupational specialization as this process of in-migration and informal sector expansion has gathered momentum.

HOUSING CONDITIONS

Urbanization which occurs without adequate industrialization, sufficient formal employment or secure wages, has condemned burgeoning urban populations in the Third World to poor quality housing. The problem has been compounded by a lack of government funds for housing subsidies, by inflated land prices boosted by housing needs and speculation, and by real-estate profiteering on the part of the upper and middle classes. The operation of the class structure of Third World cities is nowhere more geographically explicit than in the composition and working of the housing market.

Only the small upper and middle classes in Third World cities have the income, job security and credit worthiness to purchase or rent houses in properly surveyed, serviced and legally conveyanced developments. Even if mortgages are available (and they are emphatically not in most countries experiencing rapid inflation), down-payments tend to be large, and the sum borrowed must be paid back over a short time period. Middle-income housing schemes are often organized by the state for its own bureaucrats, by trade unions or by specific occupational groups, and thus automatically exclude those in informal jobs. Elites, of course, have the greatest capacity to choose building sites, house styles and ambience. Upper- and middle-class housing, especially that of recent construction, tends to be confined to formally employed workers and to match, or even exceed, standards among equivalent groups in Europe or the US.

Whereas the middle and upper classes are usually satisfactorily, even luxuriously, housed, the emergent working class and marginal population are confined to rented rooms in centrally located, high-density slums – *tugurios* in Latin America, compounds in the Middle East, *bazaars* in India – or to peripheral squatter settlements, often called informal settlements.

The renting of housing is a common feature of the Third World city, with colonial antecedents. At the beginning of this century 90 per cent of the households in Britain, then the world's most urbanized country, paid some form of rent; in 1945 80 per cent of the households in Kingston, Jamaica were in rented accommodation. In the cities of Mexico between 30 and 50 per cent of homes are rented. In Mexico City, a swathe of older, stone and concrete rented tenements – within which have been rehoused, as owners, some 28,000 victims of the 1985 earthquake – surrounds the commercial centre and occupies an area of almost 5 square miles (Plate 7.4). However, while analogous rented slum areas in Caribbean, West African, Middle Eastern and Indian cities are invariably dilapidated and grossly overcrowded, they, too, usually have rudimentary, though often inadequate, shared services, such as piped water, toilets, sewerage systems and electricity.

The demand for low-cost shelter, in the context of rapid urban population growth, has frequently been met by housing developments on plots of illegal tenure. Known as *colonias proletarias* in Mexico, *favelas* in Brazil, squatter settlements in the English-speaking Caribbean and Africa, *bidonvilles* in former French colonies, and *bustees* in India, visually they are the most outstanding feature of Third World urbanization (Plates 7.5 and 7.6). Squatters locate

Plate 7.4 Tenement, Mexico City A stark block of rented residences located close to the city centre

Plate 7.5 Squatter shanties, Kingston, Jamaica These shacks in 'Moonlight City' typify the rudimentary construction of a squatter settlement, being built of pieces of wood, cardboard and tin

on any site that is outside the market for urban land. Abandoned lots, ravines, swamps and precipitous slopes are frequently colonized; settlements of houseboats or dwellings on stilts are established in harbours and creeks.

The oldest of these squatter settlements are often adjacent to the city centre, as in Baghdad, but the most common location for squatters is the edge of the city, where land can be most easily expropriated. During the British colonial period in Nairobi, Africans were not allowed to settle within white residential areas and black peri-urban squatter settlements grew out of this proscription. In these ways the people who are economically and socially marginal become residentially and spatially marginal too. Initially, at least, squatter settlements are cut off from all types of public service, are prone to disease, and are usually located far from sources of formal employment.

A distinction must be made between squatter settlements and shanty towns. Illegality of tenure is the hallmark of the squatter settlement, but shanties – huts or mean dwellings – are defined by their fabric. Some squatters occupy shanty towns, but shanties are more usually located on small, rented plots. In Mexico City, rented shack yards make a limited contribution to urban housing, but in the Caribbean their residents vastly outnumber squatters, in the strict sense of the term.

Squatting is more likely to be tolerated by the government than by private landowners, though it is possible to find cities, such as Rio de Janeiro, where *favelas* are equally divided between public and private land. Squatter settlements are created in several ways. In Asia, the Middle East, the Caribbean and Africa, the process is usually gradual, essentially unorganized, and based on individual initiative. But in Mexico City, as in many other Latin American cities, *colonias proletarias* are normally started either through illegal subdivisions, where estate agents market plots for which they lack or fail to provide title, and/or fail to comply with planning requirements for zoning or service provision; or they may result from illegal land captures organized to place thousands of house-

Plate 7.6 Squatter shanties, Oaxaca, Mexico The dwellings in this informal settlement also use simple materials of reed, but are rather more carefully constructed

holds on the disputed property, literally overnight (Gilbert and Ward, 1985). In the vicinity of Mexico City, land reform units (*ejidos*) often accept 'illegal' settlers, who pay for titles and rapidly upgrade their homes (Varley, 1985). Indeed, the precise nature of the illegality which is involved may be variable and multiple; so the settlements are increasingly grouped together under the catch-all title 'informal settlements'.

Though squatter settlements were recorded in several Third World cities before 1914, it is only since 1945 that they have become a significant aspect of urbanization. By the 1970s more than 40 per cent of the inhabitants of Ankara, Lima and Casablanca were squatters. In contrast, squatters in West African towns rarely exceed a few per cent of the population (Peil, 1991), though dwellings erected without any control by public authorities have been reported to house 30 per cent of the population of Dakar and Abidjan. Numerous cities from a range of Third World countries record between 30 and 60 per cent of their populations living in informal settlements, and Mexico City is not only the largest city in the world, but also its informal housing 'capital'.

The spectacular growth of squatting, especially in the period 1950 to 1980, was the result of the inability of the poor to pay rent or meet house-purchase prices, and a measure of the lack of public funds to rehouse slum dwellers or provide additional accommodation for new households created by incoming waves of migrants. Despite government and international aid, long-term credit for housing is difficult to obtain and even white-collar workers find it hard to pay their rent and save enough money to purchase a plot outright – unless they benefit from mortgage facilities provided by occupational organizations or unions. In some Third World countries a substantial proportion of the public money earmarked for housing has been diverted into the banking system and used to finance private housing.

The necessity to provide some public housing is well understood by Third World government officials and planners, despite the lack of sufficient funds to finance anything more than token schemes. Projects are expensive and they are often strongly opposed by the middle and upper classes, especially if they involve the rehousing of squatters. Hong Kong has the largest and best-publicized public housing schemes in the Third World – two-thirds of the population live in government estates. But in Hong Kong the land is Crown property and the clearance of spontaneous settlements has been a response to the demand for sites for commercial and industrial purposes, and minimal attention has been given to improvement of amenities among squatters. Oil-rich Venezuela, arguably the most urbanized country in Latin America, provides a cautionary tale. The *superbloques* of Caracas, which were constructed by the dictator, Jimenez, in 1954–8, housed 159,000 people in 85 buildings and soon presented more problems of social control than the squatter settlements (*colonias*) they partially replaced.

Politics and patronage are often deeply involved in the siting of housing schemes and the selection of occupants. Even the so-called beneficiaries encounter problems, as evidence from cities as diverse as Baghdad and Rio de Janeiro shows. Most schemes are located on the urban periphery: the journey to the city centre is long and costly; shops and other facilities in the immediate neighbourhood are often inadequate; and new residents are isolated from the various community supports, pastimes and informal occupations that were once readily available in their old haunts. Despite modern design and layout, those accommodated in public housing remain as socially deprived as inhabitants of the inner sections of the city.

For most Third World governments, public housing schemes on a vast scale are too expensive to contemplate. Housing policy, with World Bank promptings, has therefore increasingly shifted to self-help schemes, using as a model the method of 'squatter **autoconstruction**'. What are the positive and negative aspects of squatter settlements and what implications do they have for self-help solutions to the housing problem?

SQUATTER SETTLEMENTS: PROBLEM OR SOLUTION?

Squatter settlements were once stereotyped as collections of insanitary, flimsy dwellings, constructed by recent migrants who created a rural environment in the towns. Only by immediate eradication could Third World countries control this 'urban cancer' (Plate 7.7). Yet in the last 25 years squatting and various other forms of illegal or informal housing have been recognized – first by social scientists then by administrators – as a potentially positive feature of Third World urbanization. Largely through the work of Mangin and Turner in Lima, Peru, in the late 1960s, squatter settlements are no longer viewed as rural slums created by new arrivals from the country, but residential areas that migrants of long standing, and many of the urban-born population, occupy as part of their adjustment to city life (Figure 7.5).

According to Mangin (1967), overcrowded tenements or established squatter settlements usually acted as initial footholds for recent migrants, and it was from them that the new waves of squatters came. While living in rented accommodation, migrants learnt how to manipulate the housing system; but it was in the squatter settlements that they could gradually build their own houses by investing money that otherwise would have been paid in rent. The move to peripheral squatter settlements from rented accommodation near the city centre frequently coincided with growing parental

Plate 7.7 Contrast in housing styles, Bombay, India In the shanty town in the foreground, some dwellings have only cloth or tarpaulin roofs, while others have tiles or corrugated iron. In the background are high-density apartment blocks

responsibilities; the availability of cheaper, spacious housing more than offset the increased cost of the journey to work. As squatter settlements aged, however, so their demographic composition and social role changed. The number of city-born increased, and the squatter neighbourhoods themselves became reception areas for migrants who were joining friends or kin.

The largest squatter settlements were rarely single-class communities. On the contrary, they were well integrated with the employment structure of the city. In Lima in the 1960s, for example, almost 60 per cent of the labour force of the squatter settlements (*barriadas*) were artisans or labourers, 16 per cent domestic servants, 14 per cent street pedlars, shopkeepers or stallholders, and the remainder (10 per cent) were white-collar workers or factory employees. This pattern was explained by the impoverished condition of the entire lower class, the magnitude of the housing shortage, and the low pay received, even by white-collar workers. In the Caribbean, however, where squatting accommodates less than 10 per cent of the population, the occupants remained uniformly lower class, impoverished, and drawn from the racially and ethically most denigrated section of the population.

Latin American squatter settlements (and illegal subdivisions) have a well-documented history of dynamic change and improvement, which is paralleled, though to a lesser degree, in the rest of the Third World, notably India. The creation of a squatter settlement leads to the establishment of community organizations whose object is to defend the community and intercede for it with the government. Demands are made for recognition, title to land, water, sewerage, electricity and roads, and if these are

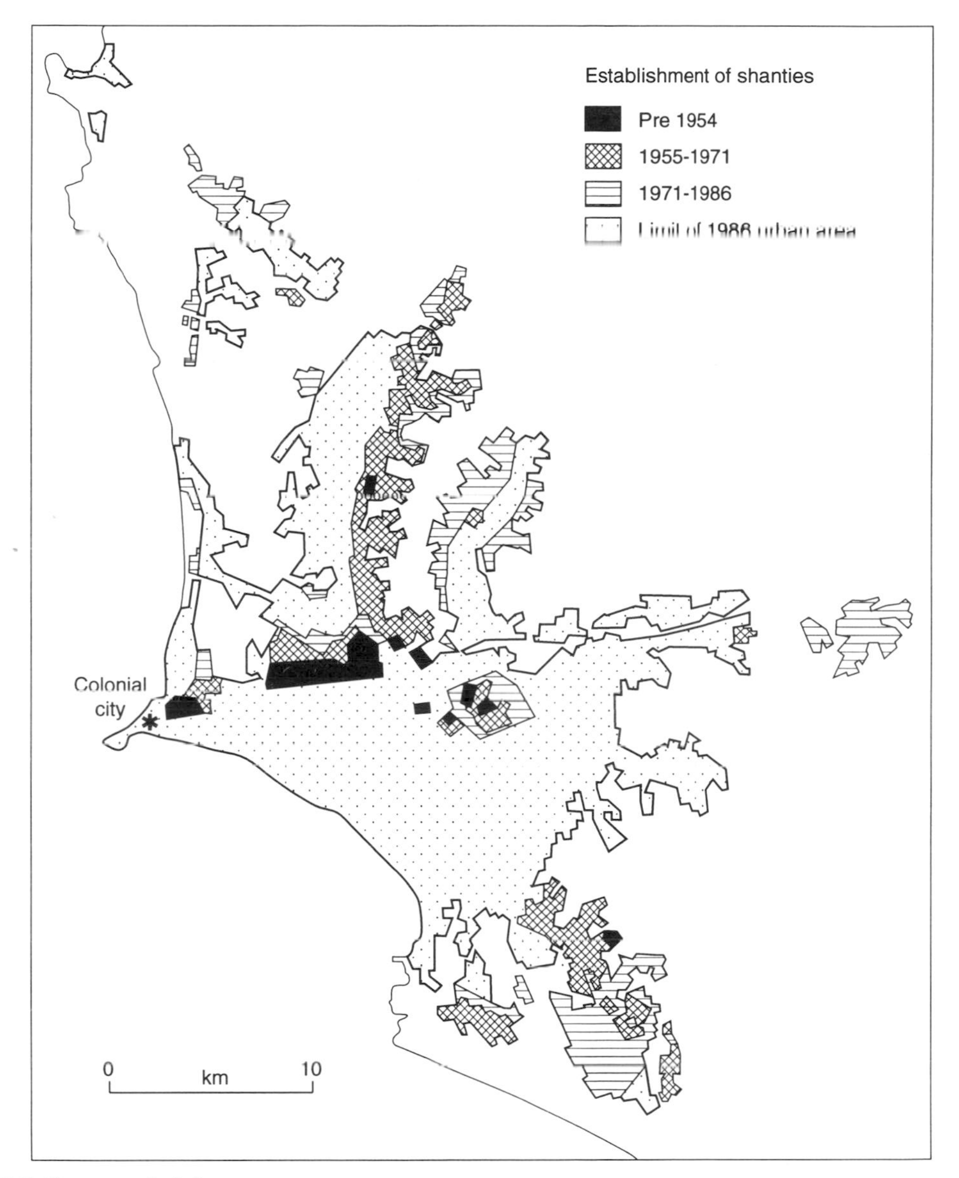

Figure 7.5 The spread of shanty towns in Lima, Peru
Source: After Driant, 1991

successful, requests may eventually extend to more sophisticated amenities – schools, clinics, community halls. Community and household improvement go hand in hand, and when title to land or some assurance of permanence is given, the original temporary dwellings are gradually replaced (consolidated) – using a mixture of family and hired labour (Figure 7.6) – by more substantial home-made structures (Plates 7.8 and 7.9).

This sequence of house and settlement development was described by Turner as a trajectory, divided into incipient, consolidating and consolidated stages (Turner and Fichter, 1972). Women are vital in negotiating all phases of this trajectory, notably in Latin America, where they account for upwards of 20 per cent of household heads in many cities. A particularly heavy burden is carried by mothers – victims of the double shift at work and at home – during the incipient settlement stage, when infants have to be reared in dwellings lacking the most basic

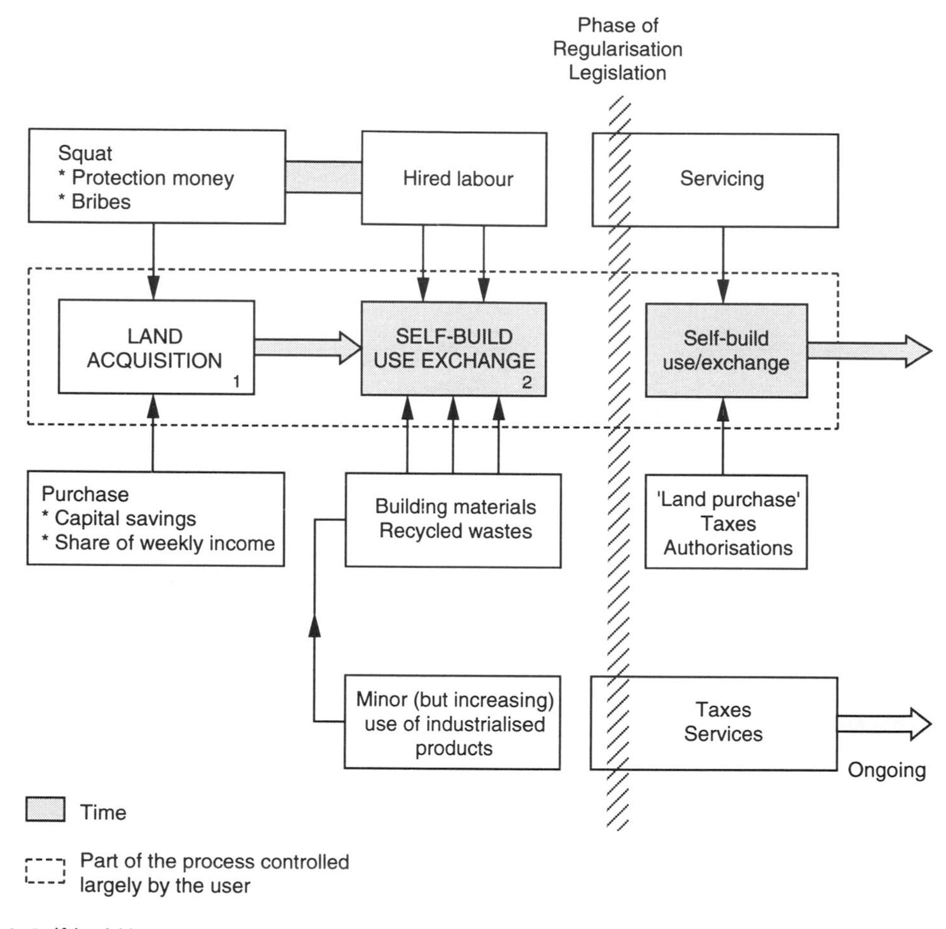

Figure 7.6 Self-build housing
Source: After Ward, 1990

amenities (Chant, 1991).

Successful upgrading of informal housing depends to a large extent upon a positive government response. But government reactions to squatters' demands range from helpfulness to hostility. In Peru in the late 1960s and 1970s, some self-help settlements (*pueblos jovenes*) were directed to specific locations where services were easy to install. Calcutta has encouraged *bustee* improvement and embarked upon massive site-and-service schemes to enable households to build on lots supplied with rudimentary amenities. Elsewhere in the Third World, government attitudes have been less encouraging to site-and-services (which are often too expensive for the poorest of the poor) and to outright squatting. The Venezuelan government, for example, has been ambivalent in its treatment of the *colonias* in Caracas, which house 60 per cent of the capital's population: water and electricity have been supplied to squatter communities, but titling of plots remains rare and a major bone of contention. Nevertheless, oil wealth has, until recently, created sufficient household surpluses to fund substantial upgrading of homes, even without the security of titles (Gilbert, 1993). A similar feature also characterizes the rent yards of neighbouring Port of Spain, Trinidad, where the economy has been oil propelled.

In Mexico, some presidential regimes have acceded to informal housing sector demands for

Plate 7.8 ***Upgraded housing, Mexico City*** Although this is an informal settlement, it has been improved and consolidated. The houses, though small and simple, are brick-built, the sidewalks are paved, and there is a water supply, electricity and public transport. A plaza has been laid out, surrounded by shops and a church

Plate 7.9 ***New squatter area, Oaxaca, Mexico*** The informal settlement of Linda Vista is located on the periphery of Oaxaca City, but though the housing is scattered and there are few facilities, some properties are brick-built, and the 'se-vende' (for sale) sign suggests there is even a housing market

security of tenure and sought to co-opt local leaderships; others (more recently) have refused to recognize land grabs by squatters (Ward, 1990). Generally, the tendency over time has been for the Mexican government to become increasingly hostile, though Gilbert and Varley's work (1991) on urban Mexico shows a marked upswing in home ownership over the period 1950–80, presumably through squatter upgrading. In Jamaica, periodic eradication of insanitary small-scale squatter camps punctuated Kingston's history of illegal settlement from the 1930s until the early 1970s, and bulldozing shanty towns had been a common response in Africa until the 1980s, especially in the former European settler colonies such as Zimbabwe, and into the 1990s, until the ending of apartheid, in South Africa.

Even if a government is positively disposed to encouraging the development trajectory of squatter settlements, micro-ecological and macro-economic circumstances may impede the process. The failure of shanty towns to improve over time is due to two different reasons, depending upon the tenurial category in which they fall. If they are based on renting, then that fact alone is usually sufficient to impede change and produce permanent shacks – as Jamaican and Haitian experience illustrates. Squatter settlements that fail to develop are usually located on congested sites where hazards threaten, or where redevelopment is impossible: in Jakarta, for example, tens of thousands of people squat on the city's pavements, as do countless millions in India's major cities.

Of course, non-recognition and threats of eradication impose equal barriers to improvement, irrespective of ecology. But even where a government's attitude is benign and other circumstances are generally favourable to autoconstruction, improvement may be impeded by lack of macro-economic development and the absence of household funds for investment. Ward's evidence for Mexico City (1982) suggested that even when squatter settlements consolidate through autoconstruction, sometimes reaching a point where it is impossible to distinguish them from middle-class subdivisions, the original occupants have not been the ultimate beneficiaries. As squatter settlements consolidate, the cost of paying for a legal title, water and electricity installations, taxes and bills, coupled to protection money for local bosses, squeezes out the poorest marginals who then have to move on to new land captures (Figure 7.6). In this way they are used by the system as a means of turning peripheral non-urban land into valuable real estate!

In addition to changes in residential composition, the Mexican urban development trajectory often involves tenurial change, in which squatting gives way to ownership and then to rental. Indeed, sub-letting is also present in the first stage of occupation, since speculators place families on plots and pay their outgoings as the settlement develops – only to remove them as the economic rent increases. Similar processes of crude capital accumulation threaten to undermine the development of site-and-service schemes. It is impossible to escape the conclusion that, in the economic circumstances of most Third World cities, only the humblest shack is viable for the poor, and even that dwelling is quite probably minuscule, lacking in furnishings and all basic services, and tenurially insecure.

There is a further aspect of the transition from informal systems of tenure (such as squatting) to renting: Third World governments, especially those that are either indebted or experiencing structural adjustment or both, are cautious about allowing further informal settlement. Although Marxist critics have argued that self-build is little more than self-exploitation, giving leverage over the masses to government without commensurate commitment of funds, Third World regimes are unwilling to pay for additions to the already extensive system of urban infrastructure in massive cities. Indeed, there may be less need to do so. Whereas the population increases of the 1950s took place in cities with a small housing stock, which could be expanded only by informal housing, the modern Third World metropolis is ripe for sub-letting; petty landlordism and 'densification' are the order of the day, according to Gilbert and Varley (1991).

SOCIAL STRUCTURE AND THE URBAN MOSAIC

Poverty and poor housing in Third World cities are closely related and – socialist countries such as China and Cuba excepted – inherent in the social structure. In Latin America, the urban social stratification comprises a European or Europeanized elite, a small but growing middle class, and a mass of lower-class persons who are set apart from the bourgeoisie by an enormous social and economic gulf. In other Third World regions, notably the Middle East, the social structure is often more traditional and reflects non-class inequalities, such as those of race, religion, culture, tribe, language or ethnicity; and even if class distinctions are emerging, they may remain subordinate to racial and cultural differences. But whatever the criteria of social differentiation, it is usual for them to be produced and expressed in the urban mosaic. During the colonial period in West Africa, indigenous city dwellers were often segregated from 'stranger towns' which housed the non-white migrants; since independence Africans who have succeeded to elite positions have taken over white housing areas and created an incipient class structure, which is reflected in the urban morphology.

In origin, Third World cities and towns – both pre-colonial and colonial – are quintessentially pre-industrial. Sjoberg (1960) identified three main aspects of land use which, he argued, distinguish the pre-industrial city from E.W. Burgess's industrial model. These are: the pre-eminence of the central area over the periphery, particularly as portrayed in the distribution of social classes; certain small-scale spatial differences according to guild or family ties; and low levels of functional differentiation in other land-use patterns. A major aspect of the pre-industrial city is the absence of a central business district at the heart of the settlement: instead, religious and government buildings provide the focus. In contrast to the industrial city, so typical of the US, homes of elite citizens cluster around the centre and disadvantaged members of the community fan out towards the periphery.

Sjoberg suggested that these social patterns in pre-industrial cities may be explained by cultural values that define residence in the historic core as most prestigious. He also emphasizes that pre-industrial technology entailed foot and animal transport and placed a premium on face-to-face contact – or spatial proximity – for interpersonal communication. Many Third World cities, such as Ibadan and Beijing, are pre-industrial in origin because they have pre-European foundations (see Figure 7.4: Box 7.2), while Caribbean and Latin American settlements, though new foundations, at their inception incorporated the pre-industrial values and the systems of motive power which were available in Europe in the sixteenth, seventeenth and eighteenth centuries.

Indigenous settlements of the Muslim world, whether planned or irregular, focus on the central mosque, citadel and bazaar (Figure 7.7); cosmo-magical symbolism is expressed in the orientation of ceremonial cities in the pre-Colombian civilizations of Middle and South America and in pre-European Asia; European colonial settlements are usually laid out to conform to a simple grid pattern. Traditionally, the small, pre-industrial elite lived in houses adjacent to the centre, while persons of low status were confined to the edge of the settlement. This declining social gradient from centre to periphery is still observable in many small Third World towns.

But large Third World cities, which have experienced rapid growth, have adopted modern technology and architecture since the mid-nineteenth century and now enjoy rapid communication by car and telephone, exemplify a partial reversal of the original, pre-industrial sequence. Elites have left their old-fashioned homes as city centre land values have rocketed, and now live in ostentatious modern properties that reveal more conspicuously their superior status and lifestyle (Plate 7.10). As in North America or Western Europe, the urban elite and middle class are suburban in location, though often concentrated in a specific security-guaranteed sector or wedge, and the pattern holds good for cities as far apart as Port-au-Prince, the capital of Haiti, and Calcutta. In general, the poor are concentrated in over-crowded compounds or tenements located near

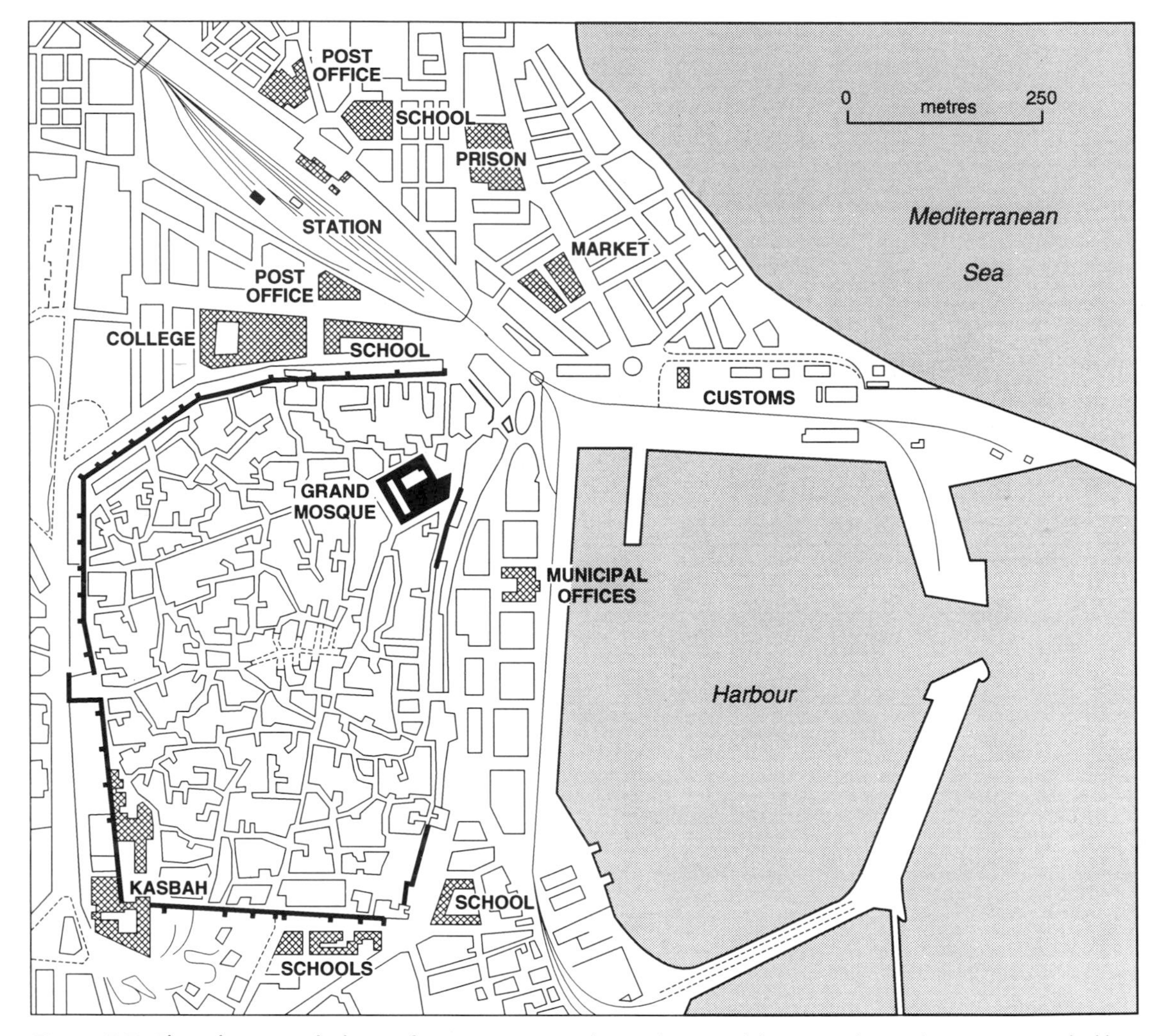

Figure 7.7 The urban morphology of Sousse, Tunisia The Arab core of the town, the medina, is surrounded by a wall, enclosing the mosque, the shopping area of the souks, a maze of narrow streets and alleys and secluded dwellings. Outside the walls the more regular and spacious layout of the modern commercial and residential areas was created by the French colonial administration after damage in the Second World War
Source: Société Tunisienne de Topographie, Environs de Sousse, Flle 4, 1:5000

the city centre or in peripheral squatter settlements and government housing schemes.

The establishment of multiple stores in the central business districts of Third World cities, the replacement of the old urban fabric by high-rise buildings, and the rapid growth of shopping precincts in the suburbs, are signs of the importation of the American style of life into Third World cities as diverse as Bangkok and Rio de Janeiro. But, in small towns, shops are often no more than lock-up sheds, and concrete construction and neon signs are appearing only slowly. There is, therefore, a continuum from pre-industrial to industrial, linking small traditional communities and large urban centres, which is reflected in economy, fabric and ecology.

To date, re-colonization of the central city through gentrification, in emulation of some North American and European cities, is rare in the Third World. The refurbishment of old San Juan, Puerto Rico, and the redevelopment of the waterfront in Kingston, Jamaica, for a container terminal and a revamped financial and con-

Plate 7.10 Affluent housing, Belo Horizonte, Brazil This district of upper-class housing is built on the mountainside overlooking the city of Belo Horizonte. Its residents can afford to pay for the provision of urban services to such a site, and the architecture echoes the style of colonial Ouro Preto (Plate 2.1)

ference centre, are contrasting, but still unusual, examples. The debt crisis of the last decade has drastically curtailed public expenditure on major urban infrastructure, so that even Latin America has responded less fully than might have been expected to North American innovation.

However, the detailed residential morphology of each city depends upon the social structure of the country in which it is located, the proportions of different classes and races and on topography. In Caracas the steep slopes which hem in the central business district are occupied by squatters, while the middle class form a buffer between the elite suburbs in the east and the original state housing schemes – now infiltrated by the middle class – adjacent to the old city (Figure 7.8); in Port-au-Prince city-centre residence is associated with high population densities, low social status, black skin colour and the practice of voodoo; in India caste distinctions give rise to residential segregation in which high-caste Brahmins and Kshatriyas take the prime locations.

MEXICO CITY: A CASE STUDY OF URBAN GROWTH AND MARGINALITY

Mexico City, laid out by the Spaniards over the ruins of Tenochitlan, the capital of the Aztec empire, is now a mega-city (approaching 20 million population), and one of the largest and most rapidly growing settlements in the Third World. Since the beginning of this century its population has increased more than 40 times and now approximates one-quarter of the nation's total. The census recorded 370,000 inhabitants in 1900, 1 million in 1930, 5 million in 1960 and 8.5 million in 1970; estimates put the city's current population at above 16 million, but the precise figure is uncertain because the built-up area does not coincide with the census limits. Guadalajara, Mexico's second

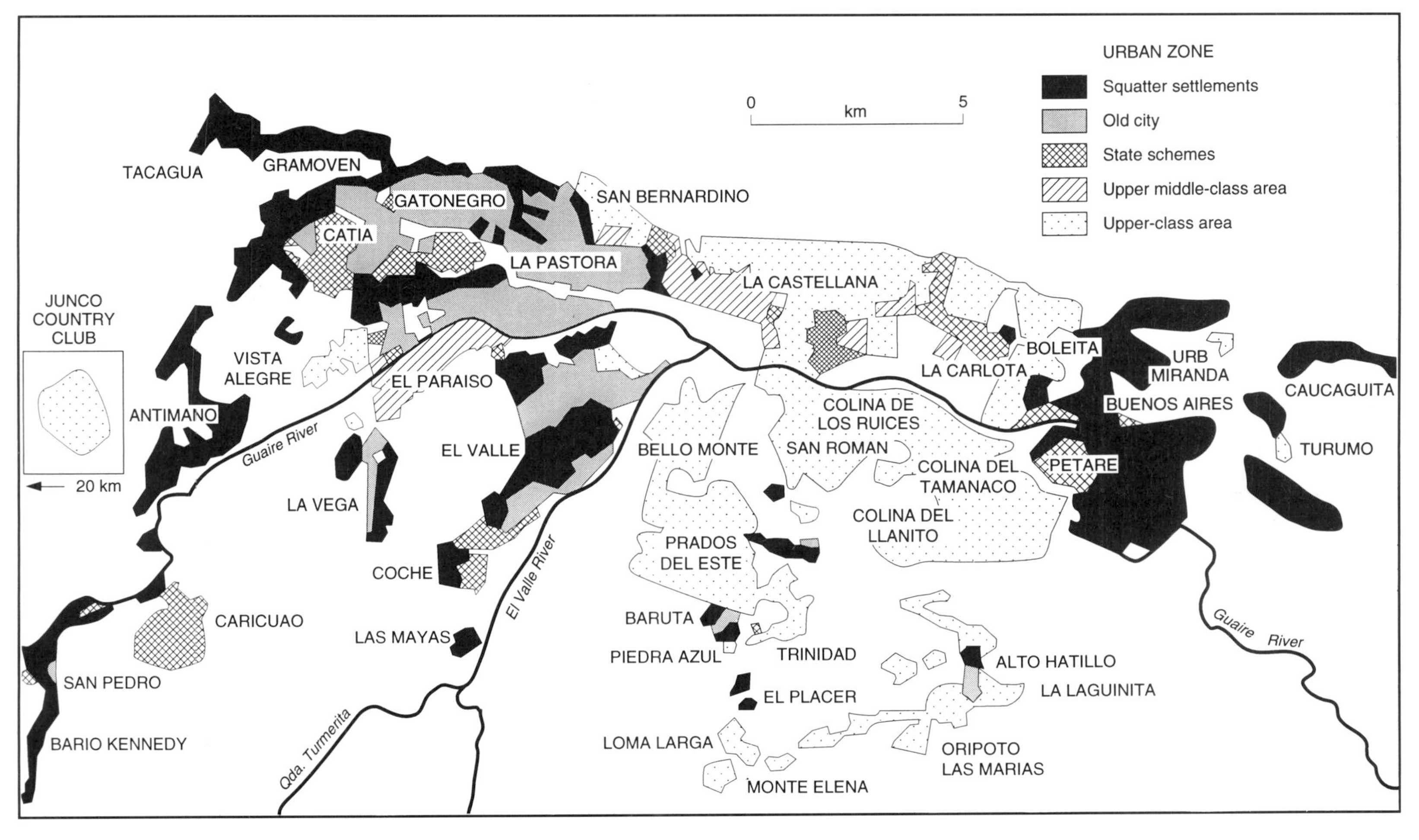

Figure 7.8 Housing and social class in Caracas, Venezuela
Source: After Franklin, 1979

biggest city, lags way behind with approximately 2 million inhabitants.

For many years Mexico City's annual growth rate has remained fairly constant at about 5 per cent, but each decade since the Second World War the contribution made by migration has fallen. During the 1970s provincial newcomers accounted for less than one-third of the addition to the population, the remainder being the result of natural increase – itself a consequence of the city's youthful age structure and the sharp decline in mortality, especially among children. Unless the demographic transition to low levels of fertility is rapidly achieved or state projects for decentralization are accelerated, the city is likely to have over 30 million inhabitants by 2010.

Migration to Mexico City has declined in relative terms, but the absolute number of cityward movers continues to grow. Three features characterize the migrants: they are young (mostly 15 to 40); the majority are women (1210 per 1000 males); and large numbers are drawn from states adjacent to the capital – Mexico, Puebla, Morelos, Hidalgo, Tlaxcala – or from impoverished regions in the southern states of Guerrero, Oaxaca and Chiapas. Most migrants, once established in homes and jobs through kin networks, remain in the capital, yet retain close ties with their villages or towns of origin. Remittances are sent to support rural dependants, and organizations are created to stimulate and finance projects in the home regions.

Massive population concentration in Mexico City has generated spatial growth. As late as 1900 the capital still clustered around the Spanish-created plaza (*zocalo*), which contained the country's principal administrative and ecclesiastical buildings, though there had already been modest suburban expansion to the west and north in the direction of Guadalajara, and small satellite communities clustered in the south, some of which had been occupied since colonial times (Figure 7.9). By the end of the Mexican Revolution in 1917, *colonias* had been added to the east and south; developments in the 1920s took place essentially to the south; and in the 1930s they concentrated in the west. Massive encroachments in all directions were made into the surrounding countryside after the Second World War. They were particularly large to the east of the city, and caused settlement on the dry bed of Lake Texcoco to spill over the Federal District boundary. By 1970 the district of Nazahualcoyotl, originally an informal settlement, recorded more than 500,000 inhabitants; by the mid-1970s its population had already exceeded 1 million.

Mexico City represents the largest single labour market in the world, with more than 7 million economically active people. Industrial development – mostly for consumer goods – was at the outset unusually successful by Third World standards: in 1980, 27 per cent were workers and artisans, 17 per cent in administration, 15 per cent service workers, 11 per cent professional, 10 per cent in commerce and sales, and 16 per cent unspecified. However, industrial growth in Mexico started to flag during the 1970s, and unemployment, underemployment and marginality have developed into immense and increasing problems. Access to jobs in the coveted industrial sector has become more and more selective and almost completely dependent on contacts – usually through kin who will 'speak for' a job applicant – and sound credentials – usually in the shape of a school-leaving certificate.

According to one study, 37 per cent of the city's working population is engaged in the informal sector, oscillating from a low of 34 per cent in 1981 to almost 40 per cent in 1987, as unserviceable oil-loans drove the economy into debt and IMF restructuring took its toll on manufacturing and the public sector. To enable households to survive the crisis, women have entered the workforce, some of them entering US-financed industry attracted by Mexican tax concessions. A major way in which the poor have accommodated has been to share accommodation with kin, and to use kin both as a social security net and to facilitate child care when women work (Ward, 1990).

Income distribution is markedly skewed, so that the majority receive less than the minimum wage. This pattern of purchasing power correlates with the occupational and class structure of the city, has a direct bearing on the quality of dwelling units and services, influences house-

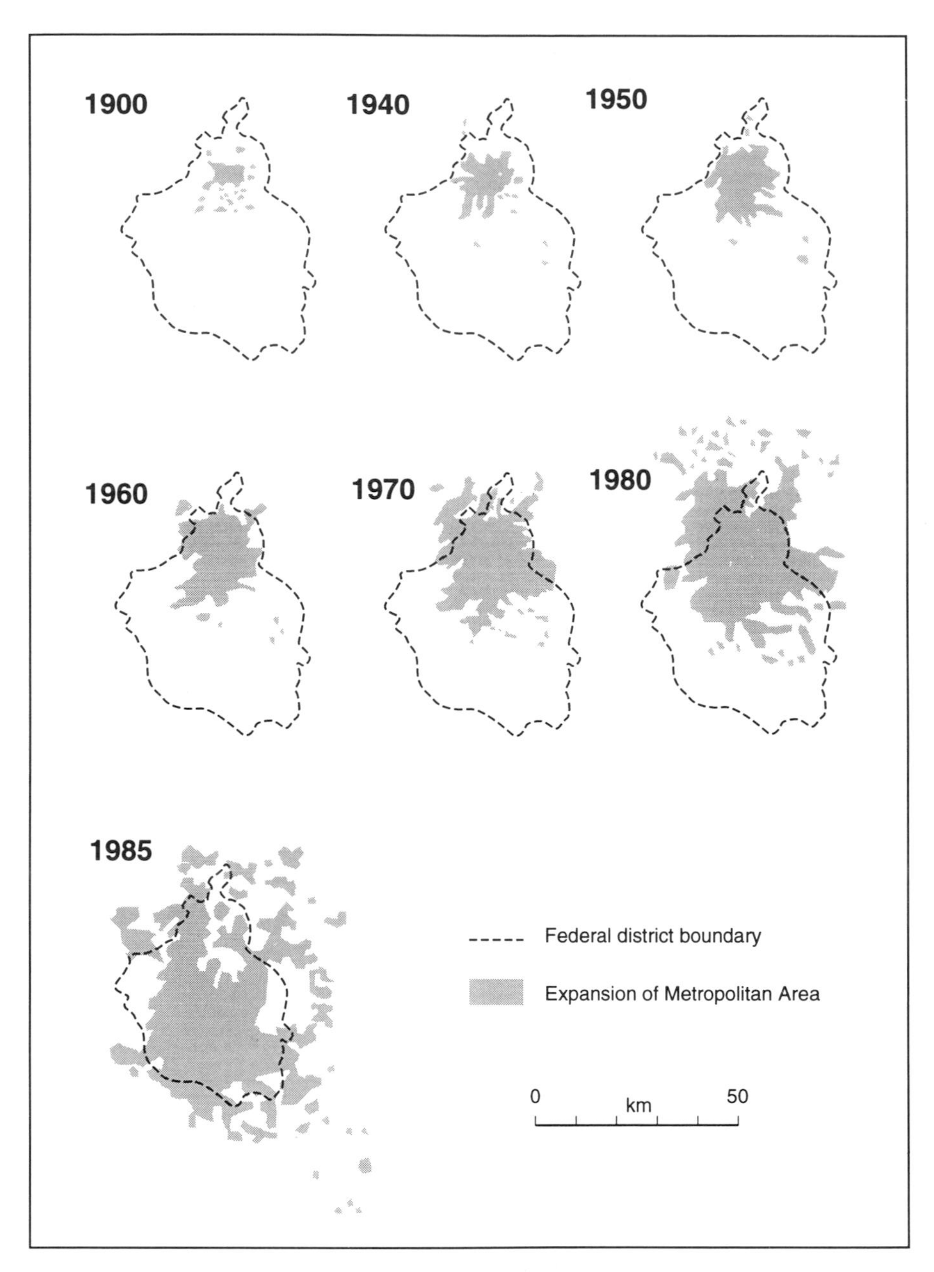

Figure 7.9 The expansion of Mexico City, 1900–85
Source: After Ward, 1990

holds' access to the housing market, and underpins the spatial structure of the city. A survey carried out by the National Mortgage Bank in 1952 revealed that over one-third of Mexico City's population lived in rented *tugurios* (slums), 10 per cent in *jacales* (shacks) in rented yards, and 14 per cent in *colonias proletarias* (low-income neighbourhoods). The remaining 42 per cent of the population was divided unequally between decaying property near the city centre (28 per cent) and new housing (14 per cent) of elite or middle-class status located near major routeways, especially those leading to the south.

Subsequently the low-income housing sector has been inflated to accommodate the growth of the city's population: tenements have been built and subdivided; enormous squatter settlements have been established through autoconstruction; and some government housing projects have been completed.

Tenements (*vecindades*) are located in dilapidated (often rent-controlled) colonial buildings, in overcrowded purpose-built apartments, or in smaller suburban speculations. *Colonias proletarias* fall into two categories: *fraccionamientos clandestinos* (illegal subdivisions) occur where the developer lacks legal title to the land or permission to sell it for housing, or more commonly where he lays out the lots but fails to provide the infrastructure that is required and thereby makes the project illegal. *Colonias paracaidistas* (parachutist settlements) are true squatter developments and are usually formed through invasion from *vecindades* or consolidating *colonias proletarias*. The third major low-income housing system is that of the *ciudades perdidas* (lost cities) or shack yards. They consist of concentrations of rough huts located on rented – usually privately owned – plots. A similar number to those in shack yards lived in government-funded projects. Government housing schemes had stipulated a minimum income level for entry which was beyond the earning capacity of at least one-third of the city's households, so that they failed to eradicate *ciudades perdidas* and made no impression on popular housing demand as represented by the proliferation of *colonias paracaidistas*.

In view of the massive growth of Mexico City's population, it is to be expected that housing deprivation would be widespread. In 1980, 30 per cent of homes lacked an internal water supply, 14 per cent were without adequate sewerage, and 3 per cent had no electricity. These deficiencies persist despite the processes of legalization, service provision, fabric improvement and autoconstruction for which Mexico City has become justly famous (Ward, 1986). Moreover, they illustrate the ever-present problems that characterize the *vecindades*, *ciudades perdidas* and those squatter settlements whose development trajectory has been cut short by government fiat or economic stagnation.

Processes of population growth, urban expansion, social deprivation and segregation have created a complex urban mosaic (Figure 7.10). However, it is possible to discern some pattern to Mexico City's urban ecology and to decompose it into zones, sectors and nuclei that echo the various models of the industrial city (Figure 7.11). Inner and outer zones in the north and east are devoted to proletarian settlements. The middle class, in its various social fractions, concentrates – with status increasing towards the periphery – in the south central wedge which follows Insurgentes Sur and incorporates the former independent colonial settlement of Coyoacan. Elite housing is confined to a western sector stretching via Polanco to the Lomas de Chapultepec and to an outer south-western sector incorporating San Angel and the Jardines de Pedregal. Finally, the outer *colonias*, Ciudad Satelite, Lindavista and Tlalpan, are set in vast expanses of *colonias proletarias*. Hence each of these three nuclei contains class stratifications organized in a pre-industrial fashion with a centralized elite or upper middle class and a peripheral proletariat. As a generalization, the poor dominate the north and east; the south-west quadrant alone is unmistakably the preserve of the elite and middle class. However, the colonial palaces in the city centre and the peripheral poor illustrate the persistence of pre-industrial features.

URBAN REGIONS AND PLANNING

Planning has had only a superficial impact on urban problems in the Third World, largely because the regulations enacted by governments have been waived or ignored in the face of the social, economic and political realities. Often planning has been for non-planning reasons – to justify a foreign loan, or to validate a decision already taken at ministerial level. Moreover, the scale and complexity of urban problems has been beyond the scope of the planning system, which has often been preoccupied with land-use zoning and transport. Above all, the discordance between the built-up areas of cities – especially the mega-cities – and city administration and planning frameworks has made urban management nigh impossible.

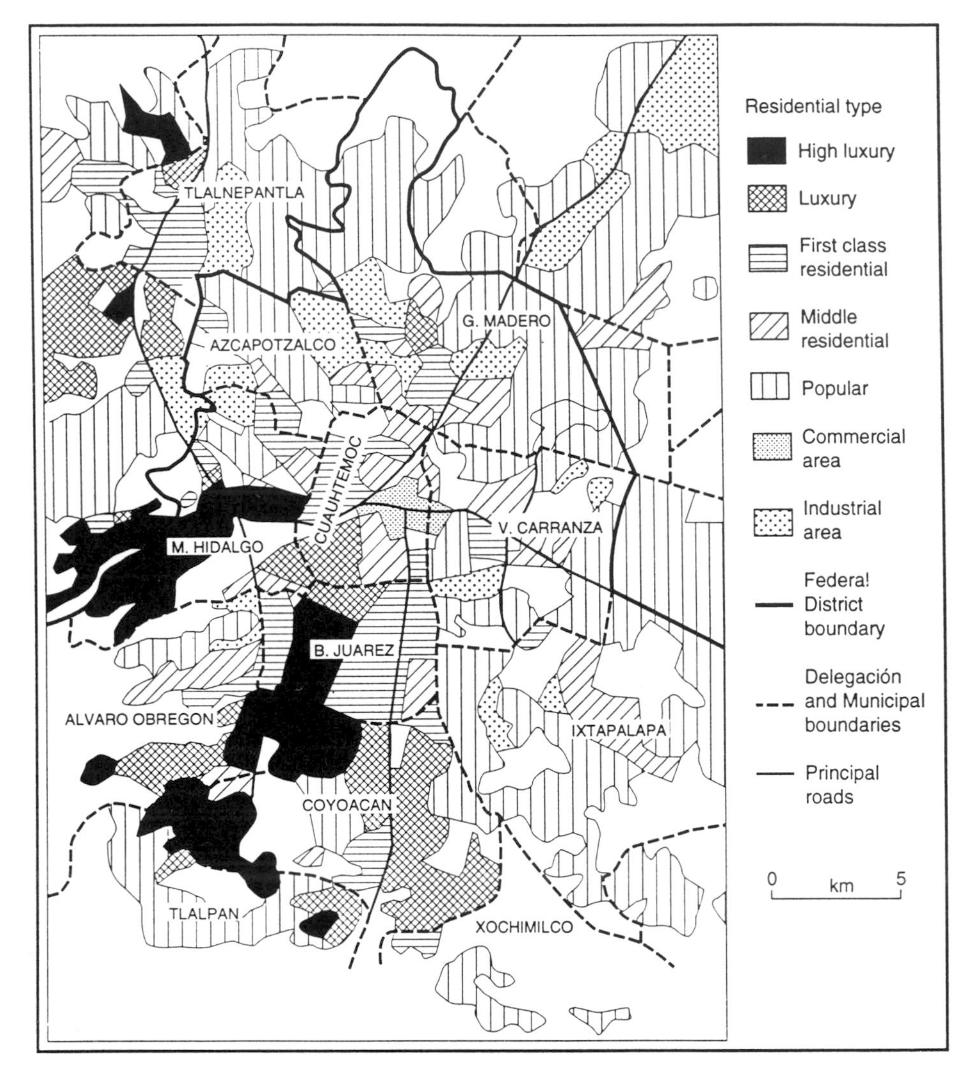

Figure 7.10 Distribution of population in Mexico City by income and residential type c. 1980
Source: Ward, 1990

Cities are places as well as spaces. Third World cities articulate the economic system, and changes in economic conditions result in alterations to the urban hierarchy. Primacy, as we have seen, is indicative of urban imbalance: states in Africa and South America, for example, are dominated by coastal core regions surrounded by narrow belts of partially integrated territory. Since the Second World War attempts have been made to establish poles of development in the interiors of countries in the Third World, and an excellent example is provided by Brasília, located in the interior of Brazil (Box 7.4).

Socialist China and Cuba provide interesting contrasts to the urban growth policy adopted by peripheral capitalist societies in Latin America, Africa and Asia, most of which have continuously reproduced marginality since the early 1960s. Since the 1959 revolution, urban and rural planning in Cuba has been shaped by socialist principles. The artificially high price received for sugar from the ex-Soviet Union, combined with cheap oil imports from the same source, enabled it to carry out ambitious programmes that emphasized rural development at the expense of urban growth, in sharp contrast to the rest of Latin America. For example, while

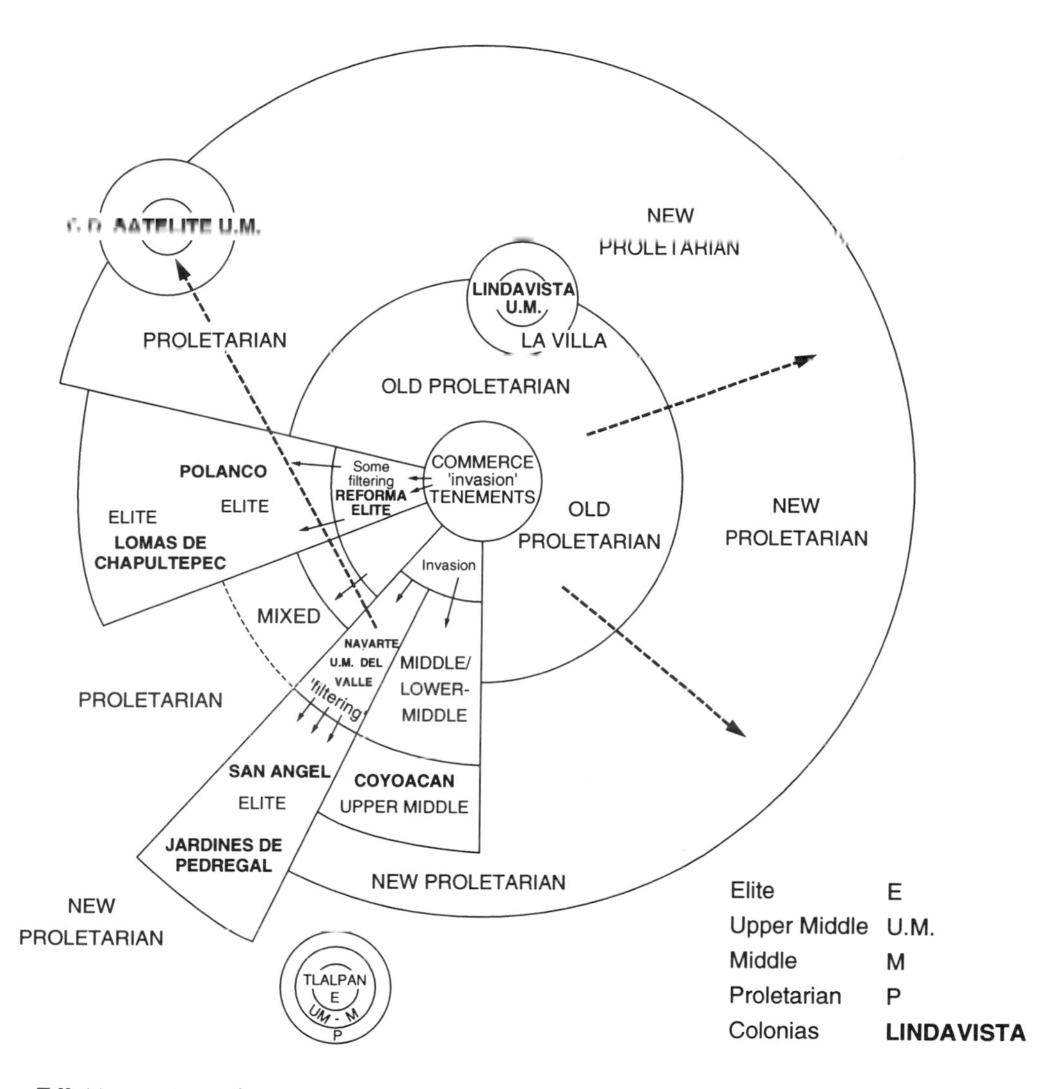

Figure 7.11 Mexico City: urban-ecological areas
Source: After Ward, 1990

Havana's population has doubled to 2 million in the last 30 years, the population of Mexico City has increased six-fold.

Cuba's public health programmes account for the lowest fertility rate in Latin America. Migration to towns and cities has been strictly controlled and forbidden without security of employment. Castro's government has decentralized the economy and run down Havana, to the advantage of the provincial cities. Several hundred *pequeñas ciudades* (small towns – in reality, little more than housing schemes) have been established in the countryside to house landless labourers, supply improved health and educational services, and deflect potential migrants from the capital. Throughout the urban system there is provision of state-funded housing, with rents fixed as a proportion of income. New housing developments have been built and old residential areas upgraded by state-organized micro-brigades. Many urban properties belonging to the white elite who abandoned Havana after the revolution have been reallocated to public services, such as hospitals and schools.

Government strategy to build up the rural sector has been underpinned by central planning and by the emphasis placed on the sugar

BOX 7.4 BRASÍLIA

Brasília's function in the national planning strategy has been to shift the political and psychological focus of Brazil away from the colonial centres on or near the coast, and to create a socially undifferentiated city – a real pioneer zone in Brazil's empty interior. Like the old, colonial towns, this new capital city was imposed upon the landscape, although the scheme was based on modern planning principles that segregate pedestrians and vehicular traffic. Lucio Costa's design resembles an aeroplane (Figure 7.12), and the city is greatly enhanced by Oscar Niemeyer's modernist public buildings (Plate 7.11).

Civil servants were loath to leave Rio de Janeiro for the isolation of Brasília, yet the new city has gradually developed into a fully fledged capital. A substantial amount of road building has been completed around Brasília and has created a wave of rural colonization, much of it highly speculative. Within the new city a major planning task was to harmonize housing, employment, shops, transport links and recreational facilities in combinations appropriate to

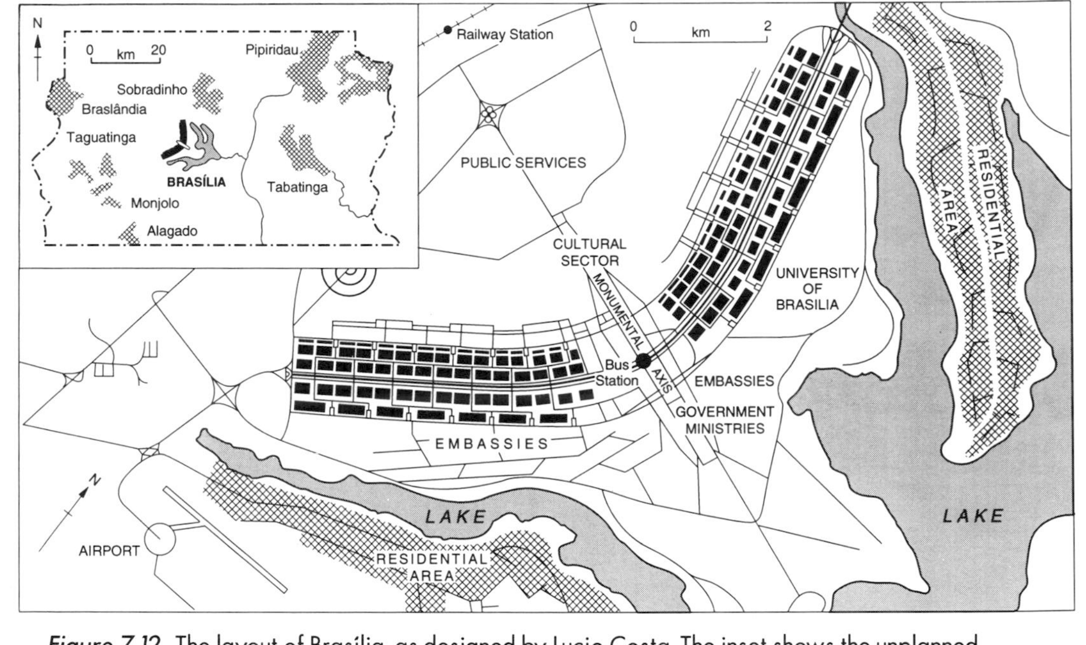

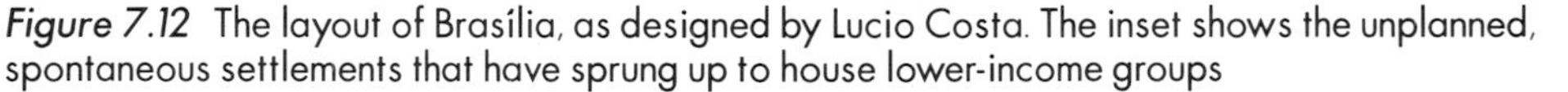

Figure 7.12 The layout of Brasília, as designed by Lucio Costa. The inset shows the unplanned, spontaneous settlements that have sprung up to house lower-income groups

industry. This strategy was underwritten by the Soviet Union's purchase of at least half of the sugar crop on terms highly favourable to Cuba, so it is unlikely that this model can be repeated elsewhere – assuming it continues to be viable in the aftermath of the collapse of communism in Eastern Europe. Indeed, in late 1995 the US was still maintaining its 35-year blockade of Cuba, and, as the economic crisis exacerbated by this stranglehold has mounted, the living standards of Cuba's citizens, both urban and rural, were plummeting.

The Chinese, too, have renounced urbanization and migration and have based their economic development on regionally decentralized systems of peasant cultivation and rural industrialization. Big cities like Shanghai, though they persist, were originally rejected by the regime as hangovers from capitalism. Millions of people living in urban slums and shanty towns (2 million in Shanghai alone) were rehoused in vacated middle-class homes and in government-inspired schemes, or encouraged to renovate the fabric of their dwellings in

BOX 7.4 *continued*

the constantly changing needs of the rapidly growing city. It was intended to be a city without class divisions, but the affluent preferred to live in single family dwellings, rather than the planned 'super-blocks', and the blocks were too expensive for the working class. In consequence, Costa's 'monumental city' is primarily middle class, the rich live in villas around the lakeside, and workers and the urban poor in unplanned and underserviced satellite cities which have grown up in the Federal District. Over 70 per cent of the population of 1.2 million now lives in such satellite communities, where squatting and lack of water, light and sewage disposal are as problematical as in the coastal cities of colonial foundation (Holston, 1989). However utopian in conception, Brasília has merely reproduced the inequalities of pre-Brasília Brazilian society.

Plate 7.11 Parliament buildings, Brasília The National Congress building forms the monumental core of Brazil's planned capital. The city was intended to shift the country's administration from peripheral Rio de Janeiro to a more central location in the interior. Its spectacular architecture has also become a symbol for the country

response to government provision of services and subsidized materials.

However, regional disparities as between a developed coast and underdeveloped interior persist, and urban growth has continued; Shanghai now has a population of 12 million compared to 5 million in 1949. But it is no longer a city in the traditional sense because one-third of the active population is engaged in agriculture, and the increased area covered by the city includes a considerable number of communes that produce food for local consumption. Nevertheless, even China has a long way to go to solve its urban problems, though they are dwarfed by the difficulty of marrying new-found capitalism with communism and convincing a population of over 1000 million of the need to reduce population increase.

CONCLUSION

Third World cities and towns confront problems that are more elementary than those

Plates 7.12 and 7.13 Urban growth, Belo Horizonte, Brazil These two photographs were taken in 1982 and 1993, respectively, from roughly the same spot. The lower one shows the continuing growth of mainly middle-class high-rise apartment blocks and of the shanty town over the decade. In the *favela*, consolidation of the site can be seen, with many of the earlier shacks of timber, tin, and wattle and daub now replaced by brick dwellings, and power lines supplying the settlement. These physical improvements reflect a reduction of the threat of clearance by the authorities, and a greater security of tenure

experienced by settlements in the economically advanced countries. Yet some of the biggest Third World cities have literally the worst of both worlds. Mexico City, for example, has the biggest informal housing sector in the world, yet with about 2.5 million cars on its roads suffers from appalling smog – factory pollution limits were set as recently as 1983 – and traffic jams. Many Third World cities combine pre-industrial and industrial morphologies and technologies, mansions and shanties, the richest and most impoverished people (Plates 7.12 and 7.13).

If urbanization has occurred without adequate industrial employment, it has also been achieved without 'breakdown'. As the cities have expanded, employment has been created by informal means, and self-help housing has been married to self-employment to create do-it-yourself urbanization. Governments are gradually recognizing that it is impossible to deploy sufficient funds to build public housing for the urban poor. Most countries have attempted to eradicate squatting at one time or another only to find their efforts met by violent community resistance. Eradication may be worthwhile if the number of squatters is small and alternative accommodation can be supplied. But to make squatters or rent-yard dwellers homeless by destroying their communities only shifts the problem elsewhere.

Nevertheless, governments have considerable power to influence the urban situation. Encouragement of the positive aspects of illegal land occupation by the active planning of site and service schemes could become part of an urban strategy for the Third World, provided the loss of indigent settlers from improving communities can be prevented. Cuba exemplifies the wide-ranging impact that a government can have on the urban system. In addition to shifting the focus of economic growth down the urban hierarchy, the communist government eradicated unemployment in the 1960s by expanding the bureaucracy, socializing the labour force (and expatriating dissidents). Street sellers were branded petty capitalists and banned. Cuba's achievements, reinforced in the mid-1980s by the anti-capitalist policy of 'the rectification of errors', are evaporating in the face of the dissolution of the East European communist world; and they are now unrepeatable in the larger, less sophisticated and more capitalistic countries of the Third World.

Marginality – the domination of the informal sector by the formal economy – is the hallmark of dependent urbanization, but its interpretation depends upon the model of development that is being used. Is it a temporary phenomenon, transitional to the absorption of surplus labour by the formal sector? Or is the informal sector a permanent feature of dependent development, the urban manifestation of the Third World's persistent poverty? Evidence suggests that marginality – informal employment and informal housing such as precarious squatting and renting – is growing not declining, that it is functionally related to dependent capitalism, and helps to sustain it by keeping wages low.

SUMMARY

Although the proportion of the national population living in cities in the Third World varies considerably, highest in Latin America and lowest in Africa and South Asia, in all countries there have been sharp increases in the levels of urbanization in recent decades, mostly as a result of migration from the countryside in search of jobs and the new ways of life that the cities seem to offer. Jobs are sought in the formal sector, but since the number of urban job-seekers greatly exceeds the formal employment available, there has been an explosion of informal sector activity, especially in services, which absorb the great mass of the marginalized urban poor. Rapid growth of the numbers of mainly very poor people has placed great strain on existing housing stock, and new forms of low-cost autoconstruction and community involvement in that provision have been characteristic of the struggle to establish sustainable urban livelihoods. As in the developed world, patterns of residential segregation and integration that are evident in Third World cities reflect the broader structures of and contradictions in society, and attempts by governments to manage the urban fabric have often heightened these tensions.

8

Internal Interaction

INTERACTION AND SPATIAL STRUCTURE

Although the rural and urban areas of Third World countries and the agricultural and industrial sectors of their economies have been examined in their own right in the preceding chapters, they are not independent of each other, but interact as part of a spatially and economically cohesive whole. Incomes from crop production stimulate rural demand for industrial and consumer products, which in turn have an effect on the urban economy by creating jobs and stimulating commerce, while industrial growth in the towns may stimulate the rural sector. Expansion of cotton textile production, for example, will mean not only more jobs in urban areas and increased rural–urban migration, but will stimulate an increase in cotton production by local farmers. Separately, town and country, industry and agriculture, comprise the essential building blocks of the human geography of any country. Together they are linked in the development process through a network of urban regions, arranged hierarchically, both internally and with respect to the international economy. Exchange or interaction between them is integral to the process of development or underdevelopment. In this chapter we consider why and how distinctive patterns of interaction have developed within Third World countries, and with what effect on the movement of goods, services and people.

The essential features of any spatial structure are *nodes*, *linkages* and *hierarchies*. Nodes are points of exchange and control, and emerge at points of accessibility through linkages (roads, telecommunications, etc.) with their hinterlands. These nodes are not only related to each other through other linkages, but also tend to be arranged in hierarchies. The levels of the hierarchies are themselves dependent on linkages with levels above and below. The most obvious type of interaction occurs through the linkages of the transportation system, by road, rail, air, river, lake and sea. In the case of transport, the *linkage* is the road or railway (or the 'invisible' sea route), and the *nodes* are the railway stations, ports and airports; the *hierarchy* may be the range from rural path to dirt road to superhighway, or from a simple mode, such as the mule track to the railhead to the main port. Other networks of movement are not related to transport in a formal sense, but to modes of communication such as electricity transmission grids or telecommunications networks. Even less formally they comprise person-to-person or village-to-village communication of information and ideas. Interaction between places at a similar level of the urban hierarchy involves mainly short-distance, local exchange between local markets. Long-distance, inter-regional and international exchanges are mostly between different levels of the urban hierarchy, either moving upwards through it from rural areas to the capital, as in the case of export crops, or moving downwards, as in the case of inputs into agriculture, such as fertilizers or pesticides (manufactured overseas or in the capital city).

The mere existence of linkages and inter-

◀ Women working on a building site, India

action in the Third World offers nothing that is unexpected or distinctive, for interactions of the kind identified above are global. Certainly, in the Third World there are fewer roads or railways per head or per unit of area than in more developed countries; there are fewer goods to be moved when subsistence agriculture remains important; low income levels generate a low overall demand for manufactured goods; newspapers are less widely read where levels of literacy are low. The limited development of interaction is both a cause and an effect of low levels of economic activity and technology, and in all Third World countries the expansion and intensification of existing networks and linkages has been a central feature of development efforts at both the national and local scales. An expanding system of transport and communications has been viewed both as a necessary condition for, and as a symptom of, economic advance and national integration.

However it is the nature of the interaction, its growth and its effect on the overall regional pattern of economic activity that merits particular attention in this chapter. Patterns of interaction reflect the wider context of rural/urban and inter-regional relationships. Their distinctiveness in the Third World is part of the wider process of integration of the country as a whole, and of each part of it into the national and international systems of unequal exchange. The essence of the patterns of interaction in Third World countries is captured by Figure 8.1. The national space economy is subdivided into the rural and urban sectors, and each of these is further subdivided into a formal and informal component, as discussed for the rural economy in Chapter 4 and for the urban economy in Chapter 7. Both these sectors, separately and together, have links through trade and other exchanges with the international economy (as previously elaborated in Chapter 2 and also later in Chapter 10). There are also the linkages within each of the main sectors between the formal and informal components. However, crucially in the context of this chapter, urban and rural sectors interact with each other through exchanges of goods and services, of capital and ideas, and of people. These exchanges are in both directions – rural–urban and urban–rural – and also upwards and downwards through the urban hierarchy. The

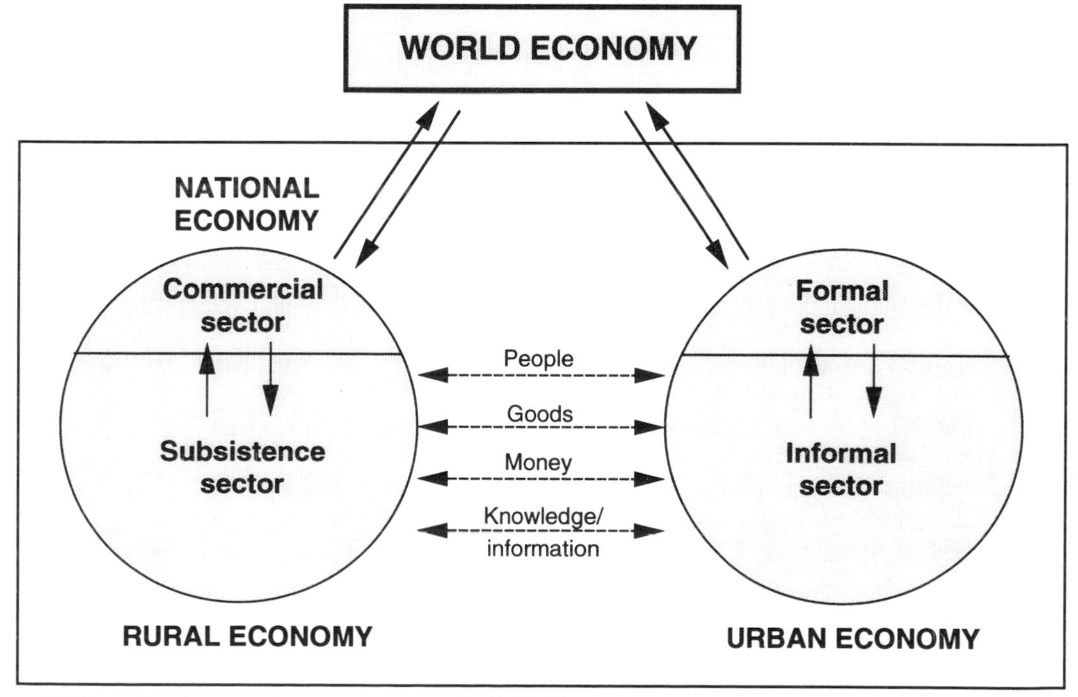

Figure 8.1 Spatial and sectoral interactions in a Third World economy
Source: After Taylor, 1981

exchanges are often unequal, having a differential effect on the areas involved – some areas benefit more than others, and some areas may even be adversely affected by the exchanges. The following sections of this chapter elaborate some of the major forms of interaction and exchange, and how these affect regional and rural/urban inequalities in economic and social development.

THE GROWTH OF NATIONAL TRANSPORT NETWORKS

In many pre-European subsistence societies movement was short distance and on foot, with limited nodes of exchange. There were some longer distance movements: in the savanna belt of West Africa, for example, horses and camels were used to carry trade across the Sahara Desert to the Mediterranean. Market places were everywhere the nodes for local exchange, and these were linked to wider trading hierarchies and networks carrying long-distance trade, particularly in essential goods such as salt or iron, or in luxuries such as gold, spices or cloth. Some pre-colonial states developed what were essentially trading empires with elaborate networks of routes over land (for example, Songhai in interior West Africa, in what is now Mali, which was at its height in the fourteenth century AD) and sea (Sri Vijaya, for example, trading from the north-east Indian Ocean east to the China Sea in the sixth to the ninth centuries AD).

Most of these ancient route networks were undermined by the new trading patterns associated with European expansion. Penetration was mostly from the sea, and the colonial presence was first felt at seaports. Established ports and traditional routes inland by foot and animal transport were used initially, but once colonial control was established and new types of economic activity began to be introduced, the new technologies of rail (in the later nineteenth century), then road (in the twentieth century) transport developed. A network and hierarchy of settlement then developed round them.

The evolution of this superimposed network was classically described by geographers in the 'ideal-typical' sequence model of Taaffe *et al.* (1963). This model considered the growth of transport routes in a colonial setting, from their origins in the ports of entry through their expansion into a national route network. In the earliest period transport is mainly in the coastal area with very limited communication inland. New routes are built inland, but only from one or two of the original ports and, as a result, these ports become larger at the expense of those with no new links to the interior, for the growing trade is concentrated at a few transshipment points. The links inland are to areas of economic or strategic importance, whether preexistent or of potential value for mineral, cash-crop or plantation development. These areas of early attraction then develop as major nodes for further expansion of transport routes, and as the basis for the newly emerging urban hierarchy. A more intensive pattern of feeder routes is then built, especially into areas of emerging export potential, eventually linking separate coast/interior routes into an integrated national network with increasing differentiation of settlement size and traffic volume to emphasize the main channels of import and export. There is a diffusion of development downwards through the urban hierarchy and outwards from urban centres to rural areas.

This diffusionist, ideal-typical sequence has been an important model in the geographical study of the Third World. It is based on the assumptions of those, like Rostow, who see development inevitably proceeding through stages of increasing incorporation into the world economy, which proceeds through the extension of transportation networks until the whole, previously undeveloped, area becomes part of the national and international system. It is seen as a process of 'modernization', with its assumptions of diffusion of the benefits of the new economic order into the backward areas through the transportation systems.

The sequence was seen to be 'ideal' in the sense that it was consistent with the history and goals of colonial development. It assumed that the networks were created primarily for the purpose of external trade, to facilitate the establishment, expansion and collection of exports of agricultural and mineral primary produce, and to facilitate the import and distribution of manufactures, as well as for strategic and military control. It therefore assumes that the area being affected has little or no economic structure of its own on which the new economic and political order may be successfully built, but that the new modernized order can start afresh without a substantial legacy of the past. The sequence was 'typical' in the sense that its essential structure could be traced historically in most countries of the Third World. Taaffe *et al.* used Ghana and Nigeria to develop and exemplify the sequence, but it can be identified in almost every country of the Third World. In Africa, in particular, the radial pattern of railways emphasizes their close links with seaports and, by implication, with external trade (Box 8.1).

Over the last half-century rail has been giving way to road transport for long-distance interaction and export routes in most countries. The railways that were built remain important, but major highways have been constructed to run parallel with them. These have tended to confirm the networks and hierarchies established in the earlier period – though they may considerably extend them into previously unserved areas, notably in the development of Amazonia in Brazil. As with railways, highways provide fast inter-regional links that are well-suited to the movement of major products. Despite efforts to restructure the inherited colonial transportation network and associated urban hierarchy, as in the development of new non-coastal capitals such as Brasília to replace Rio de Janeiro in Brazil, Islamabad to replace Karachi in Pakistan and Dodoma to replace Dar-es-Salaam in Tanzania, national transportation networks continue to reflect the dependent exchange and productive relationships established at an earlier period. Despite these politically symbolic, and in many respects successful gestures, the former capitals remain the chief ports and commercial centres of their countries.

BOX 8.1 RAILWAYS AND DEVELOPMENT IN EAST AFRICA

In colonial East Africa, railways were built inland from four ports, Mombasa in Kenya (1896), and Tanga (1893), Dar-es-Salaam (1905) and Lindi (1949) in Tanzania, each to link the coast with specific well-developed and economically attractive inland destinations (Figure 8.2). Branch lines were built to serve areas of export production in Kenya or, as in western Uganda, to serve the copper deposits of the Ruwenzori Mountains. Lines were not, as a rule, built into areas of African settlement and subsistence farming unless there was a potential cash crop to generate traffic (such as the northern Uganda extension in the 1960s associated with the expansion of cotton production). Eventually the separate rail lines were linked by transverse routes (Mombasa–Moshi–Dar-es-Salaam), but these have never achieved the importance of the original axes. In the 1960s, at the time of independence for Tanzania, Uganda and Kenya, there was great optimism about the emergence of an East African Community, building on the common services created by the colonial powers to create a unified common market with free movement of people and goods and services between the countries. In these circumstances the benefits of linking the separated rail systems was very apparent, but the impetus collapsed with the collapse of the common market objectives in the face of political and economic rivalry between the partners.

Even in the post-independence period the role of railways for external rather than inter-regional trade was confirmed by the building of the Tan-Zam railway from Dar-es-Salaam to land-locked Zambia, to permit the export of copper, Zambia's main economic support, and the importation of oil and other essentials for the mining economy. The link was of great importance following the closure in 1965 of Zambia's traditional routes through Rhodesia to Beira in Mozambique and to South Africa during a period of political disruption. With the independence of Rhodesia (as Zimbabwe) in 1980, Zambia's shorter and more reliable rail routes to the south were restored and the strategic rationale for the Tan-Zam railway faded. The railroad continues to have a relatively limited local impact on south-western Tanzania.

BOX 8.1 *continued*

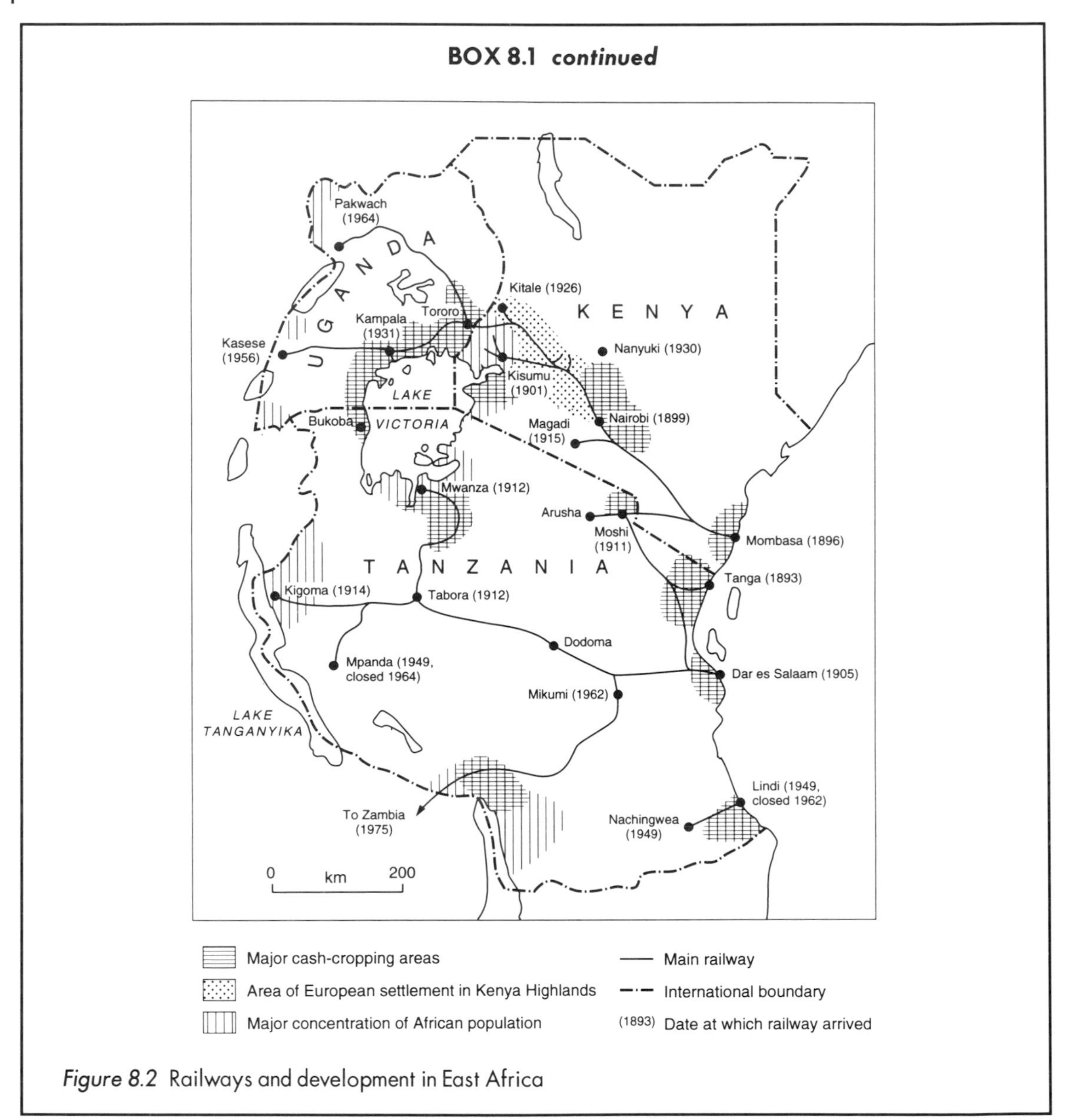

Figure 8.2 Railways and development in East Africa

Transportation systems remain far from 'ideal', operating with the effect of creating and exacerbating internal and external inequalities (Browett, 1980).

LOCAL TRANSPORT

Alongside fundamental changes in the scale of inter-regional transportation linkages, there has been considerable continuity in local patterns of transport. Exchange in the subsistence or non-commercial sector remains dependent on movement by foot, by boat, by animal or bicycle, and even where lorries and vans have been introduced, spatial patterns continue to follow long-established links at the lowest level of the urban hierarchy (Plates 8.1 and 8.2). Though there is often a functional relationship between local and national interactions, they are not necessarily dependent on each other.

The quality of local roads can vary from a bush track, suitable only for foot traffic, to all-weather roads for heavy vehicles. However,

Plate 8.1 Human porterage, Nepal In mountainous terrain where road transport is impossible, many goods are still transported as headloads, as here, by men and women

Plate 8.2 Loading a bus, North-West Frontier Province, Pakistan Passengers and goods are loaded into and on top of an ornately painted bus

Plate 8.3 Road building, Goa, India Much road construction is still carried out by manual labour. In this case there is a division of labour, in which the men do the digging and the women carry away the excavated material

Plate 8.4 Road transport, Northern Pakistan Even a modern jeep and a made road are still vulnerable means of transport. In mountain areas intermittent floods and serious erosion create problems for road building and maintenance. The jeep is inching its way over a much weakened bridge, temporarily repaired after storm damage

many areas of the Third World are well beyond reach of a motorable track, even seasonally. Not only is there a lack of funds for road building, but also in many areas of mountainous or swampy terrain, or of highly seasonal rainfall, the physical difficulties of building and maintaining roads are enormous. Alternatives may be available, notably rivers, but for many communities lack of access to adequate transport imposes severe limits on the possibility of producing goods for other than a very local market. Physical and economic isolation are thus closely related. Efforts are made, sometimes by government but often by the isolated communities themselves using communal labour, to build links to allow access to larger commercial markets, with acceptable and competitive prices, without the risk of goods' excessive deterioration (in the case of fresh fruit and vegetables) or complete loss (Plates 8.3 and 8.4). The opening up of sparsely settled but potentially productive areas has been hampered by the absence of local 'feeder' roads from the inter-regional highways, but major programmes have been launched to improve accessibility to and from such areas. The introduction of new cash crops has often been associated with the development of a network of feeder roads to give farmers improved access to collecting or processing centres (Box 8.2).

Improvements in local mobility may also bring disadvantages. Just as goods and people can more easily move in or out to benefit rural communities, they can also do so to their disadvantage. Many goods produced more cheaply elsewhere, even in another country, can be more competitive in the local market, and can undermine the traditional local production of such groups as potters, weavers, tinsmiths or ironworkers with the availability of plastic containers and factory-produced textiles. Imported foodstuffs may equally undermine the traditional balance of agricultural activity. Roads may also facilitate considerable out-migration, to the overall detriment of the local community. Although the development of local roads has undoubtedly brought some unwelcome effects, the desire both within and beyond local

BOX 8.2 ROAD LINKS IN EAST AFRICA

In Kenya new roads have extended the transportation network into African farming areas neglected in the colonial period, and were associated with a strategy for rural development that sought to expand peasant production of the major cash crops, tea and coffee, and to extend the market economy into these remoter areas. In addition out-grower schemes for sugar-cane in western Kenya, as part of government measures to restructure the industry away from a plantation base to a peasant base, have involved considerable investment in roads within 13 km of central sugar mills, so that farmers, for the first time, have effective access to a processing plant and through it to the national market, and to advice and other benefits such as credit facilities. Using the roads network to extend the national market into rural areas has been a recurring theme of successive government development plans and policy statements in Kenya, consistent with its belief in the assumptions of a diffusionist, trickle-down development model.

In socialist Tanzania expanding the road network was also important, but for rather different reasons. Tanzania is a large country with a few foci of population and development, widely separated by large expanses of areas of low economic potential. Linking these foci is expensive but necessary for creating and maintaining a national economic system and ensuring the urban food supply. Also, at the local level roads have played an important part in economic and social restructuring. Each of the *ujamaa* villages created in the early 1970s was sited in order to have access to a road to facilitate its integration into the national economy (Box 8.4). This gives access to the market for any production surplus, but also allows medicines, school textbooks and technological innovation to be available in the village, and encourages teachers, health workers and other professionals to work in the villages without experiencing or anticipating isolation from the national mainstream. While in Kenya roads have had an economic function primarily, in Tanzania they have also been seen to be part of the wider processes of social and political transformation to a socialist state.

communities to improve local transport suggests that the advantages considerably outweigh the disadvantages as the local community becomes more fully integrated into the national system.

MOVEMENT OF GOODS

As levels of economic activity and income rise, an increasing proportion of agricultural produce leaves the farm in exchange for a rising volume of non-agricultural products. This exchange is at the heart of the development sequence implied by the diffusionist modernization model. At a national level exchange focuses on the port as the final collection point for exports and the initial entry point of imports. Rural products are brought to the chief port through the urban hierarchy via a series of collection centres and, where appropriate, processing points; imports are distributed from it by that same hierarchy. The main, sometimes the only, economic activity in many smaller centres is collection and distribution, as would be the case in any market town in any country of the world. In most countries commodity flows are largely a matter of internal rather than external movement of produce, and the larger and more economically complex the country, the more important internal trade must be. In large countries such as India, China or Nigeria, there are major flows of inter-regional trade that developed because of ecological differences between one part of the country and another, or as a result of the location of manufacturing industries in some areas but not in others.

The main flows of produce from rural to urban areas supply their food markets. With rapid population growth and rising incomes, the urban demand for food has risen rapidly. Some of that demand is associated with changing taste preferences, in particular the substitution of wheat-flour products for maize-flour products, and has been met by food imports, of which cheap North American wheat is the most important. In some countries, such as Ghana and Nigeria, the level of import of basic commodities like wheat and rice has reached economically damaging levels, even though there is often falling local food production and therefore the capacity for its expansion. Internally, the sources of the urban food supply follow a familiar pattern anticipated by Von Thunen's model of intensity of land use, with more produce coming from the area immediately surrounding a focal settlement and declining in proportion with distance. Larger cities tend to have larger food hinterlands and capital cities are supplied from most parts of the country. In return, some goods manufactured in the towns are taken out by the producers themselves or by middlemen to provide the small but expanding range of purchases from rural incomes, principal of which are kerosene for fuel, some food products, plastic utensils and clothing. The collection and distribution networks tend to be organized by small entrepreneurs each with a lorry or truck and operating as a general haulier.

Most local trade continues to be channelled through rural marketing systems, which can be very elaborate, and based on movement of traders between markets in regular cycles of varying length. The major Sunday market near a church depending on the sales and purchases of the congregation is a widespread feature in Latin America, and in West Africa there are particularly complex indigenous structures of local and long-distance marketing. Almost everywhere, however, there are daily markets with small hinterlands exchanging a small range of products. These are the most common, but least noticed, means of exchange of goods.

PROVISION OF ECONOMIC AND SOCIAL INFRASTRUCTURE

The transportation system is also the means through which government and other agencies can provide many of the services that are thought to be necessary for the running of any modern state. These services are both economic, including power supply, agricultural advice and telecommunications (telegram, telephone, fax – particularly important in China with its written symbols); and social, including health care, water supply, education and media services (radio, television, newspapers). Ideally, each of these services is made available as widely as possible to the population, but in the conditions of scarcity of financial resources and qualified personnel that are all too familiar in the Third World, the services are normally not as widely available as governments and the population might wish. The distribution of service points such as schools, clinics and cattle dips, is highly uneven, and closely follows the spatial priorities implied by the urban hierarchy.

In many countries the most widely available directly economic service provided by government is the supply of information and advice to farmers. The nature and quality of the service will vary according to the historical, economic and cultural needs of any country, but most services are centrally controlled and hierarchically organized through a central agency that extends outwards and downwards through the urban system to provide a chain of communication between the centre and the farmer. In practice the benefits of such a service are felt disproportionately in areas near towns or in areas of already higher productivity, and by farmers who are already better off, by male farmers more than women farmers in the same area, and its impact is strongly related to the farmer's access to the transportation network.

Ideally, provision points would be well distributed in the area to be served and arranged hierarchically according to central place theory, with a regular hexagonal spatial pattern of provision at the lowest level, and building up to a regular pattern of provision at each of the higher levels. Idealized central place theory has been the basis of an Indian experiment to introduce services appropriate to the rank size of particular growth centres. The success of such mechanistic attempts to create an effective but artificial hierarchy has been limited, even in former socialist countries, and is likely to have any impact only where population densities are high enough to sustain them, and where traditional marketing systems have implied their potential, as in China. Plans to introduce growth centres to rural Zambia for example, where population densities rarely exceed five or six persons per square mile, did not succeed, despite more than 40 years of colonial and postcolonial planning.

The supply of electricity is one of the most directly important of the higher economic services that can be provided, whether by government or by private generating companies. Power is distributed through its own communications network: the electricity grid. The principal flows in the grid are to the capital primate cities, which have the heaviest demands, and subsequently to the smaller consuming centres, such that the pattern of the grid and of flows within it closely follow the national urban and industrial hierarchy. In Latin America, for example, much of the earliest capacity was close to the capital cities, but supply has subsequently extended quite considerably into some rural areas, particularly to the richer agricultural districts surrounding the major urban centres (see the example of Minas Gerais, Brazil, in Chapter 6). In order to provide electricity many Third World countries have exploited major hydroelectric dam sites, with large generating capacity deriving from favourable physical conditions: for example, Cabora Bassa in Mozambique, Akosombo in Ghana, Aswan in Egypt, the Three Gorges project on the Yangtse in China. However, very often the need to generate power at such sites has had serious and controversial environmental consequences (Box 8.3).

Social services, in particular health care and education, existed in traditional forms in all societies in the Third World. Now, they are provided in Westernized institutions, in medical facilities and schools, each following the models provided by richer countries, and introduced in a fairly unsystematic way, often by nongovernmental agencies, notably Christian missions in Latin America and Africa. Here too there is a hierarchical structure of provision: medical facilities range from the large hospital with sophisticated medical technology to the small, rural one-room dispensary; for education the range is from university to the rural primary school. Inevitably, these facilities have been unevenly distributed with a very strong bias towards urban rather than rural provision, and to the richer districts rather than to the poorer ones. The provision of services further confirms the overall pattern of internal disparities within each country of the Third World – high levels of health care and school enrolments are more

BOX 8.3 CONTRADICTIONS IN PROVIDING ELECTRIC POWER

The provision of electricity is an integral element in the development strategies of Third World countries, providing power for new industry, and illumination and heating for growing cities. For many such countries, abundant hydroelectric potential provides a major, and renewable, source of energy. In recent years the construction of large dams, such as Kariba, Aswan and Itaipú to generate massive amounts of power, supplying major cities and large tracts of countries, has been a striking feature of the development process. Indeed, such massive projects are seen as symbols of 'progress', and multi-purpose dams, providing electricity, flood control, nagivation and irrigation water have been significant contributors to regional development.

However, they do serve to generate conflicts at a number of levels, particularly the impact of

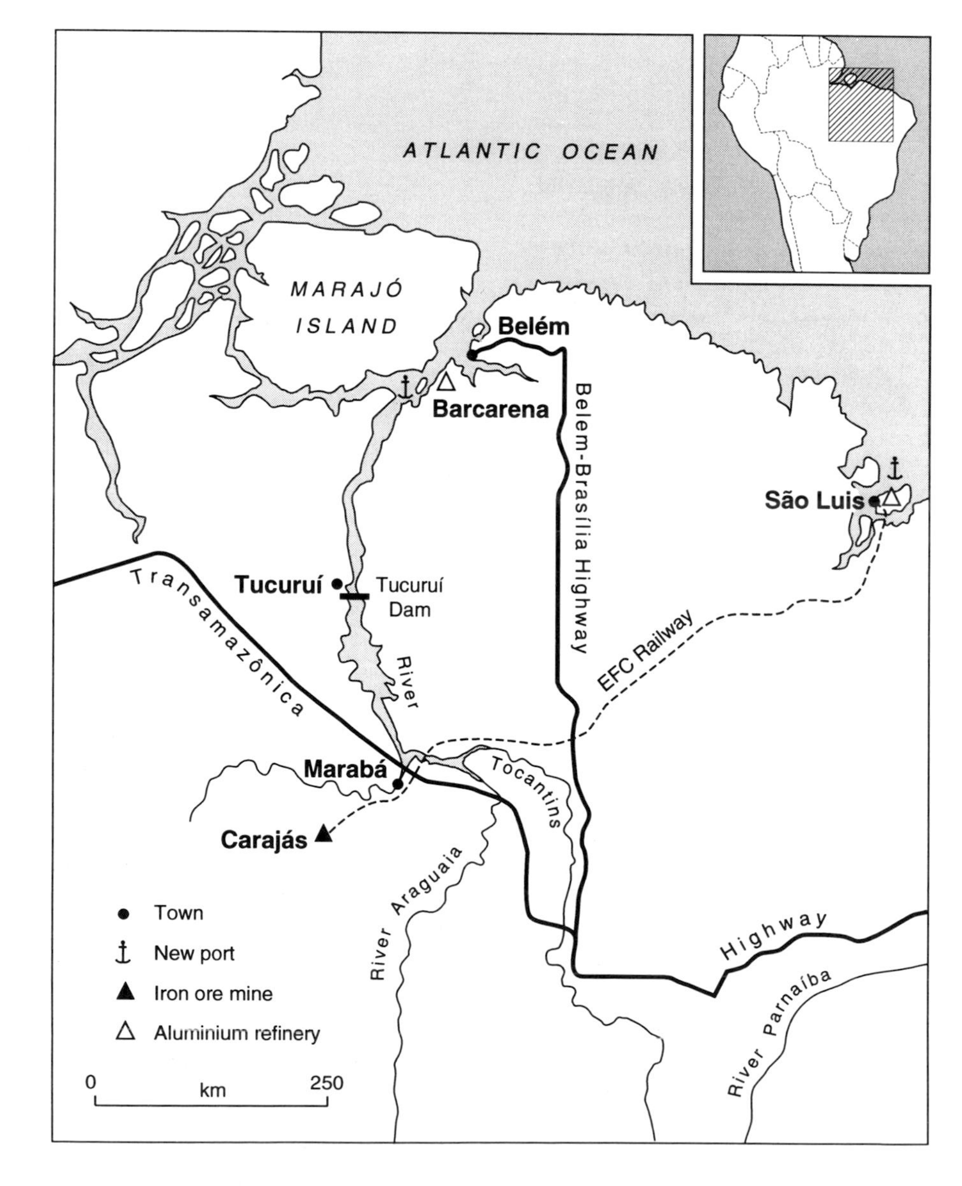

Figure 8.3 The Tucuruí scheme

BOX 8.3 ***continued***

their flooded areas on the environment, native peoples, farmers and small settlements. The Tucuruí dam, on the Tocantins River in Brazilian Amazonia, was developed to provide cheap, bulk electricity for factories producing aluminium from the region's bauxite (Figure 8.3). The companies were joint ventures between Japanese, American and state-owned Brazilian firms. They could be seen therefore as attracting foreign capital, to utilize local resources, and produce a manufactured good for the domestic market and export. The plant, the largest to be built in the rainforest, and the fourth largest in the world, was to have a capacity of 3900 MW, rising to a possible total of 7900 MW, as part of a larger project along the Tocantins–Araguaia which would involve 27 dams.

Critics argued that Tucuruí produces power that goes to plants away from the dam, without supplying the local population, and that it provides power to foreign companies at much lower cost than it is available to residential consumers. The dam also flooded some 2400 square km and affected six tribal groups, and possibly 30,000 rural dwellers. As well as flooding farm land, the reservoir also disrupted the complex subsistence economy involving hunting, fishing, and the gathering of firewood and wild fruits. There was also concern for the creation of water-borne health risks such as schistosomiasis. Conversely the project suffered from the growth of aquatic weeds, corrosion of turbines because of acidic water, and sedimentation due to rapid run-off following forest clearance.

In the initial conception Tucuruí was one of over 60 dams which would tap the enormous hydroelectric potential of Amazonia, estimated to be close to 100,000 MW. Schemes in southern Amazonia alone were projected to flood 81,000 square km and displace over 150,000 people. In environmental and human terms, this implies considerable cost. On the other hand, such schemes provide electric power for national and regional development. More broadly, the World Bank has estimated that the Third World's hydroelectric capacity, where less than 10 per cent of potential had been tapped in Asia, Africa or Latin America in 1980, would increase by 100,000 MW in the 1980s and the 1990s; most of this would come from large dams.

likely to be found in the urban and most economically productive districts. The areas with low levels of provision tend to be those further from the core area of the state or of advanced economic activity, or have populations that are politically, economically or socially apart from the rest of the state (Gould, 1993, Chapter 3).

The similarities in the pattern and organization of provision of economic and social services suggest that they face similar problems in their attempts to achieve the goal of equality of access. The cost of providing places for all children, i.e. universal access, is high and resources are limited, so priorities in allocation need to be ordered. Since economic criteria generally dictate such ordering, services have been installed disproportionately in those areas where direct cost of provision is lowest: in urban and densely populated rural areas, where costs of transport are lower; and where large facilities with considerable economies of scale can be provided. Although these are areas of highest total and per capita demand, because they are the richest areas, they are not necessarily the areas of greatest need.

Recent policies in many countries of the Third World, motivated by political criteria of equality of opportunity and access, have sought to spread services and economic infrastructure more evenly to permit better access to services in the remoter areas. Additional facilities have been allocated on the basis of criteria other than economic demand, often to achieve a more overtly equitable distribution. However, the economic basis of service provision has not been and cannot be ignored; so policies must seek to restructure the type of provision, to promote basic low-cost facilities rather than high-cost and technically sophisticated ones. Major examples of this trend are the shift to the development of rural health posts in Tanzania, and polyclinics in Cuba, rather than urban-based hospitals – part of overall national polices

This Christian Aid cartoon points to the problems of distance and access to health care faced by rural dwellers

promoting rural self-sufficiency. Policies stressing primary health care have been strongly promoted by the World Health Organization throughout the Third World as part of its strategy for achieving 'Health Care for All by 2000'. Such an approach seeks to further reduce rural/urban disparities and to provide access to basic health care for more people, as a priority over raising the quality of care for those already within reasonable access to a hospital.

INTERNAL MIGRATION

The range of population movements in the Third World was discussed in Chapter 3, emphasizing the importance of distinguishing between essentially short-term and repetitive movements (circulation) and permanent changes of residence (migrations) that result in changes in the overall distribution of population in the long term. While circulation occurs in all countries and is not necessarily directly related to the wider patterns of interaction and exchange that have been discussed above, migration has a much more obviously structured pattern which resembles that of the movement of goods and services at both local and national levels.

The principal, but by no means only, motivation for migration is the expectation of economic benefit. Where there is rapid economic change with the introduction of new crops, industries and technologies for rural and urban production, it is likely that the geographical distribution of economic opportunities will change. Some areas decline and others expand. Changing population distribution is an expected and normal response to economic change and itself contributes to that change. People will be attracted to newly developed or expanding areas; conversely, where areas are stagnating or declining, this will result in out-migration, either spontaneously or under environmental or economic pressure. The 'pull' of some areas and the 'push' from others creates an overall pattern of migration. The 'push' is generally from poor rural areas, and the 'pull' is to urban and rural areas with expanding land or job opportunities (Plate 8.5). The scale and intensity of the movement will normally be influenced by the size of the differences in income and opportunity

Plate 8.5 Migrants, China A migrant arrives in the Chinese city of Kunming, Yunnan Province, with his family and possessions piled on a bicycle trailer

between the source and destination areas – the larger the differential, the greater the volume of movement.

The principal migrations are rural/rural and rural/urban. In the rural sector they mostly comprise migrations from overcrowded areas to newly developed land, mostly spontaneously but also as part of planned rural development. Such migrations have been widely encouraged to spread population and development into relatively empty areas, relieving pressure on land at source and raising the productivity of destination areas and the incomes of the settlers in them. Among the largest movements of this type are the so-called 'transmigrations' that have been a major feature of Indonesian development strategies since the mid-1950s, involving some 6 million people (see Box 3.3). Similar inter-regional migrations, but with much smaller numbers, have taken place, for example, in peninsular Malaysia to promote the development of the eastern side of the peninsula with in-migrants from the relatively overpopulated western side; and, on a much larger scale, in Brazil to develop parts of the Amazon Basin with migrants from areas of severe poverty and environmental difficulty in north-east Brazil. Other large-scale, long-distance rural/rural migrations have been associated with irrigation projects, bringing water and therefore greatly increased economic opportunity to previously dry areas. These include the Gezira Scheme in Sudan, developed in stages since the 1920s with water from the Blue Nile, and the Lower Indus Project in Pakistan. In other cases resettlement has been politically motivated, as in Kenya in the 1960s and early 1970s, and in Zimbabwe since independence in 1980 with resettlement of African farmers on former European-owned land, sometimes merely creating a change in ownership, but with many families involved in a change of land use from an extensive European-style farming system to a more intensive African style (Box 8.4). Politically necessary population redistribution of African farmers in resettlement schemes on some former White-owned land is

BOX 8.4 VILLAGIZATION IN TANZANIA

More localized rural migrations have been part of planned settlement restructuring patterns, as in the villagization programme in Tanzania. This was designed to concentrate the population in nucleated *ujamaa* villages, each with 250–1500 families, in a country where the traditional settlement pattern is that of dispersed homesteads and low overall population densities. The villagization policy was implemented in the mid-1970s as an essential component of a rural development strategy to improve agricultural productivity through increased communal effort, an overtly socialist objective, and, as indicated previously in the context of rural health facilities, to facilitate the delivery of economic and social services, for each village is ideally on or near a road and is large enough to justify a primary school, a health post and a clean water supply. While progress has been made in meeting the social provision objectives, the production objectives have seriously faltered, partly as a result of peasant farmers' indifference to becoming involved in communal production and partly because of the serious environmental degradation.

Environmental problems are also increasingly evident round many nucleated settlements. Intensification in fragile environmental settings has brought over-use of land and pasture, with widespread soil erosion and declining yields in many villages. Away from the villages there has been abandonment of some plots and a resurgence of bush, and associated with that has been a resurgence of some pests and diseases. The traditional dispersed settlement pattern was integral to a farming strategy that minimized environmental hazard, but villagization has made the farming system more vulnerable to environmental risk.

also likely to be a feature in South Africa from the mid-1990s.

Rural–urban migration is more obviously significant in the changing production structure of any country, even though numerically it may be less than migration between rural areas. Massive differences in income between rural and urban areas generate temporary or permanent migrations to towns, even where there is overt urban unemployment, for a precarious living in the urban informal sector may often be preferable to low incomes and few opportunities for improvement in rural areas (see Chapter 7). The strength of the urban hierarchy as a channel for migration paths is very marked, and flows are much more markedly in one direction – upwards to larger centres – than is the case of the movement of goods and services. There is some reverse movement of individuals at all levels, and circulation between town and country remains important, especially in tropical Africa and throughout the Pacific islands. But the overwhelming net migration trend is city-ward, and in particular to the larger cities (Parnwell, 1993).

INTERNAL INTERACTION AND REGIONAL DEVELOPMENT

These broad overviews of the nature, forms and impacts of internal interaction are now brought together in the elaboration of two extended examples from very different contexts in the Third World. One, northern Pakistan, is set at the regional scale, while the other considers Kenya as a whole. The Pakistan case begins by considering a road, but quickly broadens to its impact on economic development, population movement and social change, and the interaction of the region with the rest of Pakistan; the other begins by considering patterns of migration, but links this with larger issues of regional exchanges and widening inter-regional inequalities. What they have in common, however, is an elaboration of how the distinctive geography of Third World countries is moulded by the high level and wide range of exchanges of people, goods and services between regions.

The Karakoram Highway, northern Pakistan

The Karakoram Highway is a major engineering feat in extremely difficult terrain, with steep slopes and constant seismic activity, through the Karakoram mountains in northern Pakistan to link the country with China (Figure 8.4). It was built in the 1970s primarily as a strategic link between the two countries then in recurring disputes with India, but that objective has also had local consequences on the remote, socially conservative and severely impoverished mountain communities through which it passed. It opened up these areas to the commercial markets in the densely populated plains of Punjab Province, with their substantial demand for fruit and vegetable products, such as apples, apricots and potatoes, previously grown mainly for subsistence in the mountains. The harnessing of commercial opportunities associated with the highway became a focus for development initiatives that sought to take advantage of the new trading opportunities to raise levels of income and overall standards of living of local farmers and their families.

The principal developments were begun in 1982 by the Aga Khan Rural Support Programme (AKRSP), an NGO financed primarily by the Aga Khan Foundation, the Aga Khan

Figure 8.4 The Karakoram Highway

Plate 8.6 Aid project billboard, Northern Pakistan The Aga Khan Rural Support Programme is an indigenous NGO, with a focus on community projects in rural development and environmental management

being the spiritual leader of the majority Ismaili Moslem population of the area (Plate 8.6). The primary objective of AKRSP is to raise the standards of well-being and incomes of the population of the Northern Areas by establishing locally managed Village Organizations through which loans would be channelled for productive physical infrastructure projects (PPIs) that would allow local communities to produce more for the new markets now available to the south. Production had in the past been severely limited by environmental constraints, notably lack of water in a very low rainfall area, and very fragile and immature soils developed on steep and unstable slopes. However, the glaciers of the area are a potential store of water which could be channelled down to settlements in the valley bottoms. Where a water supply could be guaranteed, then production for the market could rapidly expand. Most PPI loans have been for irrigation canals, with feeder roads and soil conservation schemes also prominent. The Village Organizations operate as marketing cooperatives, with local training for some members in marketing and technical skills (e.g. fertilizer use, pest management, fruit preserving), and also in sustainable environmental management, critical for such an environmentally difficult area to ensure that increases in production can be sustainable. Incomes have risen as more agricultural goods can now be produced for and sold in the national market.

With the highway people also have much greater opportunity to move out. Though this was a poor area in the past, its economic and physical isolation meant that it was not greatly involved in out-migration to the rest of Pakistan. Now the job opportunities in the large towns and major commercial farming areas are only an overnight lorry journey away, and the potential for out-migration is substantial. However, since the migration opportunities have become available at the same time as rising rural incomes from increased marketed produce, the incentive to remain in Northern Areas has been considerable. There has been no short-term evidence of increased permanent out-migration, but short-term out-migration has increased, especially of young people in search of opportunities for higher education available elsewhere. The enhanced opportunities for trained and educated personnel in the local expanding market economy have persuaded the majority of them to return, and those who have earned money elsewhere to invest in the rural areas.

The road has facilitated the out-movement of goods and people, but has also had the effect of bringing intensive impacts from outside into this once isolated area. The general effects of modernization and the market economy will inevitably bring social change, but the organization of AKRSP, and in particular its approach of strengthening the 'grassroots' through the Village Organizations, should allow the communities to be more aware of and to take advantage of potential benefits rather than to suffer as a result of potential problems. More schools and health facilities are being built, and the quality of the human resource base is being substantially raised – people are better educated and healthier. Many of the economic and social initiatives are targeted directly at women and girls in an area where they have been systematically excluded from non-domestic activity. The coming of international tourists as trekkers in what is an incredibly beautiful and 'unspoiled' landscape is one new potential problem, as it has been elsewhere in the Third World, but it also offers a further opportunity to raise incomes that needs to be carefully managed.

Migration in Kenya

The distribution of population in Kenya is very uneven. The majority of the population lives in the south-western quarter of the country and along the coastal strip (Figure 8.5a). The bulk of the country, in the north and in the east, is dry savanna or semi-desert and occupied mainly by pastoralist groups at low population densities. The south-western quarter is higher land, mostly above 1000 metres, well watered and with relatively rich soils for the most part, such that it can support large rural populations. However, there is a sharp contrast in rural population densities within these highlands between the traditional areas of African farming, especially in Central, Nyanza and Western Provinces, with very high population densities, and the area of former European settlement, mostly in what is now Rift Valley Province, where densities are rather lower and farming systems still reflect the European extensive farming model, even though most farms are now owned by individual Kenyans or Kenyan companies. With the exception of Mombasa, the chief port, most of the major towns, including Nairobi the capital, are in the highlands.

The most comprehensive data on migration in Kenya have been obtained from the 'place of birth' data in national censuses, the most recently available of which is for 1979. While 'place of birth' is far from ideal as a question on migration in a country where many people move frequently and where circular migration is particularly common, it does allow the recognition of overall patterns of movement. Figure 8.5a identifies net migration (the balance between in- and out-migration) for each of the 41 districts of the country, and Figures 8.5b and 8.5c identify the rates of out-movement and in-movement respectively. However, since most of the population lives in the south-western quarter of the country, these rates need to be set against the actual size of the flows. Figure 8.5d identifies inter-provincial flows of over 25,000. The densely populated districts of the highland provinces dominate the areas of out-movement. Nairobi, Rift Valley and Eastern Provinces are the main areas of in-movement.

Most of the large flows are rural–rural between neighbouring provinces. In particular there have been flows into rural areas of Rift Valley Province from the more densely populated Central, Nyanza and Western Provinces. Much of this movement was associated with resettlement in the 1960s and 1970s on former European-owned land in the Million Acre Resettlement Scheme. Most settlers were allocated land in high-density schemes, and there they established peasant, small farm systems. Most land in the former European areas, however, remains at low densities with extensive commercial farming or ranching, and with African labourers rather than peasant farmers. The resettlement satisfied the political needs of land

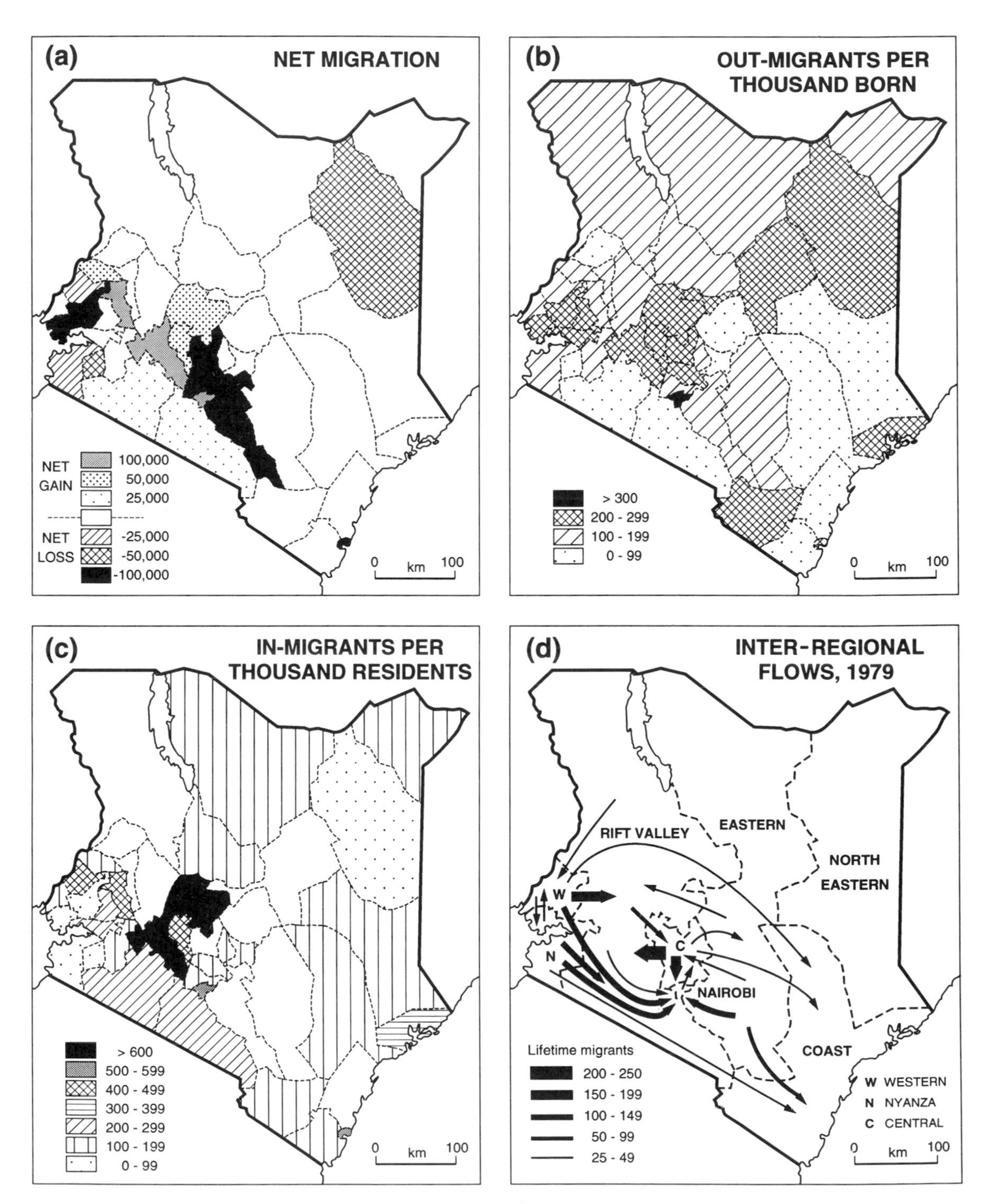

Figure 8.5 Migration in Kenya

redistribution at independence, but also relieved some of the intense population pressures in the densely populated source regions. In addition to permanent resettlement, circular movements of estate and plantation workers established in the pre-independence period between their rural 'homes' and place of work on tea, coffee or other estates persists.

By the mid-1970s the formal resettlement in the highlands had almost stopped, but population pressures continued to mount in the high-density areas as a result of very high rates of population growth. There was therefore pressure for spontaneous migration to the drier lands on the margins of the highlands, or to government irrigation schemes. Eastern Province experienced substantial inflows from several provinces, and it is in these environmentally marginal areas, not in the areas of highest population densities, that there have been the most serious environmental consequences of resettlement. There has been extensive soil erosion associated with over-cultivation and overstocking, and destruction of vegetation without adequate conservation measures and problems of human disease, notably a resurgence of malaria in the irrigation schemes.

Rural–urban migration is dominated by the importance of Nairobi, which attracts migrants from all parts of the country. It grew from a population of 267,000 in 1962, just before independence, to 835,000 in 1979, and an estimated 1 million by 1990, more than twice the size of Mombasa, the second largest town. Its rate of growth was most rapid in the 1960s, and fed mostly by migration, but it remains high in absolute terms. Recent urban growth has been more rapid in smaller provincial and district centres, such as Nakuru and Eldoret in the relatively prosperous Rift Valley Province, attracting migrants from more local sources. Nairobi, however, still has over 50 per cent of all modern sector employment, and continues to attract migrants, highly skilled as well as the unskilled, from all regions of the country. Although the net effect has been for rapid urban growth, much migration to towns continues to be circular, dominated by young men who leave their wives and families to work the family land at the rural source while they work or look for work, often unsuccessfully, in town, periodically returning to the rural area during periods of unemployment or during leave from their employment.

The continuing interaction between town and country through circular labour migration has been a critical factor in increasing regional differentiation in Kenya. Through the remittances of the workers there is a substantial transfer of cash and goods from urban areas to rural areas, and some estimates have suggested that 15 per cent of incomes in Nairobi may be remitted to rural areas. Many rural households rely on urban remittances for their survival, for in the poorest and highest density rural areas, with high rates of out-migration, rural subsistence and cash incomes are insufficient to maintain the household. Circular migration of some household members is essential to mere survival at very low levels of income and well-being. In other areas, however, and particularly in the more prosperous areas of Central and Rift Valley Provinces that are near the large and relatively affluent Nairobi market, and even have relatively easy access to the international market for air-freighted vegetables to Europe, migrant remittances from urban jobs have been crucial for financing farm investments to allow enhanced participation in the commercial economy. For the more prosperous areas the migration system seems to have brought additional benefits. For the poorest provinces, and particularly Nyanza and Western, the migration system has seemed to perpetuate rural impoverishment, and to sustain a spatial structure of sharp variation between rural and urban sectors and among rural areas (Gould, 1994a).

INTERNAL INTERACTION, REGIONAL INEQUALITIES AND 'URBAN BIAS'

We have considered the movement of goods, services and people within Third World countries through national and local networks of transport and communications and the linkages developed through the urban hierarchy. Through these interactions the several regions of a country and several sectors of an economy are integrated into a cohesive national economic and political entity. However, they also allow regional specializations to emerge, facilitated by the exchanges that take place. The geographical distribution of economic activity is thus affected by the strength and directions of movements and exchanges, and is the net result of conflicting pressures within each state that act either to spread development throughout from the original foci or, conversely, to concentrate it in a few historically or ecologically favoured regions.

Spread effects are involved where there is a diffusion of developments and innovations from the national foci to all parts of the country. The development of an urban hierarchy and the transportation system associated with it is assumed in the Taaffe *et al.* 'ideal-typical' sequence to facilitate the spread of goods and services initially provided in and organized from the core. However, the evidence presented in this chapter suggests that the hierarchy also generates flows in the opposite direction ('backwash effects') to attract goods, people and services from the periphery to the core, and widen any previous gap between them. Spread effects are evident everywhere and operate with varying strengths, but have not generally been sufficiently strong to counter the opposite forces. The dominance of backwash effects would be expected, indeed was essential, for a colonial economy as the metropolitan countries sought to dominate internal production and external trade and integrate it with the world economy.

The continuing strength of the backwash effects in the face of a widespread political desire to spread development and the benefits of development more equitably is due to two major reasons. First, the legacy of the colonial structures has ensured that, despite attempts to restructure transportation networks and the urban hierarchy and generally to strengthen the spread effects in the economy, there is still unequal inter-regional exchange, and this is linked to the persistence of the structures of export trade associated with colonialism. A process of cumulative causation maintains and strengthens the dominant position of the more advantaged areas. Rich areas get richer faster than the poorest regions, even where these poorer areas become richer at all. Second, there is in Third World countries the operation of a process of 'urban bias', a term associated with the economist Michael Lipton (1977). He has argued, initially from the Indian experience but subsequently generalized to all parts of the Third World, in centrally controlled as well as market economies, that there is a systematic bias in decision-making by governments and by individual economic decision-makers in favour of urban areas in the allocation of economic and social expenditures. It is urban areas and their immediate hinterlands, the areas where most decision-makers themselves live and come from, that attract the most productive investments, that are allocated highest per capita expenditures on health and education and are generally politically favoured. Systematic though not normally explicit urban bias compounds the inherited processes of spatial inequality that are characteristic of Third World countries.

SUMMARY

The geography of any Third World country is continuously moulded by the exchanges of people, goods, capital and ideas that occur between different parts of the country, between rural and urban areas, but also between one rural area and another and between one urban area and another. The exchanges occur as a result of differences in levels of development between source and destination, and, because there tend to be large regional differences in these, they are particularly evident in the Third World. These interactions affect source and destination areas in different ways, generally bringing benefit to both source and destination as a result of the exchange, but very often bringing rather greater benefit to urban areas than to rural. This may occur where the urban area is a source of the flow, as in the investment of urban capital in agricultural enterprises in rural areas with returns of income and interest back to the urban area. It tends also to occur where the urban area is the destination of the flow, as in the case of urban migration where the benefits of the additional labour accrue in the town itself. There is therefore a tendency in Third World countries for internal interaction to be a critical factor for maintaining and widening spatial inequalities.

"Some ventures require a rather unconstrained environment"

Freedom to operate in a free market economy – Freedom to enjoy the same legal treatment as nationals – Freedom to invest in any business, economic sector or activity – Freedom to transfer abroad in hard currency capital gains, profits, and royalties – Freedom to trade stocks and obtain tax-free returns in one of the most profitable bourses in the world – Freedom to apply for concessions to build and manage public facilities and supply public utility services – Freedom to participate in the privatization of state-owned mines, ports, banks, telecom, electricity and oil companies – Freedom. We really mean it.

LOOK AGAIN AT PERU. A BREATHTAKING OPPORTUNITY.

National Economic Management

Development planning and the market

As the earlier chapters have shown, the countries of the Third World are faced with a great variety of problems: rapidly growing and impoverished populations, inadequate medical and social provision, an agriculture often unable to meet basic food needs, limited and sometimes inappropriate industry, restricted and dependent trade, uncontrolled and rapidly growing cities, widening regional inequalities, and inadequate access to the dynamic economic system of the developed world. Colonialism left many Third World countries with economies that have seemed incapable of meeting the needs and aspirations of their inhabitants. Although there are many and diverse problems to be faced, Third World countries, by definition, have only limited financial resources with which to tackle this multitude of problems. In addressing them, countries have sought to put in place mechanisms to bring about change and provoke economic growth. It is therefore not surprising that they have invoked some form of development planning in an attempt to identify an approach that will yield economic and social advance; but it has now become clear that development planning has not been sufficient to achieve the desired objectives.

Development plans have been formulated in an effort to use scarce domestic resources carefully and productively. In addition, in seeking to develop, many Third World countries require external resources in addition to their own, and the drafting of precise, detailed plans outlining specific objectives makes it easier to secure foreign aid than a blanket request for funds for unspecified purposes. Development plans are also seen as symbols of political independence, through which Third World countries will make economic progress and secure economic independence from colonial and neo-colonial powers. They may also be the means of focusing the national consciousness of newly independent states, and securing the unity of new nation states which may contain populations divided on ethnic, religious or social grounds.

The objective of economic planning has been described as seeking to use national resources in the best interests of the nation as a whole. This involves some decisions as to what these best interests are and what the goals of development are to be. Once these have been determined, the process of economic planning tries to direct development along appropriate paths to achieve them. Planning therefore involves the choice of desired objectives, decisions about the best strategy by which these objectives may be secured, and the allocation of available resources to implement the strategies and fulfil the goals set.

Such a diverse and complex process inevitably means that the state and the political system have a crucial role in the planning process. The early association of the USSR with economic planning tends to equate planning and socialism, but this overlooks the role which the state plays in economic activity, even in market capitalist economies, and in the provision of basic services. Even in the seemingly laissez-faire economies of industrializing nineteenth-century Europe, the state had a positive role in

 Invitation to invest in Peru

"I'm looking forward to being independent and having an airline, a Hilton and a nuclear reactor."

the development process. In consequence, the role of the state may vary from those countries in which, as in socialist states, the government controls all, or most, of the means of production, to others where state control is restricted to a declining and limited number of sectors. In the former, economic planning embraces the whole economy; in the latter it may be restricted to certain sectors with the government in an enabling role, formulating guidelines or providing incentives for the objectives deemed desirable in the private sector.

While comprehensive development planning was a familiar approach to Third World problems in the 1960s and 1970s, there has been a distinct shift in approach in the late 1980s and into the 1990s, with increasing positive use of market forces to shape the development process. This reflects a strong commitment to market forces and a 'rolling back' of the state in First World economies such as those of Thatcherite Britain and Reaganite America, and a similar view in major international agencies such as the World Bank. The break-up of the Soviet bloc and moves to create capitalist economies in its constituent states has also undermined the 'planning model' offered by the Second World.

DEVELOPMENT PLANS, PROGRAMMES AND MODELS

Within the planning process a number of levels may be distinguished. A 'plan' relates to the economy as a whole, divided into major sectors – that is, sectoral planning – and possibly to areas within the country – regional planning. A 'programme' relates to the more detailed determination of specific objectives to be achieved within the various sectors or areas. 'Projects' are the individual components that together make up a programme.

The form of development plans and the techniques of planning are very variable from one Third World country to another, but most tend to derive from a limited number of more or less formalized economic models depending on differing interpretations of the factors that influence the rate and direction of development.

Three major categories have been distinguished. The first, using quantitative or aggregate models, emphasizes the role of saving and investment as the determining element in development. Such models deal with the entire economy in terms of macro-economic variables such as savings, investment, production and consumption. They assume that lack of savings and investment are a major check to development, and seek to identify the amount of saving and investment necessary to secure a given rate of growth in GNP. Where domestic sources are inadequate such models give an indication of the amount of foreign assistance needed to secure the desired growth rate. Such an approach sets only the broadest objectives and may provide only background for a more comprehensive and operational plan which is concerned with economic activity *per se*.

Second, sectoral models emphasize the generative role of specific activities or sectors. Here the economy is divided into component sectors such as agriculture and industry, so that the prospects and targets for each and both can be identified, and an effort made to ensure that rates and types of growth in one are consistent with growth in another and will contribute to the overall development of the economy. In some cases concern may be with only a single sector, thought to be crucial to the economy or where data to facilitate comprehensive planning are unavailable. This gives rise to what is in essence a partial plan concerned with a single sector, or with a number of distinct sectors which lack any internal consistency. Such partial plans were characteristic of the pioneer schemes of a number of ex-British territories in Africa and the Caribbean in the early 1960s, but more characteristic have been overarching national plans within which sub-sectoral plans could be formulated.

Third, there are models that are intrinsically geographical, in that they focus on regions and seek to promote regional differentiation through region-specific plans. The regional planning approach complements the sectoral approach, but they both need to consider the interactions between the components, regions in this case (as examined in general terms in Chapter 8).

Changes in an economy are not independent of one another. Change in one sector may require or provoke change in another. Even the seemingly simple economies of poor countries are often complex so that, because of this interdependency, consistency is required between the various sectors. Planning at this level therefore seeks to take account of targets, resources and investment of production and consumption for the various sectors, and to inter-relate the activities of all the productive sectors of the economy. All activities are viewed as producers of outputs and users of inputs. For example, an increase in agricultural output may require inputs from industry in the form of machinery and fertilizers. Thus the changes which might ensue from changing output in one activity, both direct and indirect, are sought as inputs, investment and employment in another using multi-sectoral or comprehensive input–output models. Given the selected targets of a development plan, these models can be used to trace all the implications for the economy and to produce a comprehensive plan with mutually consistent production levels and resource inputs. Complex models of this type require a considerable amount of data and sophisticated analysis, so that they tend to be more characteristic of more advanced countries. However, Third World plans do seek both comprehensive goals and internal consistency.

In the formulation of any planning strategy a number of basic conditions have to be taken into consideration. The stage of development is clearly significant. In a very poor country that is largely dependent on subsistence agriculture, with a weakly developed commercial economy and few natural or financial resources, there is only a small possibility of formulating a realistic complex multi-sectoral plan. Government priority would probably be to provoke some initial change to lay the foundation for later, more substantial development. On the other hand, a country with a prosperous, well-developed commercial, agricultural and mining economy may wish to begin a transformation to industrialization, and will have the skills and resources to do so. These more sophisticated plans may require large inputs of demographic and economic information for analysis and prediction. In some Third World countries such data may be unavailable or unreliable, so that less sophisticated planning structures may initially be more appropriate.

Resource endowments may impose particular constraints on development, or shape its form. Deficiencies in particular resources, such as fuel or water, may require particular strategies or major efforts to provide them. Shortage of labour skills requires an emphasis on education, while limited finances require careful assessment of what might be obtained from trade or aid sources.

In the choice of a planning path to be followed a major influence, of course, is the political and economic context within which planning is to take place – capitalist, socialist or some indigenous alternative. Within such a context a country needs to decide which approaches are most suited to its particular needs and objectives. Having done so, the planning strategy decided upon should, within the political context of the country, define the goals to be achieved by the plan, then set out a strategy by which these goals are to be achieved, and endeavour to ensure that the strategies for the various sectors are consistent with one another and with the overall objectives of the plan. Commonly a plan has some time-scale for its completion, typically 5 or 7 years, though there may be a longer-term and more generalized perspective plan which sets out broader goals for the country and its development aspirations, and also shorter-term targets to be met, perhaps on an annual basis.

THE ROLE OF THE MARKET AND THE ENABLING ROLE OF THE STATE

In formulating their development strategies, Third World countries are obliged to make choices between a wide range of alternative paths, and these choices are conditioned by the political and economic frameworks in which these strategies are set. A fundamental choice exists between planning and development under socialism and the dominance of the state in allocating productive resources on the one hand, and under capitalism and the predominance of the market on the other. These are not necessarily polarized choices, as all sorts of gradations between the two exist. The choice, however, will influence the nature of the economy and its organization, for under a socialist command economy virtually all economic activity will be owned and controlled by the state, which can therefore determine and shape priorities. In such an economy decision-making may be centralized and planning essentially coercive. In a capitalist economy direct control by the state may be restricted to a few sectors, and for the remainder the government may seek to influence the direction of development by indicating its priorities and trying to persuade the private sector to fulfil them.

As the previous section has shown, the most characteristic approach in Third World countries has been through some form of centralized planning. From the 1950s to the early 1980s planned development through centrally managed strategies and a dominant role for government in defining these strategies was largely unquestioned. By the 1980s, however, the role of the state in the development process was being seriously questioned, and not only in Third World countries. That period was what has been described as a 'counter-revolution' in development economics, with a shift in emphasis to the importance of the market rather than the state as the prime instrument of economic management. This was associated in Western states with the rise of monetarism. The political dominance of President Reagan and Prime Minister Margaret Thatcher, both vigorously pursuing monetarist policies in their own domestic economies, had repercussions for the international economy in the 1980s in creating a climate of opinion that saw planning and state intervention as creating 'distortions' in the workings of the market. Economic development, especially in the Third World, was being held back, so they argued, by states intervening too much to manage markets. There were too many state enterprises, mostly subsidized and operating inefficiently, and many of these needed to be privatized and freed from government restriction. International trade was being prevented by tariffs and other barriers, such as artificially controlled exchange rates, and a freer trade and floating exchange rates with freely convertible currencies would aid the developing as well as the developed countries. The state was too involved in social expenditures at the expense of production, and needed to divert its scarce resources to production and to providing an enabling role for private enterprise to create wealth, with a general reduction of the direct role of the state in welfare provision.

In most countries the role of the state in economic management and social provision was cut back during the late 1980s. World Bank data show that in the industrial countries of Europe, North America and Japan central government expenditures as a proportion of national GNP fell from 29.2 per cent in 1985 to 27.7 per cent in 1989. The equivalent figures for the Third World as a whole were from 29.2 per cent to 23.4 per cent, an even larger fall. The fall was sharpest in the Asian Third World countries (from 22.8 per cent to 19.4 per cent), where the leading role in rapid development in the NICs was taken by the market rather than the state. In Sub-Saharan Africa, however, the proportion of GNP spent by government actually rose (from 24.1 per cent in 1985 to 29.7 per cent in 1990) in a period of severe economic crisis when there were substantial reductions in government expenditures, but even more rapid falls in national wealth.

The role of government had changed from being the leading direct provider of economic and social investment, to one of providing an enabling economic and political infrastructure, ensuring the rule of law and the quality of governance, and encouraging local and international companies to invest in productive enterprises. In Kenya, for example, a country that sustained an explicitly capitalist approach to development in the earlier period when more centralized planning was the norm in Sub-Saharan Africa, the government offers incentives to certain types of industries and in certain parts of the country in the form of investment allowances and tax relief for each new job created. In addition, direct grants to small-scale enterprises permit their expansion where feasible. These measures had been successively extended as part of Kenya's Development Plans, from the first in 1965 to the fifth, 1984–88, but even these have become less important in the 1990s.

A second area of apparent conflict involves the priority to be given to investment in public works or productive activity. These two sectors are sometimes referred to as social overhead capital (SOC) and directly productive activity (DPA). The former refers to basic services such as water and electricity supply, transport, law and order and education, without which productive activities cannot function. DPA refers to the directly productive primary, secondary and tertiary sectors. The debate here has something of a 'chicken and the egg' pattern. Is the provision of SOC essential before any DPA can develop, or will the creation of DPA stimulate provision of SOC?

In many African countries more than one-quarter of government expenditure is on education, most of which is teachers' salaries. Politically motivated policies to expand education, to make it more widely available, have increased the direct financial burden of the education service. Providing for a literate and numerate population can be seen as investment in SOC for the future development of DPAs with a more educated labour force. There is, however, a growing feeling in many countries that the immediate benefits of such a high level of SOC investment are small, and that better use of scarce resources would be made by investing in DPAs, notably small industries, rather than in schooling, or within the education sector to invest in vocational education and training, including on-the-job training, which will have a more direct effect on production, rather than on general primary schooling for the mass of children. In East Asia, in sharp contrast to Africa, governments have successfully prioritized DPAs, offering financial incentives to technological innovation in small companies, and giving strong bias to investment in technical education. The success of Asian NICs in raising levels of economic activity and living standards has not been achieved by the absence of government and the power of the market, but because of particular forms of enabling support from the state for private development initiatives.

Given that investment capital is scarce and living standards are low, what should be the balance between consumption and investment? Should any short-term improvements in the economy be made immediately available to the population to improve their condition, or should they be asked to maintain low levels of living, such that improvements can be saved, invested and used for a higher but later level of improvement? Would a modification of the pattern of income distribution to provide greater equality or draw more people into the money economy be beneficial to the overall rate of growth? A country may also wish to consider what is the most suitable pattern of ownership for achieving its development goals. What should be the balance between the state and the private sector in ownership of the means of production? What should be the balance of domestic ownership and foreign investment?

BOX 9.1 WEALTH FOR WEAPONS

A sinister issue, noted by the Brandt Report (1980) but with growing prominence in the Third World throughout the 1980s, and culminating in the debates associated with the Gulf War in 1991, is expenditure on defence. The proportion of government expenditures allocated to defence is higher in Third World countries than industrial countries, whether in the NATO or former Warsaw Pact countries, and such military expenditures use funds that are necessarily diverted from other, more obviously development-oriented purposes. Defence expenditures, in these terms social overhead capital (SOC), were consuming a rising proportion of Third World budgets in the 1970s and 1980s – much more in some countries than in others. The average military expenditure for Third World countries in 1985 was 17 per cent of government expenditures, and Middle Eastern countries, awash with oil revenues, were spending on average 29 per cent. These levels have since fallen in all groups of countries since then, partly as a result of the 'peace dividend' with the ending of the Cold War but also as a result of severe constraints on government revenues in the majority of Third World countries. Between 1985 and 1990 the Third World average fell slightly from 17.0 per cent to 16.1 per cent of government expenditures, which were themselves falling rapidly as we have seen, but the equivalent figures were from 15.0 per cent to 11.7 per cent in Western industrial countries and between 1989 and 1994 the falls in the former Soviet bloc were even more dramatic.

The large defence spending countries are of two types. The largest spenders are those, like Saudi Arabia and other Gulf states, that have massive budget surpluses and relatively small populations, and have been able to invest in the most sophisticated defence systems, without obviously sacrificing allocations to other development expenditures. Most of the arms were supplied by Western arms manufacturers and by state-owned or, by the 1990s, privatized East European companies, but increasingly some of the larger Third World countries, including Brazil, India and China, are themselves arms manufacturers and exporters. The other group includes countries with military governments (e.g. Nigeria or Myanmar, though countries with military governments are now fewer in number than they were even in 1980), those involved in internal repression (e.g. Pinochet's Chile pre-1988, General Amin's Uganda in the 1970s or Mengistu's Ethiopia and Said Barre's Somalia pre-1990), or international conflicts (e.g. Pakistan and India, or Middle Eastern countries in their recurring disputes), where priorities are consciously ordered to favour defence over development expenditures, very often with serious implications for an impoverished population. In the Cold War years up to 1990, many of these countries were armed as client states of the superpowers. Pakistan and Somalia received massive arms shipments as part of US 'aid' as counter to the arming of India and Ethiopia respectively by the USSR. Western countries were very active in supplying arms to Saddam Hussein's Iraq during its war with Iran, 1981–89, but found, to their very considerable embarrassment, that these arms were turned against them in the Gulf War in 1991.

Should a country invest in people or materials? To what extent should priority be given to improving the physical as opposed to the material lot of the population? Should scarce resources be invested in health care and education, to improve peoples' basic well-being, on the argument that a healthy, literate population will provide a more productive labour force; or will investment in productive activity provide funds for social improvement at a later date?

Countries may wish to restructure their trading patterns. Should a country seek to become more self-sufficient in the full range of economic activities and less dependent on exporting a few primary products and importing manufactures; or should it endeavour to expand and diversify its trade pattern, producing commodities it is best able to and thus benefit from comparative advantage?

A major debate over the development process, particularly in the early days of development planning, was whether to give priority to developing agriculture or industry. If countries are dependent on traditional, low-productivity agriculture, should they give priority to developing a modern, efficient, industrial sector to give dynamism and diversity to their economies, or should they seek to transform agriculture which will, after change, be able to sustain the industrialization process? Associated with this has been a debate over the relative merits of capital-intensive and labour-intensive technology. It has been commonly argued that the development process should be easier for developing countries because, unlike Britain and the other now-developed countries which underwent economic revolution in the eighteenth and nineteenth centuries, they do not have to 'invent' a new technology; they can simply 'borrow' it from the developed countries. It has become apparent, however, that this technology, whether in agriculture or industry, developed to suit the peculiar conditions of those countries over a considerable period of time and it may not be suited to the distinct physical, cultural and economic environments of the Third World, especially where countries are anxious to achieve change in a couple of decades rather than centuries. Furthermore, the technology of the developed world has continued to change to take account of its changing factor structure, in which capital is relatively abundant and labour relatively scarce and expensive. In less developed countries the factor mix is often reversed, with capital scarce and labour cheap and abundant, so that available technology may not be appropriate. This gave rise to notions of an 'intermediate technology' more suited to the needs and resources of the Third World, but this has not secured widespread acceptance, for a variety of reasons. The developed world remains the main source of technology and has expended little research on the particular needs of Third World countries for cheaper and less complex equipment, while Third World politicians have seen this technology as inferior and second rate in relation to the modern machinery of the USA, Japan and Germany.

The debates about agriculture and industry, and about levels of technology saw different solutions from country to country and different solutions in any one country over time. In India, for example, the First National Plan, 1951–56, gave particular attention to agriculture but the second, 1956–61, moved towards industrialization as the major priority, and successive plans have oscillated in their commitment to the two sectors. However, the success of the East Asian NICs in the 1980s, through export-led industrialization, and the attainment of a high degree of technical sophistication as well as the advantages of a relatively cheap labour force, have given the debates about sectoral priorities and levels of technology a dimension that other Third World countries are anxious to incorporate into their development strategies.

In fact, most of the 'choices' of strategy identified above are not extreme and absolute; there are a whole range of compromises and intermediate positions. Nonetheless, they do indicate the very complex nature of development and development planning: elements of choice, balance and strategy have to be considered. Moreover, no single satisfactory model and plan for development has yet been formulated. In the relatively short time that development of the Third World has been a major priority (perhaps 50 years), few countries have achieved the breakthrough from developing to developed country status, and no single method of bringing about such a transformation in the short term is apparent.

It will be evident from the earlier chapters that the Third World is not a homogeneous entity but a heterogeneous collection of nations with some common features but many contrasts. In consequence, no single model or strategy is likely to be applicable to their varying circumstances. Each country and set of circumstances requires a procedure appropriate to its particular case. It has certainly come to be recognized that no single factor can account for success or failure in a process as complex as development. In addition, empirical research has not yet clearly identified all the significant factors in the development process or the means by which they operated in the now-developed nations.

STRUCTURAL ADJUSTMENT AND 'LIBERALIZATION'

The relatively disappointing performance of Third World economies by the 1970s, even during a period of relatively easy availability of finance from commercial and multilateral sources to support investment programmes, meant that by the 1980s conditions for development became even more difficult. By then global trading conditions had sharply deteriorated, and terms of trade had moved against primary producers. As the price of oil fell steeply in the 1980s from its 1979 peak, OPEC funds for direct development assistance, and to supplement World Bank and IMF funds, were substantially reduced, and interest rates rose world-wide. The result was greatly increased indebtedness of most Third World countries, and a potential collapse of the world banking system as some of the major borrowers from commercial banks, mostly in Latin America, were unable to repay even the interest due on previous loans. Debt rescheduling became a familiar practice, but with conditions imposed by the IMF in particular that fundamentally affected the economic management strategies of most Third World countries.

The IMF and its sister institution, the World Bank, had the resources to rescue the defaulting countries, to provide finance to repay interest and also to continue to make investments in development (Box 9.2). However, these institutions also had the political and economic muscle to impose conditions on the loans that they made to move economic management towards a macro-economic model that presumed the superiority of free market economics over planned development to stimulate economic growth. As in developed countries affected by monetarist economic management, this involved the rolling back of the economic and social roles of the state, the privatization of state enterprises and a reduction of government interventions and 'distortions' in the workings of economic markets (Plate 9.1). In addition, further conditions over environmental, administrative, social or population policies were

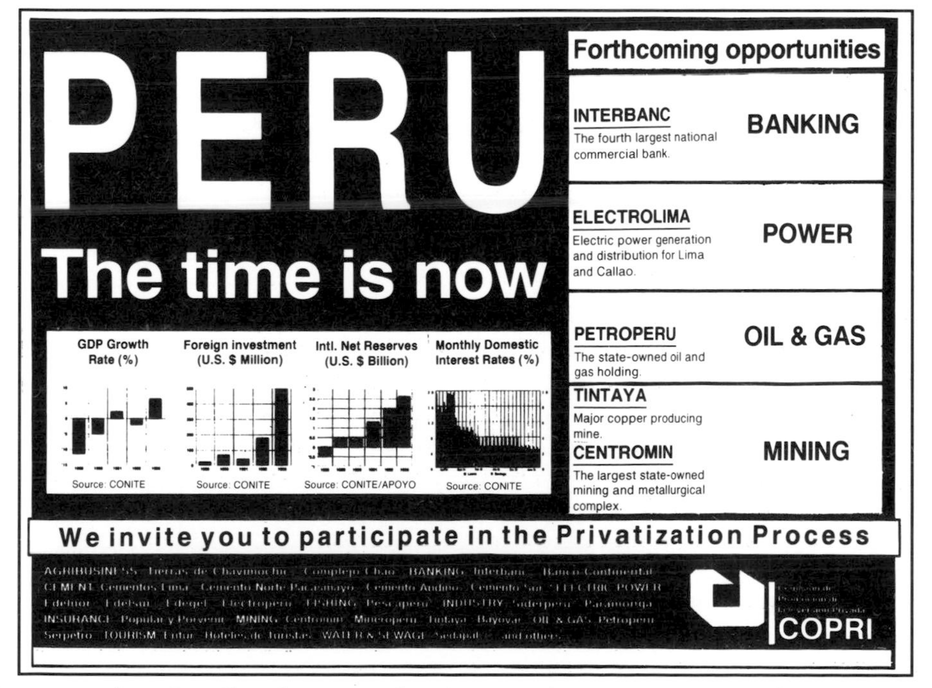

Plate 9.1 Investment advert, Peru This advertisement, from the *Financial Times*, is an invitation to foreign investors to invest in companies formerly owned by the Peruvian government, in banking, electricity, mining, oil and other sectors, as part of the process of 'opening up' the economy to the type of 'market model' preferred by the World Bank

BOX 9.2 THE WORLD BANK

The World Bank was founded in 1944, along with its sister institution, the International Monetary Fund, initially to provide finance to governments in Europe to support economic reconstruction after the Second World War, but by the 1960s its main operations were in developing countries. It is a bank, owned by its members, currently 155 countries, whose purpose is 'to assist its developing member countries further their economic and social progress so that their people may live better and fuller lives' (World Bank, 1994b, 4). This it does by lending directly to governments through its two major institutions, the International Bank for Reconstruction and Development (IBRD) and the International Development Association (IDA), and on a much smaller scale to the private sector through the International Finance Corporation (IFC). IBRD loans are given at slightly below commercial rates and go mostly to middle-income countries; IDA loans are virtually interest free and go only to the 40 low-income countries, mostly in South Asia and Africa. As part of these loans the World Bank provides economic advice and technical assistance from its international staff and consultants, and it is by far the largest agency directly involved in the development process throughout the Third World.

However, as a bank it is controlled largely by those countries that contribute funds for its lending programme. Lending policies and priorities are set by its Executive Board, with votes allocated to individual countries according to financial contributions. Table 9.1 summarizes the voting power in IBRD and IDA in 1994. It is clearly dominated by the seven richest countries, the G7 group, with nearly half the votes, while the remaining 148 countries share the rest. Only seven Third World countries have over 1 per cent of the votes, and most of the poorest countries have less than 0.1 per cent. The USA clearly dominates the institution, not only in its voting power, but also in the fact that its headquarters are in Washington DC, and that its President has always been a US citizen.

The lending priorities of the World Bank have changed over the years. In the 1970s it was the leader in developing basic needs and targeting its support to projects designed to provide economic and social services to the rural poor. By the 1990s, however, its priorities had shifted to structural adjustment lending, reducing the power of the state and enhancing the role of the market, and supporting productive investments. Not only is its influence over policies of individual countries considerable, but the World Bank also has a very considerable effect on the broader climate of thinking by geographers and others on development through its publications, notably the annual *World Development Report* (Plate 9.2).

Table 9.1 Voting strength in the International Bank for Reconstruction and Development and the International Development Association, 1994: selected countries (percentages)

Developed World	*IBRD*	*IDA*	*Third World*	*IBRD*	*IDA*
USA	17.42	15.66	China	3.15	2.03
Japan	6.58	10.39	India	3.10	3.11
Germany	5.08	6.93	Saudi Arabia	3.15	3.47
France	4.87	4.11	Brazil	1.73	1.64
United Kingdom	4.87	5.22	Iran	1.68	0.16
Canada	3.10	3.11	Mexico	1.33	0.70
Italy	3.10	2.68	Indonesia	1.06	0.98
G7 countries	45.02	48.10			

Source of data: World Bank, 1994b

BOX 9.2 *continued*

WORLD BANK INFORMATION BRIEFS

Toward a Market Economy

Privatization is a worldwide revolution still in its beginning phase. The key lesson is that the transformation is complicated and there is no single blueprint. The World Bank is heavily involved in helping governments with privatization.

Food Security

Hundreds of millions of people in developing countries suffer from hunger, although the world produces enough to eat. Food security is a development as well as a humanitarian issue. A combination of broad economic and social policies and specific interventions has reduced the incidence of hunger. But continued commitment is necessary to improve food security.

Nutrition

The immediate effects of malnutrition – poor health, mental debilitation, early death – can obscure its long-term consequences on a country's economy. The World Bank has adopted a number of strategies to help developing countries tackle this problem. These range from supplying urgently needed vitamins and minerals to helping governments to reorient their anti-poverty spending to make the most efficient use of limited resources while improving the nutrition of the most vulnerable groups.

The Gender Gap in Eduction

In developing countries girls have far less opportunity for attending school than boys. The evidence shows that raising per capita income does not automatically reduce the gender gap in education. This suggests that government commitment to girls' education through policy interventions is necessary to make a real difference in closing the gender gap in education.

Private Sector Development and Finance

The private sector is critical to development. In the 1990s many developing countries are trying to establish a market-friendly environment for a flourishing private sector. The effort is most obvious in countries moving from centrally-planned to market economies. The World Bank tries to help developing countries strengthen their private sectors.

Energy and Development

At the heart of successful economic development is the development and use of cost-effective energy. But energy almost always requires expensive investments, and many developing countries face great problems in meeting present and future energy needs.

Middle East and North Africa Region

The World Bank's Middle East and North Africa (MENA) Region extends from Morocco to Iran. The region is very diverse economically, and nearly all countries face substantial development challenges. Better prospects for peace across the region could redirect the people's energies and domestic and overseas investment into peaceful and productive activities. The World Bank is playing a key role in helping the region develop.

The Impact of AIDS on Development

AIDS has been the most serious new medical challenge of the last decade. But the epidemic is more than a health or even a social problem. The impact of AIDS on the economies of some poor countries is already significant. The epidemic could become a tragic brake on development if its spread is not halted.

War and Development

Despite increases in international trade, constantly improving worldwide communications networks, and other constructive global interactions, international and civil wars have remained a constant threat to development. While war can sometimes bring technological advances, the loss of human life and long-term productivity far outweighs any tiny benefits that may result.

Environment and Development

In recent years realization has grown that development – reducing poverty – and environmental protection are mutually dependent. Poverty is a major cause of environmental degradation, and environmental degradation exacerbates poverty. Sustainable development will emphasize two sets of policies: those, such as energy efficiency, which promote economic growth while protecting the environment; and those, such as controlling fish catches, which make acceptable tradeoffs between economic growth and environmental protection.

Debt-for-Nature Swaps

Conservationists and bankers have joined forces in an innovative technique – the debt-for-nature swap. By buying a share of a country's bank debt at a discount price, a conservation group can reduce that country's loan repayments, the bank gets some of its money back immediately, and the country protects its environment.

Plate 9.2 World Bank activities
This montage is made up from World Bank information sheets, and shows something of the range of fields in which the Bank is active, and the perspectives it has on matters such as health, finance, agriculture and the environment

invoked in lending programmes. Social sectors would be more appropriately supported by introducing user charges, euphemistically termed 'cost recovery' or 'cost sharing' programmes. These economic and social changes involved major structural adjustments to economic management of Third World countries.

In the 1980s the World Bank made 187 'structural adjustment' loans to 64 Third World countries, beginning with Turkey in 1980, and as much as 25 per cent of all Bank lending in that period was related to structural adjustment. Each loan required some or all of the sorts of major development strategy changes identified above as a condition of continuing support for development purposes as well as for repayments on previous loans. Its previously dominant approach of a strong reliance on formal development planning to define and implement national development strategies was effectively undermined as the operation of the open market dictated the spatial and sectoral patterning of investments, within and among states in Structural Adjustment Programmes (SAPs), sometimes termed Economic Recovery Programmes (ERPs). Since it was recognized that this 'liberalization' in SAPs would have adverse effects, at least in the short term, on the poorest people and in the most backward regions of each country, in order to reduce the regressive effects of structural adjustment policies funds have often been created to soften the blow. In Ghana, for example, this is called PAMSCAD (Programme of Action to Mitigate the Social Costs of Adjustment); in Uganda it is PAPSCA (Programme for the Alleviation of Poverty and the Social Costs of

Adjustment); in Guyana it is SIMAP (Social Impact Amelioration Programme). In these and other cases funds are made available, through government, to communities to support the building of schools or dispensaries or for food-for-work programmes, but these are clear palliatives relative to the more fundamental changes required by the approach of the SAPs.

The function of the state has changed to an enabling role, to ensure the establishment and maintenance of conditions that allow the free market to operate. Hence there is a concern for the maintenance of law and order and for the quality of governance, rather than the quality of education or of health care as a first priority. Entrepreneurs, local and foreign, are offered incentives for investment, rather than being constrained by government-imposed controls. National development planning is indicative rather than allocative: it identifies what might be possible or desirable and creates a climate for this to be achieved, but does not become directly involved in its implementation. The acceptance of such fundamental changes has met with considerable resistance and reluctance by many Third World governments, and in practice a negotiation between the lending agencies and most national governments has resulted in a compromise. It is clear, however, that since the 1980s approaches to and arrangements for national economic management in the Third World have been moving from a planned to a free market strategy. The state remains important in setting the framework for development strategies, but is less directly involved in setting out its detailed path.

REGIONAL PLANNING

It will be apparent from previous chapters that there is considerable imbalance in levels of development: between countryside and town, and between regions within countries. Much early planning was aspatial and took little account of the location of projected development. Without control or concern new projects tended to locate in the economically most attractive and profitable places and areas. It gave rise to, or intensified, the core–periphery pattern within countries, so that the benefits of development tended to concentrate rather than spread. Even when the problem or imbalance was identified, there was a debate over the economic benefits of allowing growth to concentrate, which would maximize economic progress, or, on grounds of social justice, of dispersing it, which might involve higher costs and lower returns for the nation as a whole. The need for some concern for the spatial aspects of development has come to be increasingly accepted, such that most countries now have some element of regional planning, either as an integral part of national planning policies or as an independent activity. The objective of such regional planning is to secure an equitable distribution of benefits from economic development for all geographic regions. SAPs have tended also to be aspatial, more concerned with macro-economic adjustment than with regional change. However, in one respect rolling back the operation of the state has had an impact on regional policies. SAPs have generally been well disposed to decentralization of government, rather than encouraging the centralized state, to give local communities more responsibility for their own affairs. Administrative decentralization certainly preceded SAPs as a positive step in enabling substantial local autonomy (e.g. in Papua New Guinea from the time of its independence in 1976), but in the context of SAPs decentralization is seen as part of the reduction in the economic role of the state.

DEVELOPMENT STRATEGIES: SOME EXAMPLES

Brazil

For a century after its independence in 1822, Brazil remained essentially a primary product exporting country, in which items such as coffee, rubber and cocoa were important elements in the economy and trade. The First World War and the inter-war depression revealed the vulnerability of dependence on such a structure, and in the 1930s the government set out to modify the economy, by widening the range of commodities exported and diversifying the internal economy by encouraging industrialization. The state sought greater influence in the production of mineral, petroleum and water resources.

Early development programmes aimed at breaking perceived 'bottlenecks' to development, primarily in the transport infrastructure, and the first formal development plan, 1949–53, had as its objective the creation of an economic environment in which development *might* take place – improving health conditions, energy and transport provision, and food supply.

The second plan, 1956–60, took a much more positive role, and was undoubtedly a great stimulus to Brazil's economic advance, laying firm foundations for the 'economic miracle' of the late 1960s. Called 'the Plan of Targets', it sought to stimulate growth in five areas: energy, transport, manufacturing, education and agriculture. It was particularly successful in the first three areas, securing considerable increase in energy production, extension of the highway system, and industrial expansion. The state was directly involved in highway building, railway improvement, electricity provision and petroleum production. It also participated directly in industrialization, particularly in the steel industry, and actively encouraged both basic industries and 'growth' industries such as automobiles, engineering and electrical goods. Progress in the schemes for agricultural expansion was limited, reflecting the low priority given to agriculture (for this was essentially an industrialization strategy) and also the difficulty of securing agricultural improvement without radical change in Brazil's patterns of land ownership and land use. Although the plan brought high rates of economic growth, it was less successful in generating new jobs for a rapidly growing population.

The early plans contained no regional perspective or locational controls, so that many of the new developments went to south-east Brazil, and economic progress was much more rapid there than in other regions of the country. The 1963–65 plan, though never fully implemented, acknowledged these emerging spatial imbalances, and thus raised the profound conflict facing most developing countries: should maximum economic growth be secured, by allowing it to concentrate at the most favourable locations; or should some attempt be made to diffuse it, at possibly higher cost, or lesser efficiency, to stimulate economic advance in other parts of the country?

Between 1964 and 1985 Brazil had a military government and there was a sequence of development plans, set in the framework of a 'market economic model' in which state, private, domestic and foreign capital had a role. The latter was of great importance in expanding industrial sectors such as the vehicle, electrical, chemical and pharmaceutical industries. The state was important both in shaping the overall strategy, and via its major participation in transport, electricity, petroleum provision, and in the steel industry. Of Brazil's twenty largest companies, thirteen are controlled by government and six by multinational corporations.

These plans tended to concentrate on infrastructure and industry; programmes for agriculture avoided the issue of reforming the pattern of land ownership, and focused on improving agricultural support services and developing new areas for farming. However, planning in this period showed greater awareness of regional imbalances and, in 1970, a Programme of National Integration (PIN) was announced, designed to promote the development of the poorer areas of Amazonia and the Northeast. Such a programme was still, however, a compromise, for much development had taken place in the 'basic nucleus' of economic activity, focused on São Paulo, Rio de Janeiro and Belo Horizonte, and the government was reluctant to compromise the high rate of national economic advance, in spite of its desire to foster the development of Amazonia.

Brazilian planning presents a striking dichotomy in its structure, for although concern for regional imbalance was not made explicit in national plans until 1963, the country has a history of concern for less developed regions which is possibly the longest established of any Third World country. Attempts to relieve the problems of the impoverished Northeast date back to 1877–79, when a major drought decimated the region, destroying crops, cattle and people. As the problem was perceived purely as one of drought, the response was to try and provide more reliable water supply by building dams. These were initially intended to support irrigated agriculture, but as that required altering existing land-tenure patterns, which reflected the established influence of powerful landowners and politicians, the reservoirs became primarily sources of water for cattle and people in the dry season and dry years.

The problems of the Northeast have been increasingly recognized as not merely the consequence of an environmental hazard, but the product of a whole complex of physical, social, historical and economic influences. In an effort to counter these, Brazil has tried a variety of regional development strategies copied from elsewhere. These include a multi-purpose river basin project, focused on the region's principal perennial river, to provide irrigation, electricity and navigation; a development bank to provide credit for the introduction of drought-resistant crops and new industries; and a more comprehensive regional development agency, with major priorities in improving infrastructure and industrialization. The Transamazônica highway colonization projects were part of the PIN programme, to decant excess population from the Northeast to new frontiers. All of these programmes have had some impact on the region. They have generally made a contribution to economic advance, but the region remains poor when compared to average national levels of well-being, or those of the developed Southeast. Within the region, totalling some 1.5 million square km and with a population of 40 million, economic progress has tended to concentrate in a few areas, particularly the coastlands and the large towns; there is pronounced intra-regional inequality. As at the national level, progress measured in economic terms has not been matched by provision of jobs for an expanding impoverished population. Nor has the region been secured against the episodic recurrence of environmental catastrophe. The failure of the Transamazônica scheme (see Box 5.3) to provide an adequate outlet for migration from the Northeast led the government to seek intra-regional solutions to the region's poverty. In the last formal development plan to date, the 1975–79 II National Development Plan sought to reduce the economic gap between the Northeast and the rest of the country by investing in industry and in transforming agriculture. The industrial programme included using the region's natural resources to develop nuclei of petrochemical and fertilizer production, and 'growth' industries of mechanical, electrical and electronic engineering. Agro-industries, again based on local products, were to be developed for cotton, cashew nuts, fruit juice and vegetable oil. New cash crops, such as soybeans, sesame and sorghum were also to be introduced. To support this, and in an attempt to provide greater security against drought, there was to be investment in the irrigable valleys and humid uplands of the region, and in the peripheral – and moister – areas in the north-west of the region (Figure 9.1).

As a result of its planning strategies, Brazil achieved very high rates of economic growth, above 8 per cent a year in the late 1950s, and above 10 per cent in the period 1968–73, the years of Brazil's 'economic miracle'. The country had implemented a major import substitution industrialization, created a significant capital and intermediate goods sector, and installed a sound infrastructure of energy and transport. Problems remained, however, in agriculture, the poverty of the Northeast, and the development of Amazonia. Over the period 1965–80 GDP increased at an average of 9 per cent, but in the 1980s this slowed to 2.7 per cent.

This slow-down can be attributed to a range of factors, internal and external. The removal of foreign capital during the deep recession of 1981–83 had checked the rate of economic progress. With the end of the military regime in 1985, the country needed to re-establish democratic institutions, a process disrupted by the death of the new democratic president, and the resignation of another. These political uncertainties, high rates of inflation and rising debt levels made the formulation of a new economic strategy difficult. Stagnation of the global economy compounded these difficulties, given the country's dependence upon it as a source of capital and as an outlet for raw material and manufactured exports. In consequence, there has been no formulation of new planning strategy; indeed the state has sought to diminish its

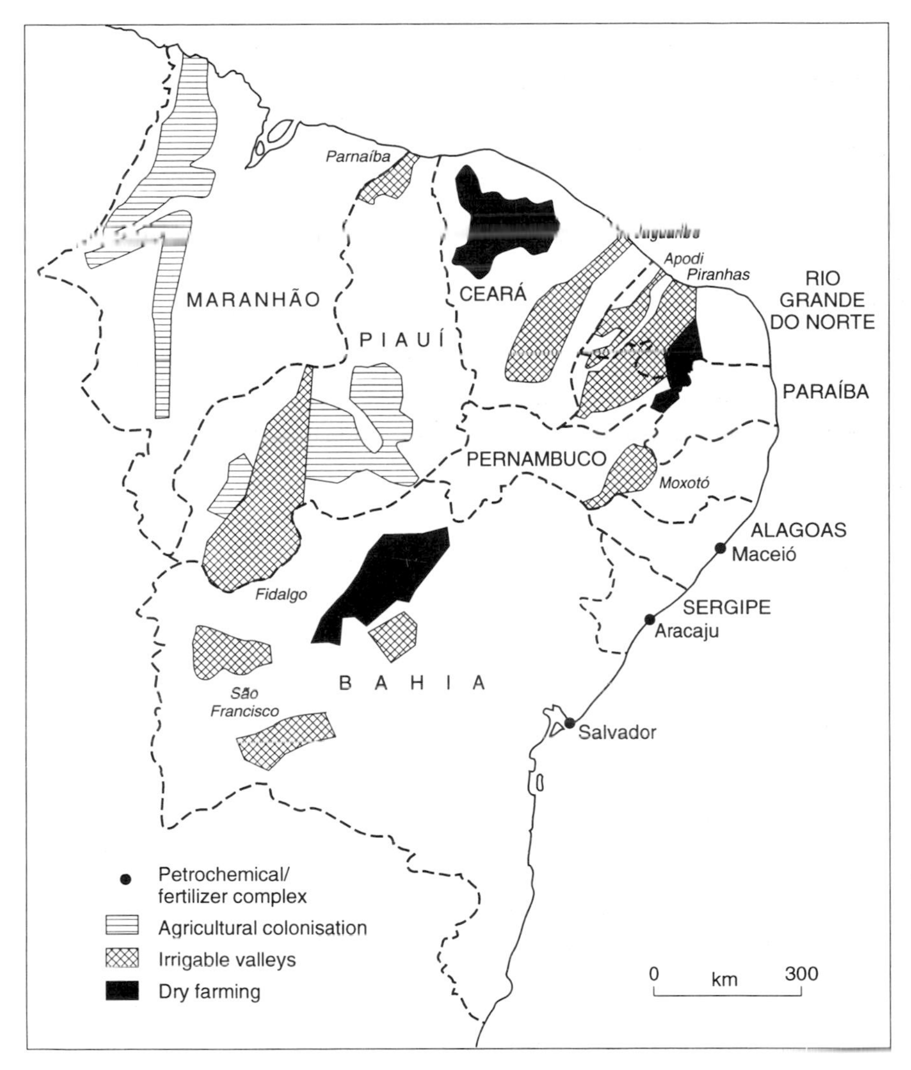

Figure 9.1 Regional planning for agriculture in Northeast Brazil
Source: After FRB, 1976

role in the economy by trying to privatize the energy and steel sectors, no easy task in such constrained circumstances. In consequence, the Brazilian experience demonstrates what *can* be achieved by a planning strategy and what sectoral and regional constraints there may be, but also that such a successful Third World economy remains vulnerable to external influences.

Ghana

Ghana provides a good example of the trends in the conditions and assumptions of national economic management of Third World countries over the last 40 years. It was, in 1957, the first Sub-Saharan African country to gain independence from colonial rule, and at that time was among the richest countries in the continent, with a per capita GNP in the mid-1950s roughly the same as that of Spain. It produced 10 per cent of the world's gold and was the world's leading producer of cocoa at a time when world prices were very high. As a result of high earning from cocoa exports, the government had amassed a surplus of over $1 billion. Economic optimism was combined with political strength and leadership, for Ghana's President, Kwame Nkrumah, voiced the aspirations of African nations not only for political independence, but also in articulating a vision of a Pan-African state with a strong economy based in industrialization and advanced technology that would be economically as well as politically free of First World and Second World dominance.

Nkrumah's economic model was strongly influenced by the Russian experience of a centralist planning strategy, with industry as the leading sector, and guided by a formal structure of national and sectoral plans. Early industrialization was to be based on hydroelectric energy, generated by the building of the Akosombo Dam on the Volta River. This was to be associated initially with alumina refining (the bauxite being imported from neighbouring African states, notably Guinea), with multipliers generating a range of other relatively technologically advanced industries, many of which were located in the new industrial deep-water port of Tema, established some 20 km east of the capital, Accra. Resources for industry and for agricultural consolidation and the development of a strong transportation infrastructure were provided from foreign sources, including from the socialist world, but also from the reserves accumulated from gold and the cocoa boom. There was substantial industrial expansion, with import substitution (i.e. local production of goods previously imported, such as building materials) and export processing (e.g. export of refined alumina) in the first few years of independence. The industrialization was mainly in state-owned companies – by 1972 the state had a majority or minority share in 235 enterprises – and the economic role of the state was paralleled by a massive expansion in its role as provider of social infrastructure, with very substantial expansions in education and health care provision, and a booming bureaucracy to administer the increasing role of the state.

In 1966, after a period of slump in cocoa prices and increasing pressure on government budgets, Nkrumah was deposed in a coup d'etat by a government less committed to centralized socialism, but lacking the political strength to systematically restructure the economic management systems that had been created. Throughout the 1970s there was a series of coups d'etat and weak governments and progressive economic decline, reaching its nadir in 1983 when the basic political weakness was exacerbated by two further factors. In that year there was a severe drought and associated famine throughout West Africa, resulting in a food crisis for Ghanaians as well as for the environmental refugees who came to Ghana from the even more affected states of the Sahel region to the north. In addition many of the migrants who had left Ghana for Nigeria in search of work in the 1970s and early 1980s, when its oil-based economy had been booming at the time of economic decline in Ghana, were suddenly expelled as scapegoats for a downturn in Nigeria's economic performance (Rimmer, 1992).

The economic structures of state ownership and state planning in Ghana had remained largely unchanged, with governments lacking either the political will or the resources to seriously implement recovery, whether through continuing commitment to a socialist transition or to some alternative strategy. Real per capita income in Ghana more than halved between 1972 and 1983, recovering subsequently, but it is still in the 1990s less than 60 per cent of the 1972 level.

In 1981, before the worst of the economic collapse, the Provisional National Redemption Council (PNDC), led by Flight Lieutenant Jerry Rawlings, had gained power in a coup with a political agenda of populist, self-sufficient, grassroots mobilization, including decentralization, but its economic strategy for recovery has been strongly associated with structural adjustment policies and the prominent role of the IMF and the World Bank in providing the financial and ideological support to enable an Economic Recovery Programme (ERP). In the late 1980s the World Bank was providing structural adjustment loans at the very high level of $40 per head per year. Much increased producer prices were paid to cocoa farmers as incentives to produce more, which they did even though the cocoa price was falling throughout the 1980s. Many government enterprises have been sold in a substantial privatization programme. There have been severe cut-backs in health and education expenditures, with the imposition of some user-charges (e.g. for prescriptions in health, and school fees in education), and there were over 50,000 fewer civil servants in 1990 than there were in 1985. The cedi, the local currency, now floats freely (mostly downwards) against the dollar, and there has been some influx of foreign commercial capital to support local resource based developments, particularly in the mineral and forestry sectors, with substantial short- and medium-term environmental consequences. Exports have risen sharply, but so have imports, particularly of vehicles and consumer durables, for some Ghanaians, mostly in the business sector, have greatly benefited from the trade liberalization. However, a necessary component of the structural adjustment programmes financed by the World Bank and associated multilateral and bilateral organizations has been a large PAMSCAD programme of community infrastructural works (e.g. building school rooms and dispensaries, installing small-scale water purification and sanitation) targeted at the rural and urban poor, whose job, housing and environmental conditions had often sharply deteriorated.

Decentralization has also been a much discussed aspect of the ERP. Administratively, 110 new districts were established in 1987 with plans for highly devolved economic and social responsibilities to them, including the local revenue-raising powers. Without a large central government compensation fund, this would be likely to have the effect of favouring the richer districts, urban areas and the more advanced rural areas in the south of the country at the expense of the poor north (Gould, 1990). In practice, however, the decentralization plans have been only partially implemented, with elected District Assemblies and associated administration in place, but without the financial or personnel resources to give the districts any effective economic power. The withdrawal of the state from economic management in favour of the workings of market forces has intensified the north–south and rural–urban differences in Ghana, and this is reflected in contemporary migration patterns, with large and continuing migrations into towns, and particularly into Accra, even though urban conditions are not buoyant for poor and poorly skilled job-seekers.

Indonesia

Indonesia has, after China and India, the third largest Third World population, 180 million in 1993. It is a very diverse and resource-rich country, with high population densities in the inner islands of Java, Madura and Bali, where over 60 per cent of the national population live, but low densities, typically less than 20 per square km, on the large and relatively underdeveloped Outer Islands. The country has experienced economic fortunes that have been quite the reverse of those of Ghana. During the first two decades after independence from the Netherlands in 1947, Indonesia remained one of the poorest countries, with GNP per capita roughly at the levels of India and Bangladesh, and with a very weak though centralized economic management that sought, without great success, to redistribute existing wealth rather than generate new production. After political changes in 1965, however, a much more dynamic economic management style was established with a central planning organization, with provincial components, and the First Five-Year Plan ran from 1969 to 1974, with successive five-year rolling updates. Though there was strong central control of the planning process in terms of levels of government expenditures and borrowings, it was technocratic rather than ideological in its implementation. It was not guided by an overarching commitment to a socialist or capitalist order, but was rather pragmatic in its ability to respond to changing local and international circumstances.

In particular it had to respond to moving from constraints of severe financial shortage before 1973 to a position of rapid economic expansion and resource surplus from 1973–85, for Indonesia, a member of OPEC and the third largest oil producer, derived massive windfall surpluses from the oil price rises of the 1970s and from development of other natural resources. Indonesia is the world's largest producer of liquefied natural gas, mostly exported to Japan, Korea and Taiwan, and a major producer of coal, manganese and other minerals (Soussan, 1988, 45–50; Figure 9.2). In addition 30 per cent of export revenues are currently derived from forest products and, after Brazil, Indonesia

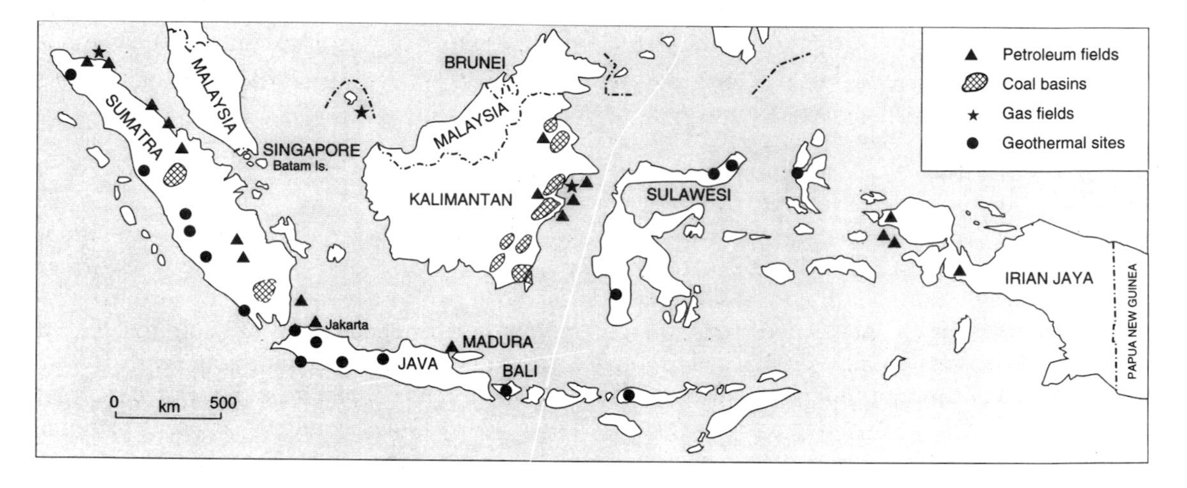

Figure 9.2 Indonesia: energy resources
Source: Soussan, 1988, 49

has the largest expanses of equatorial forest. It is the world's largest producer of plywood. As an archipelago nation it is also the source of a range of marine products, including large exports of shrimps and prawns. The revenues that have been derived from these resources raised the GNP per capita at an average annual rate of 6 per cent, 1970–90, and it has risen to the levels of Egypt. Though per capita income remains lower than that of its Pacific Rim neighbours, it is considered by many as an NIC in the same sense as Korea, Taiwan, Hong Kong or Singapore. There are, however, important differences between Indonesia and these countries, not least of which is the size of the population and the need for massive infrastructure expenditures to develop such a large and fragmented country. However, the benefits of income growth seem to have been filtering down to the population: whereas in 1976 it was estimated that 42.5 per cent of the population were officially defined as poor, that proportion had fallen to 15.2 per cent by 1990 (*The Economist*, 1993, 4).

The principal economic strategy for using the resource surplus revenues for development purposes has been to encourage resource-based industrialization (Auty, 1987). This has involved two distinct components. First, there has been the encouragement of up-stream and down-stream activities from resource industries: e.g. to manufacture drilling equipment and refined oil respectively in the oil sector. Second it has used the earnings to encourage diversification of the industrial base, in particular to develop engineering and manufacturing plants, often in association with foreign companies. Indonesia has been a pioneer in creating export-processing zones (EPZs), mostly near Jakarta and other large towns, but also in outlying centres, such as on Batam Island, adjacent to Singapore, to take advantage of the prosperous Singapore market. In 1991 manufacturing contributed 21 per cent to the national GNP (it was only 10 per cent in 1970), more than agriculture (19 per cent), which was 45 per cent in 1970. It has been a significant force in the creation of a large number of jobs, especially associated with massive urbanization and particularly on Java. Labour costs are low, lower than in neighbouring countries and very much lower than in the major Asian NICs (hourly wages in the textile industry are one-third of those in Malaysia and Thailand and 10 per cent of those in Korea), a further attraction to investment by MNCs in export production in electronics and textiles.

The national strategy has been to encourage foreign investment through joint ventures with local companies, and Indonesia has clearly been well placed to benefit from the general climate of development in the Pacific Rim. The state has invested heavily in schooling and health services and in its transport infrastructure, with central revenues being made available to provincial and local planning authorities to establish their own priorities in response to local needs. Agriculture has benefited too. Investment in miracle rice, the high yielding variety seeds, and distribution of the new technologies by government extension services have transformed the country to self-sufficiency in its staple food in most years, and there has also been expansion of area under agricultural export crops, notably oil palm, peppers and tea through Indonesian companies (Manning, 1988). The planning role of the state has been an enabling one rather than being directly involved in economic investment.

Its role in economic management has also been in controlling the levels and types of development. By the early 1990s, after the main years of resource price boom, revenues have fallen and the level of indebtedness to multilateral and commercial banks has risen. The government has imposed ceilings on foreign investment to keep indebtedness in check, but the problem has not reached the crisis levels experienced by some Latin American countries, and there has been no need for debt rescheduling. A more critical area of control has been in environmental management. It is acknowledged that the rate of removal of tropical forest products has vastly exceeded the sustainable replacement level in recent years, and deforestation has become a national as well as an international issue. There is, however, explicit recognition in the development plans that rapid development of the forest resource has been a necessary short-term expedient to generate revenues for other investments, and that the forest land might have alternative uses. The very large transmigration programme has taken people from the Inner Islands to develop land on the Outer Islands that has required the clearing of forest (see Box 3.3). However, the clearance has often brought environmental difficulties for the settlers, for the soils of the Outer Islands are generally much poorer than those of Java and prevention of rapid soil impoverishment has required extension advice and fertilizer inputs. Stronger controls on forest exploitation have been imposed by national and provincial governments, with the revoking of some licences and the imposition of stronger conditions on replanting on others (Auty, 1995, Chapter 10).

A NEW DEPENDENCE?

In the period after 1945 a development plan was regarded as an almost required part of the independence process; it would lead to economic independence from the colonial power, and would stimulate economic growth for the country and its people. It can be argued that, for some Third World countries at least, planning did bring success – high rates of economic growth were achieved; economies diversified, with a modernization of agriculture and the creation of a manufacturing sector; there was some improvement in basic conditions of living – in nutrition, health care, education and income levels. But in few countries was progress sustained, and the plan-driven impetus declined. In the mid-1990s it is difficult to point to an active and successful strategy of development planning. The reasons for this are varied, the consequence of a diversity of particular internal circumstances within individual Third World countries, and external pressures emanating from a slow-down in the global economy and uncertainties within the developed world on the role of planning in economic management.

A crucial factor for Third World countries has been that, as their economies have faltered, they have become subject once again to the strong influence of external agencies, not by individual colonial powers, but by international agencies, particularly the IMF. As part of the rescue packages offered by the IMF, Third World countries have been required to accept 'conditionalities' about the way their economies are run. Since the IMF favours the role of the market as against state intervention, the latter has been diminished in areas such as economic planning. Riddell (1992) has identified four areas where these external policies impinge on the internal experience of Third World countries. The devaluation of local currencies raises the price of imported goods, whether basic necessities such as medicine or fuel, or inputs such as machinery for agriculture and industry.

Second, the removal of agricultural subsidies raises the price of foodstuffs, particularly for the poorest segments of society. In its efforts to create a niche for these countries in the world economy, the IMF has tended to favour expansion of the primary product export sector. Whatever the global economic logic of this approach, it has the consequence of returning these economies to their earlier role of trading minerals and tropical crops for First World manufactured goods, and of undermining some of the local industries which have been established. Finally, and most crucially as far as development planning is concerned, the IMF's commitment to capitalism and the play of market forces has served to diminish the role of the state as an agent in the shaping of development strategies or in controlling sectors of the economy.

Some critics suggest that IMF policies are based upon an idealistic view of the role and efficacy of market forces, which, on the one hand, ignores the informal and non-market sectors of Third World economies and, on the other, reinforces the capabilities of monopolies, multinational corporations and powerful Northern economies. This approach, which has become pervasive in the shaping of the development process in the Third World in the 1990s, may have economic logic, but it represents external intervention in that process and, as Riddell indicates, it ignores the fact that 'Development, in the final analysis, is about humans. The debt crisis is not simply a matter of economics: health, education, employment, poverty, standard of living, diet, politics and the nature of society are also integral elements' (1992, 67).

SUMMARY

The methods and strategies of national economic management have distinctly changed from dependence on centralized national economic planning between the 1950s and 1980s to an increased reliance on market mechanisms and less direct public involvement by the 1990s. In both periods, however, the models used have been largely imported using the experience of economic management in the First and Second Worlds. Centralized economic planning, with state ownership of the commanding heights of the economy that was characteristic of socialist economies, was especially attractive to newly independent countries anxious to assert their economic independence and to re-order internal and international economic relationships. This was typically modified by integrating some of the welfarist approaches to economic policy that were characteristic of Western economies at that time. The result was typically a state with ambitious plans to achieve economic and social goals, but lacking the resources and power of implementation to achieve its bold objectives. Much was achieved by planning, but it had become clear by the 1980s that planning alone could not solve the longer-term problems of poverty, regional inequality and external dependency. By that time, however, changes in the global context of economic management, with the continuing economic weakness of socialist and former socialist states and the growing economic strength of the multilateral and commercial financial institutions in a seriously indebted world, had encouraged the emergence of market-led strategies for economic management in which the direct role of the state is considerably reduced relative to the direct role of individual and corporate economic decisions and priorities. Third World countries increasingly adopted, or were obliged as a result of that indebtedness to develop, more market-oriented styles and practices of economic management. This has resulted in a reduced concern by many governments for the spatial as well as social consequences of their development strategies. Internal regional economic inequalities have been growing with the retreat of explicitly regionally directed policies.

10

External Relationships

The foregoing chapters have sought to demonstrate the nature of the Third World. Their premise is that there is some distinctiveness to the idea of 'the Third World' which is capable of geographical recognition and analysis. The edges may be fuzzy and flexible, but there are common qualities and experiences which allow us to distinguish this macro-region from other parts of the world. Previous chapters have suggested that historical experience, demographic characteristics, economic activities, urban life and internal structures all contribute to a distinctiveness for this region and its component countries. Some commentators, such as Corbridge (1986), Friedmann (1992) and Drakakis-Smith (1993) have questioned the utility of the term, and whether a 'Third World' still exists. Yet the term remains in common currency, not least by such sceptics, and in our view a geographically distinct tract of the world continues to be recognizable, defined by characteristics of poverty, colonial experience, and contemporary social and economic structures and patterns (see also Chapter 1). If, as in the view of Norwine and Gonsalez, the Third World is not 'real' (in the sense of being concrete or tangible), it is an abstraction, but a tremendously profound and vital one. In their bald view – 'The Third World exists' (Norwine and Gonsalez, 1988, 2)!

Such a Third World exists, moreover, in the context of First and Second Worlds, at least in its original formulation. Yet, since the first edition of this book appeared in 1983, the nature of that tripartite division of the world has changed. The First World has become more clearly defined, through the more sharply market-driven economies of Europe and the USA, the activities of the Group of Seven (a grouping of the world's largest and most powerful economies: USA, Japan, Germany, UK, France, Canada, Italy), and moves towards greater coherence within the European Union. Conversely, the Second World has become less well defined. The destruction of the Berlin Wall, the 'collapse' of communism in the Soviet bloc, and the break-up of the Soviet Union – and of some of its former satellites such as Czechoslovakia and Yugoslavia – have created an ill-defined, fluid and sometimes volatile congeries of countries. The nature of their societies, political structures and economic development remains unclear. The most evident survivors of the political frame of the Second World are the remaining communist states in the Third World, such as Cuba, North Korea and China. Whatever the further evolution of Russia and the countries of Eastern Europe, since the first edition of this book the former republics of the USSR have been accepted as independent states, members of the United Nations and the World Bank, and separately identified as such in the *World Development Report*. Some, such as Ukraine and Byelorussia have much in common with Eastern Europe; but others, in Asia, might be seen by many criteria as adding to the periphery of the Third World. Uzbekistan, for example, a republic of over 21 million people located in the plains and mountains south-east of the Aral Sea in central Asia, had an estimated

◀ Unloading food aid, Sudan

per capita GNP of $850 in 1992, placing it at the lower end of the lower-middle-income countries in the World Bank's classification, at about the same level as Cameroon, Peru and Papua New Guinea. Like these three Third World countries, Uzbekistan has had a colonial past that has affected its internal economy and its new relationships with Russia, the former colonial power. It is estimated that 33 per cent of its GDP is from agriculture, 40 per cent from industry and 27 per cent from services, the equivalent proportions for Papua New Guinea being rather similar at 25 per cent, 38 per cent and 37 per cent respectively, and suggesting a substantial range of common economic characteristics.

One much-vaunted outcome of the dilution and fragmentation of the Second World was to be a 'peace dividend'. With the end of the 'Cold War' between the superpowers, expenditure on armaments, nuclear weapons and 'star wars' technology would no longer be necessary. Funds thus released from military offence and defence would be freed for other, peaceful, uses. One of these was perceived to be greater assistance for the Third World. The rich countries now had funds they could invest in the poor countries, and disburse as aid. To date such largesse has not materialized. In part this may be due to the lingering world recession of the 1990s; in part to an apparent preference for First World governments and investors to deploy resources in the rehabilitation and re-direction of Second World economies, rather than into helping those of the Third World. It may also be that First World politicians, despite their rhetoric, see the needs and demands of their domestic constituents as more pressing than those of the Third World poor.

What is evident in the shifts in the relationship between First, Second and Third Worlds over the past decade, and over the much longer period from the Age of European Discovery is that, if the Third World 'exists', it does not do so in isolation. It is shaped and defined by its relations with other, now more developed parts of the world. Inter-relatedness was, and remains, an important element in the nature and experience of the Third World. In 1984 R.J. Johnston claimed that 'The world is our oyster' (Johnston, 1984). In part this was a rather narrow argument, deriving from Johnston's view (not subscribed to by the present authors) that British Geography had been neglectful of a wider world. Nonetheless, Johnston asserted that understanding of a wider world was fundamental to world peace and ultimately to world survival. In his view, geographers (at least in Britain) needed a renewed awareness of the variability of the 'real world'. He portrayed this variable world through a series of maps not dissimilar to those in our introduction, and argued that the key to understanding the whole world was 'the economic system which links all parts of the world into a single functioning unit and which contains the dynamo for all societal activity'. He claimed that 'Economic, political, social and cultural activities are all inter-related in a world-system whose trajectory is guided through decisions made by knowing individuals, constrained but not determined by their context, and enabled by their knowledge to make decisions that will restructure that context' (1984, 443–4).

Events since the fall of the Berlin Wall would certainly sustain that view, as the First and Second World have begun to redefine their relationship; the implications for the Third World of this rapprochement are less clear. However, as previous chapters have directly and indirectly indicated, the relationships between the Third World and its constituent parts and the wider world are crucial to its definition and interpretation. External linkages through colonialism, economic exploitation, trade and aid, social, cultural or medical hegemony, are all crucial to understanding the nature of the Third World, as is the more structured approach through the world-systems model (see Figure 1.18).

Johnston's claim is that there is a 'single world-economy, to which all places are linked, to a greater or lesser degree' (1984, 446). This world-economy, he states, is capitalist in nature, and 'dependent on an increasing propensity to consume' (1984, 456). For a capitalist world-system to flourish, Johnston argues that movement is necessary – 'goods, capital and labour must all be shifted around to realise potentials as they are perceived' (1984, 457).

The Third World, almost by definition, has had a modest part to play in this world-system, and its place in the movement of goods, capital and labour may be slight in comparison to those within and between the First and Second Worlds. Nonetheless the movement of people, commodities and ideas has been, and is likely to be increasingly, important for the Third World and its role in the world-system into the twenty-first century.

Comprehensive discussion of such external linkages would necessitate another book, so the present chapter confines itself to some examples of the movement of people, goods, capital and ideas between the three worlds in order to illustrate some of the net effects of the interactions, separately and together.

PEOPLE

Labour migrants

The movement of people has been a crucial element in the linkage between the First and Third Worlds. The 'voyages of discovery' brought Europeans into contact with a 'new' world in the tropics which has evolved into the Third World. Initial movement of Europeans to these new lands was relatively modest. It is estimated that Spanish migrants to Latin America numbered less than 4000 a year between 1600 and 1750, and Portuguese migration to Brazil was similarly limited, as was the migration of other Europeans to their colonies in the Caribbean. However, the desire for labour saw substantial enforced movement of Africans to the plantations of Brazil and the West Indies. Between 1600 and 1850 an estimated 3 million slaves entered Brazil, and overall estimates of the numbers involved in the transatlantic slave trade in that 250-year period range from 10 million to 30 million, creating large populations of African origin from Brazil north through Central America and the Caribbean islands to the United States, and in the countries of the Caribbean region they are the majority.

In the nineteenth century there was substantial migration from Europe to parts of Latin America, mainly from southern and eastern Europe. Brazil and Argentina each received around 3.8 million immigrants between 1880 and 1930, with lesser numbers to Uruguay and Chile. The 'scramble for Africa' and the colonization of Asia in the nineteenth century generally involved less permanent settlement by Europeans, except in limited areas in East and South Africa. Otherwise European migration to the tropical empires was temporary, as administrators and for business reasons. There was some significant intra-Third World movement during this period, partly to offset the abolition of slavery, and a consequent need for free labour on the plantations. In the period up to 1920 some 1.05 million South Asians, mainly from India, moved as indentured labour to Mauritius, Guyana, South Africa, Trinidad, Malaysia and Fiji, with 300,000 to other countries (Figure 10.1 and Plate 10.1) (Clarke *et al.*, 1990).

In the present century, and especially since 1945, the predominant pattern of movement has been much more from the Third World to the First. This has been a movement of both 'brawn' and 'brain'. In the 1960s and 1970s there was significant movement of unskilled labour to take up basic service jobs and production line work in the conurbations and industrial towns of the metropolitan powers – for example from the Caribbean and India to Britain, North Africa to France, and Indonesia to the Netherlands, but these migrations have largely ended. In western Europe the main immigrations of unskilled workers are from southern and, particularly since 1990, from Eastern Europe. In the Americas, the principal destination of recent unskilled labour migrant streams from Puerto Rico, Mexico and Central America, and recently from Haiti and Cuba, has been the United States. Many of the migrants have been 'illegal', and therefore subject to constant harassment by authorities and forced to accept low wages paid by those who employ them.

There has also been a numerically smaller, but economically highly significant movement, of skilled labour and more educated people from Third World countries, seeking to capitalize on their education to secure higher income and better opportunities and living conditions than are available in their homelands. Such

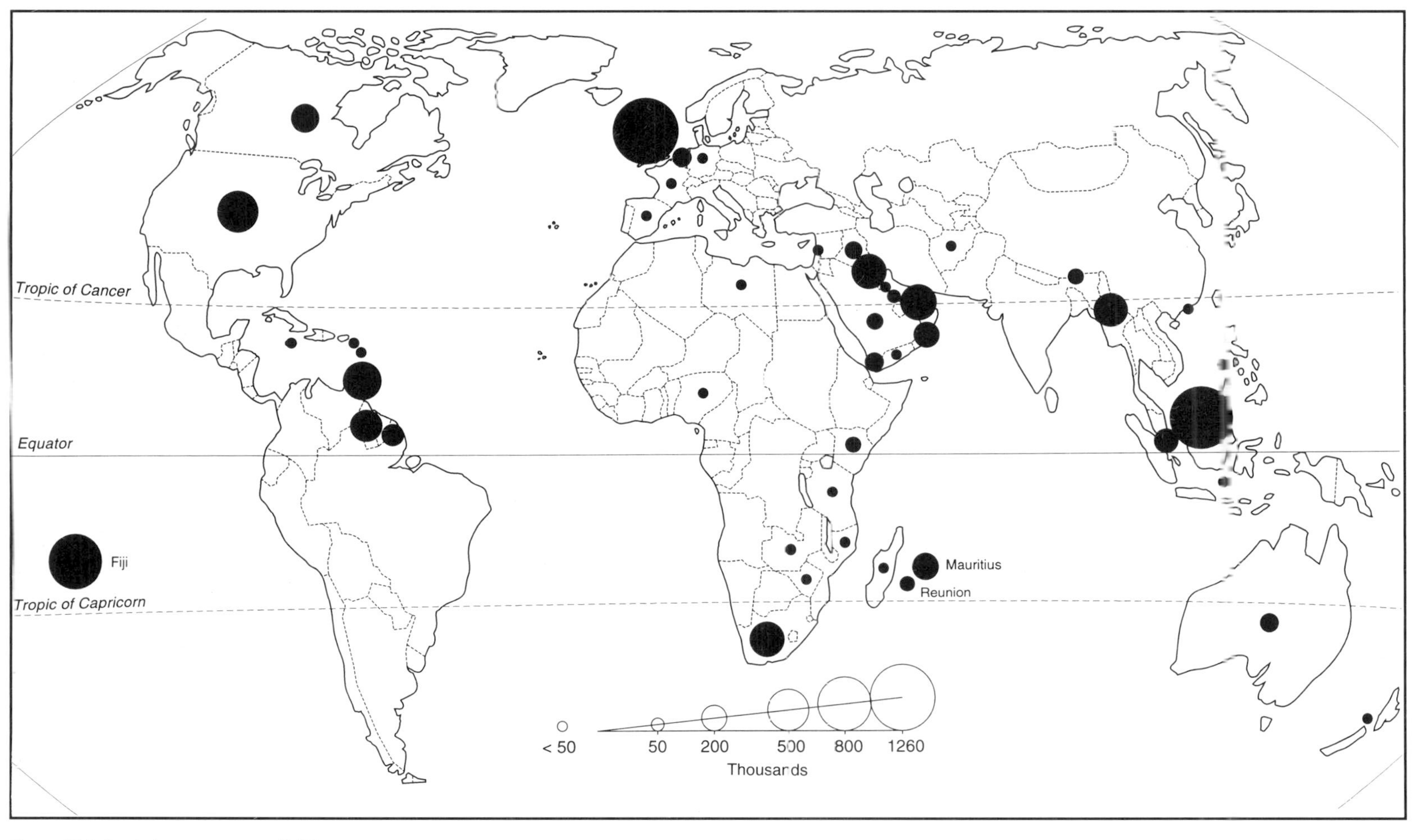

Figure 10.1 South Asians overseas 1987
Source of data: Clarke *et al.*, 1990

Plate 10.1 East Indians, San Fernando, Trinidad Between 1845 and 1917 over 140,000 East Indians were brought to Trinidad to work as indentured labour in the sugar plantations. Their descendants now comprise about 40 per cent of the island's population, and have retained many of their cultural traditions in religion, family and marriage

migration involves the movement of people such as doctors, engineers and entrepreneurs, whose skills are scarce in the countries from which they emigrate. The largest sources of skilled migrants to the United States, the most important destination, are the larger and more developed Third World countries, such as Colombia, Egypt and India, but many smaller and more severely impoverished countries lose high proportions of their skilled workers, attracted by the much better salary and professional conditions available in the USA and similar destinations. The United Nations estimated that in 1990 30 per cent of Africa's stock of skilled professionals were working abroad, and this proportion can be as high as two-thirds in countries like Ghana, Sudan and Uganda, where internal economic collapse has exacerbated the outflow (Gould, 1994b).

Since 1960 at least 35 million people have moved from the developing world to the North, legally or illegally. Prior to that date about 80 per cent of immigrants to the USA, Canada and Australia came from elsewhere in the developed world; by the late 1980s it was the Third World which provided four-fifths of the migrants, while in Europe they provided about 40 per cent. This change was aided, in the case of Australia, New Zealand, Canada and the USA, by relaxation of immigration policies which had previously discriminated on the basis of national origin.

In situations of labour shortage in the First World, and rapid population growth, poverty and unemployment in the Third, it is not surprising that people seek to migrate. In addition to economic opportunism, natural disasters and warfare also create pressures to move. Much of such movement, especially by illegal immigrants, has tended to be to low-paid, dirty and difficult jobs which were unattractive to people in the developed world. However, in the more difficult economic circumstances of the 1990s, there has been pressure to protect the job opportunities of First World citizens. In addition, receiving countries have begun to set higher standards for immigrants, favouring the highly skilled, or those who bring in capital. In

Table 10.1 International economic migrants from the Third World, 1960–69 and 1980–89

	Millions		*As % of immigrants*	
Receiving country	*1960–69*	*1980–89*	*1960–69*	*1980–89*
USA	1.6	5.5	50	87
Germany	1.5	2.6	23	48
UK	–	1.1	–	52
Canada	0.2	0.8	18	66

Source of data: UNDP, 1992, 54

1966 the proportion of migrants from the developing world to the USA classified as skilled was 46 per cent; by 1986 it was 75 per cent. In the case of Canada the proportions were 12 per cent and 46 per cent respectively.

For the labour exporting countries, there are advantages in such movements – at the simplest that a citizen is at work overseas, rather than unemployed at home. More crucially, earnings sent home as remittances contribute to family and national income. In 1989 it was estimated that the level of official remittances to the developing world totalled some $25 billion, the equivalent of $700 per worker; it is certain that such figures are on the low side, and that clandestine remittances are significantly higher. Such earnings are significant for a country's foreign exchange position; in 1989 official remittances to Bangladesh were the equivalent of 34 per cent of the import bill and were nearly half the value of exports. Bangladeshi labour migrants find work in Europe and North America, and the majority of them in settled communities (King and Knights, 1994). However, the largest number of Bangladeshi labour migrants are in the oil-rich economies of the Middle East, and it is from these economies that most remittances to Bangladesh derive. Dependence on remittances from oil-producing states is even higher for some Middle East countries, including Jordan, Egypt and Yemen, the world's most migration-dependent state, with remittances in the 1980s being four times the value of all other exports (Findlay, 1994).

There are also negative features associated with such labour emigration. The migrant worker needs to be replaced in the household economy, and in the circumstance of the tendency for the migrant to be male, this places additional burdens on the females left behind. With the increasing trend of receiving countries to be selective, there is a growing loss of skilled labour; disproportionately the emigrants from Bangladesh are skilled professionals or men (almost exclusively) with some technical skill, such as electrician or teacher. Such selectivity deprives countries of necessary skills in short supply – and is a net loss of people trained, at some cost, within the country. It may also have the consequence of necessitating the import of expensive foreign 'experts' to replace lost skills.

Asia has been a significant source of migration to the developed world, as wage labour, domestic servants or as skilled professionals, with Britain, North America and the Gulf oil states as significant destinations. In the case of North America, the loss of skills has been particularly striking, as such migrants made up half the Asian entrants to the USA and one-third of those to Canada in 1986. Table 10.2 indicates the significance of such movements for several Asian countries.

The scale of such migration is considerable, but quite varied in its sources. Clarke *et al.* (1990) estimate that there are some 8.1 million South Asians, 22 million Chinese and 30 million Africans living outside their homelands.

Not all labour migration from the Third World is to the First World, for the growth of the Middle East oil-exporting economies from the 1970s generated an international migration system of global significance, attracting unskilled migrants from South Korea and the Philippines in the east to Morocco in the west, and skilled migrants from First World as well as Third World sources. This labour migration is very different in its structure and effects from the migrations to the First World, for it is

Table 10.2 Migrants and remittances from some Asian countries, 1989

Country	*Migrants abroad (thousands)*	*Remittances (US $ billion)*	*Remittances as % GNP*
India	–	2.7	0.9
Pakistan	1200	1.9	4.7
Bangladesh	250	0.8	3.9
Philippines	650	0.4	0.8

Source of data: UNDP, 1992, 56–9

primarily temporary and circulatory rather than leading to permanent settlement. Workers are hired for specific jobs on formal labour contracts that require repatriation at the end of the contract, and thus encourage remittances as part of the contract as well as directly by the individual migrants when they return to their home country. In this case these destination countries in the Middle East are operating as countries of the semi-periphery, in Wallerstein's world-systems terminology, using the labour resources of the periphery in order to produce oil to be consumed largely in the core, and widening the already growing economic gulf between the semi-periphery and the periphery.

However, the persistent long-term trend continues to be for migration from the Third World to the First World, fed not only by the widening economic gap, but also by the reduced cost and relative ease of international movement. There are great pressures within Third World countries for more migration of skilled and unskilled workers, but these pressures are being thwarted by more restrictive immigration policies, more stringently implemented, in the countries of the First World. The build-up of pressure for migration well beyond permitted immigration levels leads to much illegal movement, a particular problem for the USA with its long land border with Mexico. The increasing pressures for, but growing resistances to, Third World to First World migration are symptomatic of the widening global economic gulf (Gould, 1994c).

International tourists

If the movement of people from the Third World is of increasing significance, and involves permanent migration or circulatory migration (Chapter 3) for protracted periods, there is a growing short-term circulatory migration of people to the Third World in the form of tourism. The rise of this kind of movement is largely a phenomenon of the period since the 1960s. Earlier, tourism to 'far-away places' had been the prerogative of the rich, with the time and money to travel the distances involved, mainly by sea. In the period after 1945, with increasing affluence and leisure time in North America and Europe, the development of the jet aircraft, and the creation of the 'package tour' industry, mass tourism overseas became possible. In the case of North America this was to the proximate opportunities of Mexico and the Caribbean. Within Europe initial development was intra-continental, with northern Europeans in particular going to the sunnier environs of the Mediterranean, but in recent years the factors of affluence and ease of organized travel have extended tourism to include parts of the Third World.

Tourism has become a 'growth industry'; between 1981 and 1990 international tourist expenditure rose from $99 billion to $238 billion, and over the period 1986–90, the number of international tourist arrivals increased from 341 million to 455 million. Nonetheless, the developed world still dominates tourism; in 1990 30 per cent of *all* tourist arrivals were to only four countries – France, Spain, Italy and the United Kingdom. Third World countries received about one-quarter of international tourists and tourist revenue, 1986–90 (United Nations, 1993, 962–9).

People take holidays to relax, to escape from their everyday lives, familiar environments (and weather!) and to find new experiences in terms

The impact of tourism in the Third World may not be entirely positive. 'Package' tours may result in traditional society and culture being 'packaged' and debased to entertain tourists

of climate, environment, civilization, people, culture, lifestyle or cuisine. The rise of international tourism reflects access to more distant lands, rather than Blackpool, Atlantic City or Deauville. For the developed world tourist the Third World offers a perhaps stereotypical 'tropical paradise' of golden beaches, palm trees, sparkling blue waters, and sunshine. It also offers exotic environments, different cultures and ancient civilizations, strange wildlife – or, for some, the 'hippie trail' to India or Nepal or 'sex tourism' in Bangkok or Kenya.

Tourism has, therefore, become a growth industry in parts of the Third World, particularly as governments, entrepreneurs and First World tour operators have recognized its potential. It has a particular significance as an economic activity compared to most other sectors – the consumer has to come to the 'exporting country', rather than having the 'product' delivered to his or her home. Tourism is seen as a potential source of revenue and foreign exchange, of overseas investment, of job creation, and of the creation of demand for local goods and services, such as foodstuffs, souvenirs and transport. It has also been seen as significant for regional development, in that demand for resort tourism can be directed to less developed and 'unspoiled' areas. There are costs involved, in terms of promoting tourism and in providing the essential infrastructure of airports, hotels, transport and basic services of water and electricity.

There are also less direct 'costs' of developing a tourist industry. Much of the cost of long-distance travel is accounted for in air travel, usually in developed country airlines, and in tourist hotels, owned or managed by MNCs specializing in leisure industries, such as the Hilton or Sheraton (USA) or Forte Crest (UK) hotel chains. Their profits are repatriated to the home country. Tourists create a demand for imported (i.e. familiar) rather than local foodstuffs. There are impacts on local culture and traditions: positively in encouraging local musicians and craftsmen and women, but also negatively in adapting their work for the international mass market in tourist trinkets. The 'demonstration effect' of First World affluence, often ostentatiously paraded, within poor societies can also bring severe social conflicts, and a clash of cultures over such issues as

Plate 10.2 Tourist rickshaw, Kathmandu, Nepal Indigenous forms of urban transport have become part of the tourist infrastructure and provide an element of 'local colour'

appropriate dress standards and behaviour in mosques, temples and churches can further inflame conflicts over the value of the tourist industry (Plate 10.2). The environment may also be affected by tourist developments due to provision of resort facilities in previously 'unspoilt' areas. Natural landscape is a 'resource' for the tourist industry that, like any resource, needs to be managed if it is to support its long-term sustainability, but intensive tourist developments may place that at risk.

Nonetheless, some countries have become very active promoters of tourism, as sources of revenue and economic growth, and as a stimulus to regional development. Figure 10.2 shows the broad pattern of revenue generated by international tourism into the Third World. In gross terms the largest movement is to Asia and to Mexico and the Caribbean, and tourism remains of lesser significance to Africa and the Pacific islands. The leading Third World tourist destinations are Mexico, China, Malaysia, Hong Kong, Thailand and Singapore. However, though the total number of arrivals may be much smaller, tourist income has become of considerable significance to island economies such as Aruba, Barbados, the Virgin Islands, Fiji, the Seychelles or French Polynesia, and it is in such locations that tourism has its greatest cultural and visual impact.

It is important to recognize that international tourism involves not only a one-way exchange between developed and developing countries. For the affluent at least of the Third World there is also tourism to Europe and the USA, and also to other parts of the Third World. At a macro-scale, most movement is, in fact, inter-regional. Thus, for the leading tourist countries in Latin America, most tourists come from the Americas; similarly for Asia and the Middle East, the main sources are within their respective continents. There are a few exceptions – for Egypt,

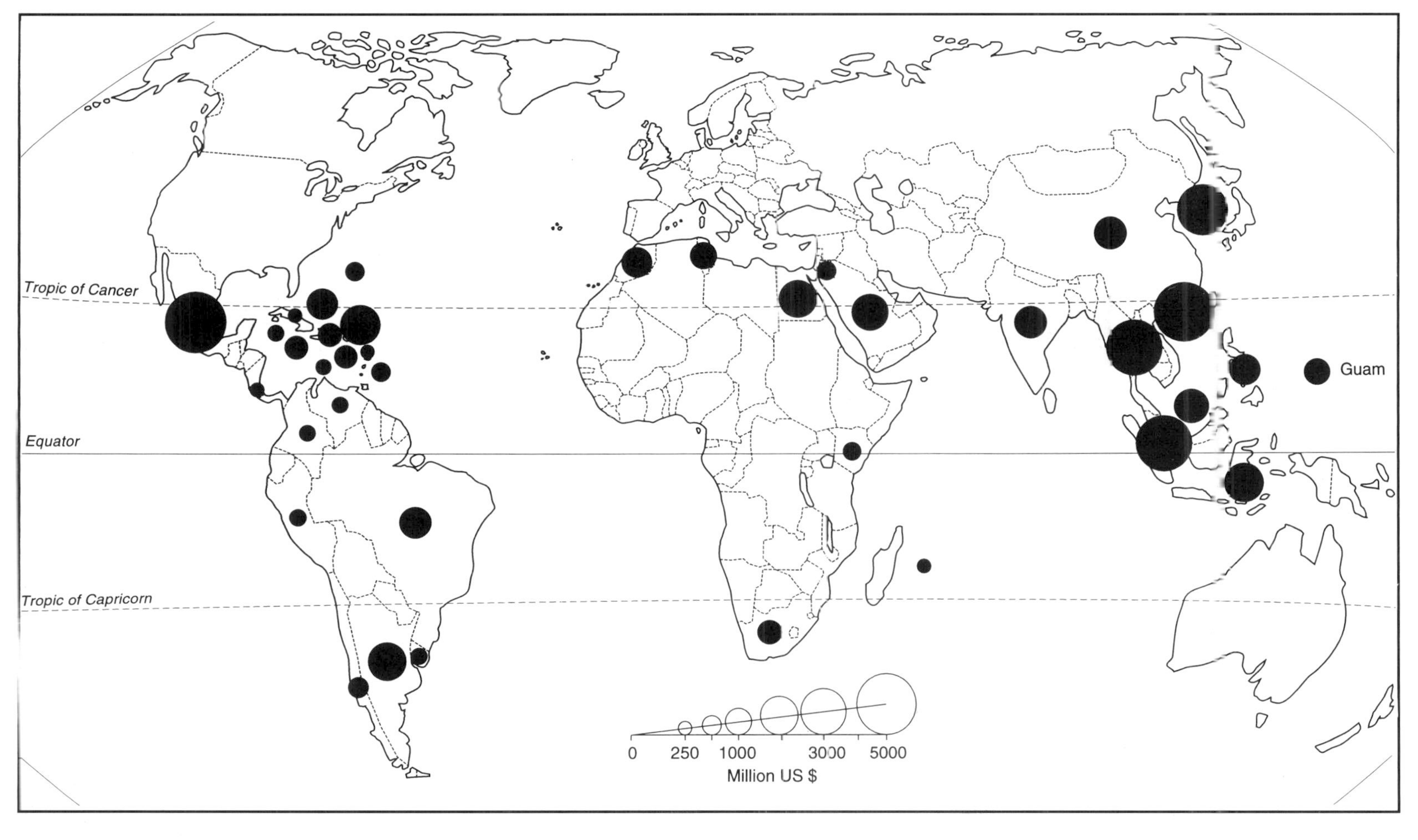

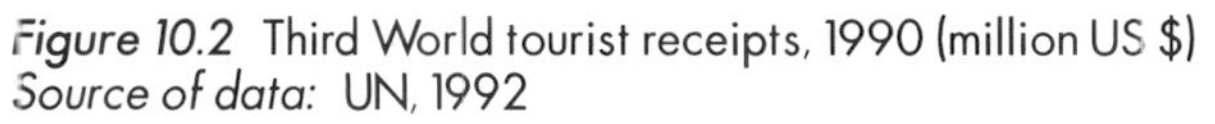

Figure 10.2 Third World tourist receipts, 1990 (million US $)
Source of data: UN, 1992

BOX 10.1 TOURISM IN THE CARIBBEAN

The thirty-two small island states of the Caribbean are comprised of a variety of geological formations. The resulting topography, waterfalls, sandy beaches, together with colourful wildlife and tropical climate, and the legacies of varied colonial histories reflected in the cultural landscape, provide valuable resources for tourism.

The industry has become an important component in the economies of a number of Caribbean countries since the 1950s and, with the notable exceptions of Haiti and Cuba, expanded rapidly after 1980. Between 1985 and 1993 the number of tourist arrivals grew by 67 per cent, and the number of hotel rooms in the region virtually doubled between 1980 and 1993. In the latter year there were 161,500 hotel rooms in the Caribbean, with half of the total in the Bahamas, Cuba and the Dominican Republic. In addition, visits by cruise ships are important; in 1993 8.88 million cruise visitors were recorded in Caribbean ports (Plate 10.3).

The United States is the most important source of tourists, though there are a few cases where former colonial ties give a slightly different pattern (Figure 10.3). In 1993, 52.4 per cent of all tourists to the region came from the USA, 17 per cent from Europe and 5 per cent from Canada.

Substantial investment has been made in tourism infrastructure and development, and in advertising campaigns. The industry has been promoted for its ability to generate employment and earn foreign exchange. In the Bahamas tourism provides two-thirds of total employment, and in Jamaica, though the

Plate 10.3 Cruise ships, Miami, USA These liners encapsulate the linkages between First World and Third. The ships are registered mainly in Europe, their passengers are mainly Americans, and their destinations the Caribbean islands and South America

BOX 10.1 *continued*

proportion is just under 5 per cent, the industry sustains 250,000 jobs. Tourism contributes significantly to Caribbean economies; in the case of Jamaica tourist receipts in 1990 accounted for 18.6 per cent of GDP and 54.9 per cent of export earnings.

There are, however, negative aspects to tourism. All-inclusive packages mean that much of the expenditure is in the country of origin of the tourist, and multinational companies control a significant proportion of the airlines, hotels, tour operations and services, and reap a significant share of the profits. The enclave nature of the industry limits interaction with other facets of the island economies and societies, particularly in the case of cruise tourism. There are also environmental costs as a consequence of the high density of resort development, the reclamation of wetlands and mangroves for hotel construction, pollution by cruise ships and yachts, and pressure on fragile environmental attractions. Tourism also contributes to an erosion of Caribbean cultures and intensifies the penetration of North American consumer values.

Though tourism is an important contributor to the economies and employment of the Caribbean islands, a crucial element for its future growth is to secure a balance between economic gain, social well-being, and the environmental integrity of the region.

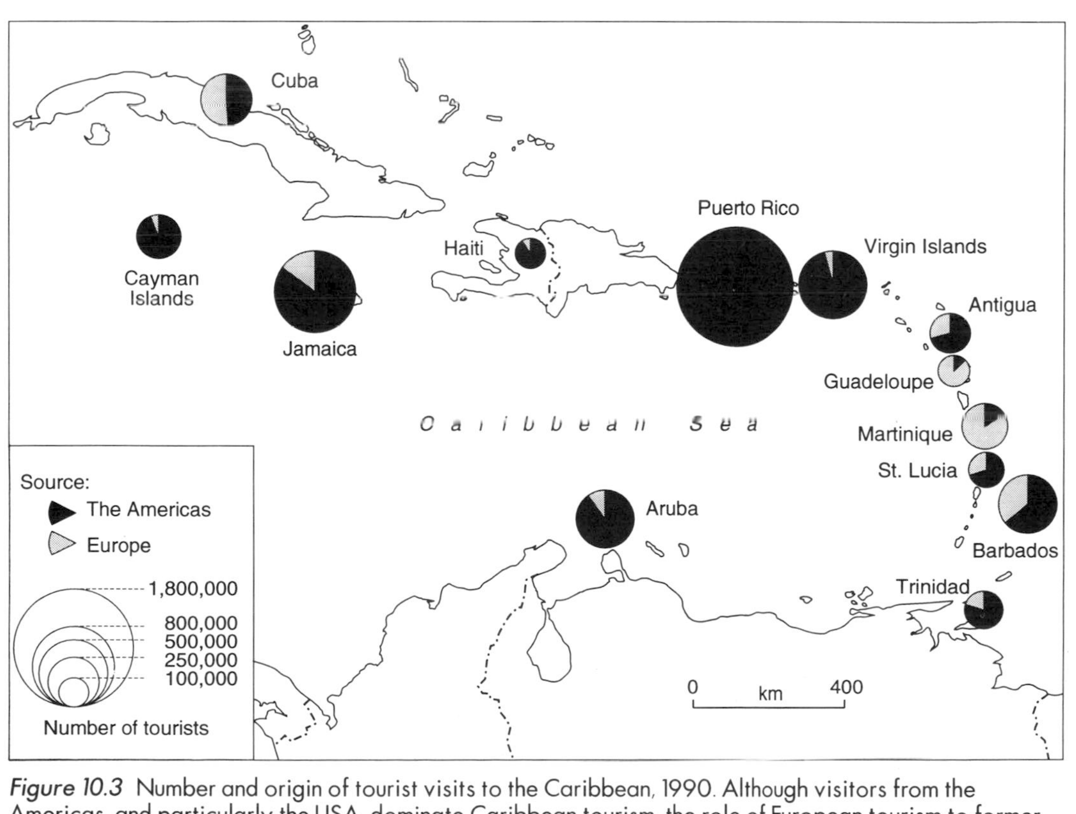

Figure 10.3 Number and origin of tourist visits to the Caribbean, 1990. Although visitors from the Americas, and particularly the USA, dominate Caribbean tourism, the role of European tourism to former French and British islands is evident
Source of data: UN, 1993

Tunisia, Kenya and India, Europe is the leading source of visitors. Europeans are also leading visitors in some of the smaller or newer destinations, such as Guadeloupe, Martinique, Gambia, the Seychelles and Maldives, where tourism, though unimportant in global terms, has great significance for the local economy. However, the role of domestic tourism – Brazilians in their many beach resorts, Indians in the former colonial hill-stations and now tourist towns – should also not be overlooked as a contributor to the development process.

Over the period 1986–90 tourism in the Third World grew relatively slowly, in terms of both visitors and revenue. This was primarily due to the world recession, but clearly, economic recovery, increased affluence and leisure, and improving transport all make it an activity of considerable potential. Its limitations are perhaps its potential impact on local societies, and the homogenization engendered by mass tourism designed for, and often controlled by, the developed world. There is also a limitation imposed by the mutual competition derived from resort tourism alone, if it depends only on sea, sun and sand, and from which individual countries attempt to escape by creating their own distinctive image through extensive media advertising in the First World.

COMMODITIES

International trade

The exchange of commodities is a major element in the linkages between the developed countries and the Third World (Plate 10.4). A search for the spices and precious minerals of the Orient was a key factor in the European exploration of the world; the exploitation of these commodities, tropical crops and base metals was an essential element of imperialism and colonial control. By the turn of the present century a pattern of international trade dominated by Europe and the United States was clearly defined, based on the assumption of free trade and theoretically dependent on the economic doctrine of comparative advantage, whereby each country produced those goods where it had a cost advantage over others, and traded them for goods for which other countries had advantage. The pattern established was one in which the developed countries produced manufactured goods most cheaply and traded them for primary products which they could not grow or mine, or could produce less efficiently. This pattern involved not only the countries we now define as the Third World, but also countries such as Canada, Australia and New Zealand, which could produce temperate crops and livestock more cheaply than Europe. Their development in the twentieth century has taken

Plate 10.4 Tea advertisement, Sousse, Tunisia The poster illustrates the multinationality of trade, in which 'Ceylon tea' from Sri Lanka is advertised in four languages in another Third World country

them out of such a dependent relationship. For the Third World, however, in both colonial and post-colonial circumstances, such a balance has largely persisted, in which many of them still have a trading relationship in which they export primary products in return for manufactured goods from the First World.

This general pattern is evident in Figure 1.9, and elements of it have recurred in other chapters. However, it is evident that the pattern is not uniform; while for much of Africa minerals and tropical crops provide over three-quarters of exports, the situation in Asia and Latin America is more varied. Some Third World countries have established a role in the export of manufactured goods. In the case of the NICs this is mainly the result of deliberate export-oriented industrialization structures, in which they have developed the export of electrical goods and similar products. In other cases the rise of manufactured exports comes from the comparative advantage they have in producing low-cost consumer goods taking advantage of local raw materials and low labour costs, as in the production of textiles, clothing and footwear. Thus, in 1990, textiles and clothing generated over half the export earnings of Pakistan and Bangladesh, and over one-fifth of those of Egypt, India, China and Sri Lanka. For many Third World countries, however, dependence on primary product exports persists, with petroleum as a distinctive case. Thus, in 1990, fuel and minerals provided over 90 per cent of the export earnings of Libya, Saudi Arabia and Iran, and over half those of Chile, Peru, Bolivia, Zaire and Papua New Guinea.

Trade is of great significance for the Third World, as a source of goods they do not produce, and more crucially as a source of revenue with which to finance their development. However, this process of exchange has numerous disadvantages for them. Under colonial regimes these resources were often 'exploited' for inadequate returns, and the terms of trade continue to be disadvantageous. In the period since the Second World War prices for primary commodities have tended to grow less rapidly than those for manufactured goods, so that Third World countries have needed to produce *more* in order to purchase the same amount of goods from the developed world. Although the oil crises of the 1970s afforded considerable benefits for the OPEC countries, as a result of greatly increased prices for oil, these trends have not been sustained, and ideas that other primary products might similarly benefit have not been realized. Instead, alternative oil sources in the developed world, such as the North Sea and the north slope of Alaska, and even in the Third World have tended to diminish this temporary advantage.

During the 1980s, a period of slow-down and recession in the world economy, some of the worst effects were felt in the Third World. For most developed countries the terms of trade in the period 1987–90 showed slight improvements or remained static; in much of the Third World they deteriorated; in some cases, particularly in Africa, by more than 10 per cent. Similarly, over the longer period of the 1980s, although most of the developed world and the Third World experienced some growth in their export trade, a number of countries, again mainly in Africa, experienced negative growth rates, or decline.

The Third World countries are, therefore, dependent on primary product exports, for which prices fluctuate and which in the medium term have moved to their disadvantage. They tend to be the weaker partners in trading relationships – and their role in global trade is relatively modest. As Table 10.3 shows, world trade is dominated by the countries of the developed world.

Perhaps even more striking, given the importance of primary products in the trading role of the Third World, is its *relative* non-importance in global trade even for these commodities. The region generates about one-fifth of world exports, and receives the same level of imports (Table 10.4).

However, as the table indicates, the developed economies dominate the export *and* import structure for most groups of commodities: *intra*-First World trade is the dominant element in world trade. Only in the case of the export of minerals and fuels is that status reduced, and the role of the Third World more evident. It is also worth noting, given the contribution mentioned above that some Third

Table 10.3 Shares of world trade, 1990 (percentage of total)

	Exports	*Imports*
EU	39.8	39.8
Rest of Europe	6.7	6.6
USA/Canada	14.7	17.3
Japan	8.5	5.9
Other developed countries	2.3	2.3
Ex-Soviet bloc	5.1	4.2
Africa	2.0	2.4
Latin America	3.9	3.9
Middle East	3.1	3.0
Asia	13.3	12.6
Other Third World	0.7	0.9

Source of data: United Nations, 1993

Table 10.4 Structure of world commodity trade, 1990 (percentages of total)

	Developed economies		Ex-Soviet bloc		Developing economies	
	Exports	*Imports*	*Exports*	*Imports*	*Exports*	*Imports*
Total	72.1	71.6	5.1	4.2	22.8	22.8
Food, drink, tobacco	69.4	71.6	3.1	6.1	27.5	21.7
Crude materials	65.1	70.5	6.3	4.5	28.6	24.4
Minerals and fuels	30.4	68.2	14.1	5.9	55.4	19.8
Chemicals	85.0	69.8	3.5	4.0	11.5	25.4
Machinery and vehicles	84.3	71.9	3.2	3.7	12.5	23.9
Other manufactures	70.4	75.1	2.9	3.1	26.7	21.4

Source of data: United Nations, 1993, 932

World countries have begun to make to world trade in manufactures, that whereas developing countries provide around one-tenth of exports of 'advanced' manufactures such as chemicals, machinery and vehicles, they generate one-quarter of 'other' manufactures, such as textiles and processed foods.

The pattern of world trade has changed in the twentieth century, with a decline in free trade, with Commonwealth preference agreements in the 1930s, the General Agreement on Tariffs and Trade (GATT) from the 1940s, and common markets, notably that of the European Economic Community (EEC), from the 1960s. The Common Agricultural Policy of the EEC, now the European Union, for example, was designed to promote agricultural self-sufficiency within the community, to the exclusion of temperate and some tropical imports, notably cane sugar in favour of European beet sugar, to the particular detriment of Caribbean sugar producers such as Jamaica and Guyana. Similar trading blocs to promote trade within the Third World countries, such as the Latin American Free Trade Area (LAFTA), have generally been less successful at stimulating intra-Third World trade. The creation of the North American Free Trade Area (NAFTA) in 1993, involving Canada, the USA and Mexico, will provide economic opportunities for Mexico in the richer markets to the north, but probably at the expense of other Latin American and Caribbean countries' access to the large North American market. Furthermore, world trade in food has shown some change, partly in response to food deficits in parts of the Third World. The USA and Canada are major sources of cereals to Third World countries (for example Nigeria,

Ghana and Papua New Guinea), which is partly a reflection of their comparative advantage in producing these crops, and as a result of new patterns of consumption, as urbanizing populations adopt a demand for wheat flour and other developed world foodstuffs, rather than traditional foods. The Third World is thus, as Figure 10.4 shows, an importer as well as an exporter of foodstuffs. Third, as noted above, for some countries at least, the export of textiles, clothing, electrical and other manufactured goods has shifted their trade structure.

An important, though far from complete, trend since the publication of the Brandt Report in 1980, has been the effort to create freer patterns of trade, particularly between North and South. It was argued by Brandt and others that 'Trade, not Aid' would be a better stimulus to development. Developed countries have accepted that an expansion of trade would benefit the Third World, though it would also benefit their exporters and so there would be mutual benefit. The developing countries, on the other hand, suspicious of the power of multinational companies, whether producers, consumers or distributors, to dominate free global markets, have pressed for greater concessions, in terms of price and quantity for specific commodities. The global recession of the 1980s, and the anxiety of advanced countries to protect their economies, trade and jobs, has meant that progress in promoting freer trade has been slow.

The General Agreement on Tariffs and Trade (GATT) was established in 1948, at about the same time as the World Bank and the IMF, as one of the institutions of global economic management, with the explicit purpose of negotiating tariff reductions to stimulate trade, and in that role GATT is supported by the joint economic muscle of the World Bank and the IMF. There have been eight rounds of GATT negotiations since its foundation and these have produced impressive results that have been seen in the steadily growing volume and changing patterns of trade, as outlined above. The eighth round of negotiations, begun in 1986 and concluded after long periods of bitter debate in

"Of course, it is our way of helping the overdeveloped nations."

The balance of trade between the First and Third World is unequal in value and volume, and usually involves an exchange of primary products for manufactured goods

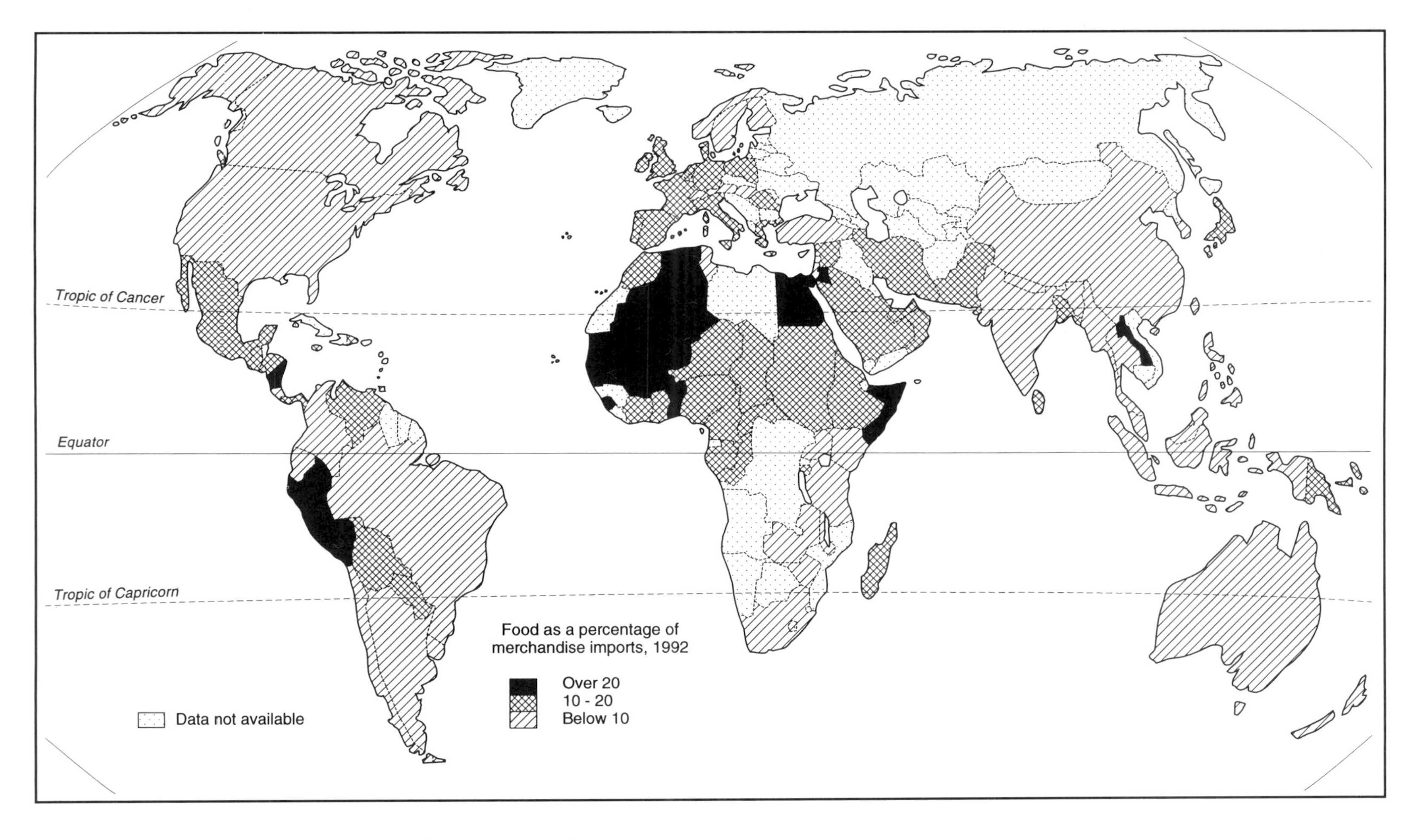

Figure 10.4 Food as a percentage of merchandise imports, 1992
Source of data: World Bank, 1994a

1993, sought not only to greatly reduce trade barriers for established products but to extend GATT's concerns into such areas as banking and insurance that by the 1980s, with the great advances in global communications, had become much more international. The principal controversies at the GATT talks, however, were those between the USA and Europe on the levels of subsidies for agriculture (there was no agreement on banking and insurance), and issues of critical concern to the Third World, such as better access to First World markets, were rather relegated in importance. A freer market in agricultural products may stimulate some Third World agricultural exports, but can be a threat to cereal production in some countries, now more exposed to cheap American imports, especially of wheat. One frequently voiced criticism of GATT and freer trade is that multinational companies, mostly from First World countries, will be the immediate beneficiaries, and there seems little prospect of the GATT negotiations affecting the growing dominance of MNCs in international trade.

Plate 10.5 Drug control advertisement, Peru Drugs are a problem in producing as well as consuming countries. This poster was part of a Peruvian campaign to discourage the cultivation of coca. This has long been grown and used in Peru but has become a major source for the international cocaine traffic. The slogan reads 'We are all part of the solution'

Illicit trade

Just as the increasing demand for, but increasing controls on, migrant labour in developed countries has created an illegal 'trade' in people, so too there has been a growing proportion of international trade that has been 'illegal'. One particular element of increasing significance in trade between the Third and First World derives from the illegal traffic in drugs. The production and movement of drugs such as opium (for heroin) in the so-called 'Golden Triangle' of the Thailand/Myanmar borders and the 'Golden Crescent' of Afghanistan and Central Asia, and of coca (for cocaine) from the 'Silver Triangle' of Andean Colombia, Bolivia and Peru, has grown rapidly since the 1970s, to fuel demand in the cities of Europe and North America. The drug market in the USA alone is estimated at over $100 billion, and drug production, smuggling and distribution has become, literally, 'big business'.

The altiplano of Peru and Bolivia is the source of 90 per cent of the world's cocaine. Until the 1970s the coca plant and its leaf had been grown and used for centuries as part of the traditional culture of the Andean peasantry, who chewed the leaf as a mild stimulant. With the emergence of demand in the USA for cocaine, derived from coca, the peasants began to plant on a large scale, generating a considerable illegal industry and trade. It is estimated that the crop generates over $750 million in foreign earnings for Peru and $600 million for Bolivia, the equivalent of at least 25 per cent of their legitimate export earnings (Garcia, 1991, 464).

The implications of this traffic in the relations between First and Third Worlds are considerable. In 1987 the estimated expenditure of North American consumers on cocaine ($80 billion) was equivalent to the combined external debt of Colombia, Peru and Bolivia of $79 million (Sage, 1991, 329). In 1990 the President of Bolivia claimed that half of the real GNP of his country was related to cocaine. In a narrowly economic sense drug crops offer considerable benefit to the producer. It expands land use and creates income for the peasantry. In the case of coca it has the advantages of being cultivable on steep slopes and soils of low fertility; it requires little in the way of fertilizer; the leaf is a low-bulk commodity for transport; and it can yield three or four crops per year. It stimulates local processing activities in the making of coca paste and in more sophisticated processing laboratories in countries such as Colombia. It brings in foreign exchange, some of which may be invested in legitimate businesses. However, it has also been associated with violence, crime and guerrilla movements in the producing areas in Peru, Bolivia and Colombia, and in staging posts for the drug traffic in the Caribbean and Panama (Plate 10.5).

Attempts to eradicate the activity pose many problems beyond those of policing the industry in the consuming countries. Collaborative efforts by the USA and Mexico to reduce production of opium and marijuana merely resulted in a shift of cultivation to Belize, Jamaica and Colombia, and Mexico still provides possibly one-third of the American market for marijuana. The use of powerful herbicides to destroy drug plantations has been discouraged because of risk to more innocent crops and fear of ecological damage. The USA's 'war on drugs' in Latin America in the late 1980s provoked nationalistic hostility to American intervention.

The rise of the inter-American drugs trade since the 1980s has been seen as one of the major dilemmas for Latin America and the Caribbean, ranking alongside debt, trade and the environment as major development concerns linking North and South (Sage, 1991, 332). Such concerns are also relevant to drug-producing areas in Asia and other parts of the Third World. The trade has provided major financial assistance during a period of economic recession, and has created a significant, though illegal, economy offering jobs and income to large numbers of otherwise impoverished people (it has been estimated that there are between 40,000 and 70,000 coca growers in Bolivia). Stamping out demand on the streets of First World cities might be one way of breaking the trade, but a key element is to provide alternative, legitimate crops for farmers in Latin America and Asia which will offer them adequate and sustainable financial returns.

CAPITAL

Countries of the Third World, generally lacking sufficient capital from internal sources but anxious to embark on ambitious development programmes, are obliged to seek support from external sources to obtain the capital resources to finance them, and in so doing incur debts that need to be repaid. The measurement of indebtedness is problematic, and can be done in several ways – as a gross figure of repayments, as a per capita figure, as a proportion of GDP, or as a proportion of export earnings. However measured, the crippling burden of indebtedness has become a major economic issue to individual countries and to the Third World taken as a whole. Indebtedness is highest in gross terms in Brazil, Mexico, India, Indonesia and Argentina, and in each of these countries repayment of interest on loans accounts for over 20 per cent of exports. This is only 3 per cent of Brazil's GDP but 9 per cent of Indonesia's, and the per capita level of repayments for Indonesia is more than twice that of Brazil, so that Indonesia has a larger problem. Overall Third World countries in 1990 paid about 15 per cent of their exports of goods and services in debt repayments, a level that is significantly higher than the average of 10 per cent in the mid-1970s, but this level varies enormously, rising to over 30 per cent in a few countries (Figure 10.5).

This debt is created by support to governments either from commercial sources or through 'official development assistance', more commonly described as 'aid'.

Commercial loans

In their borrowing from commercial banks, Third World countries are in direct competition

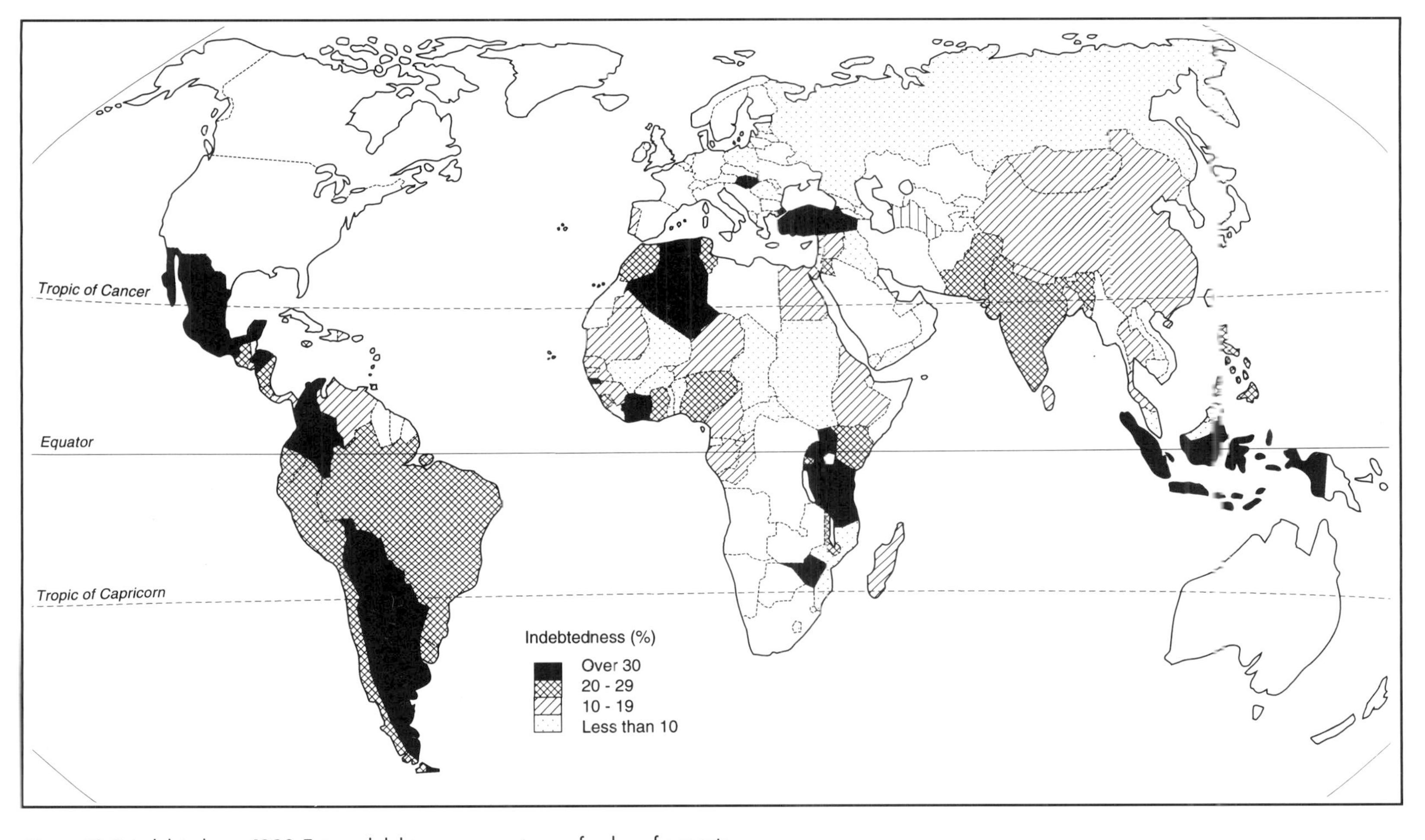

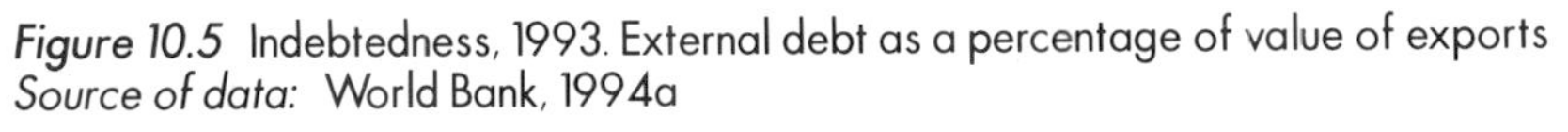

Figure 10.5 Indebtedness, 1993. External debt as a percentage of value of exports
Source of data: World Bank, 1994a

with demands from within the developed world for private sector and government loans from these same banks, for banking is very much an international business, with the world leaders located in New York, London, Tokyo and Frankfurt. Since investment risks in developed countries are perceived by banks to be less than they are in Third World countries, these countries are at a disadvantage. Banks do not wish to lend to countries with high levels of indebtedness and political instability. Not surprisingly, therefore, many countries do not find commercially raised capital an attractive prospect. Nevertheless, many countries, and particularly the middle-income countries, largely in Latin America and East Asia, have in the past been able to borrow extensively from commercial banks, and these countries have generally become severely indebted to them. The main period of borrowing from the commercial banks was in the 1970s, when there was a global capital surplus fuelled by the massive windfall profits from oil. These profits were being invested by oil producers in the world banking system, for the oil producers were not themselves able to use effectively the revenues they were accumulating. There were therefore low interest rates at this time of capital surplus, and many countries sought to borrow to finance their development plans. However, by the 1980s and into the 1990s there has been increasing capital shortage as more countries sought to borrow and the capital supply contracted in a period of global recession. The transformation of the economies of Eastern Europe and the former USSR since 1989 from socialist to market economies has created a great demand for commercial finance to support their restructuring which has raised interest rates further; and the international banking community, assisted by such new institutions as the European Bank for Reconstruction and Development, which, unlike the International Bank for Reconstruction and Development (the World Bank), lends solely to the private sector, is more attracted to investment possibilities in these countries than it is to the prospects in much of the Third World.

Despite these pressures, however, the net flow of commercial resource to developing countries has been growing in the early 1990s, but it is increasingly focused on Latin America, East Asia, as well as the countries of the former USSR, with low-income countries being much reduced in significance for the commercial sector. The debt repayment crisis was at its height in the mid-1980s when it threatened to undermine the global banking system. Major Latin American countries, notably Mexico, intimated their reluctance to continue the high level of repayments, but subsequent rescheduling of the repayments and 'softer' international conditionalities associated with structural adjustment, have made the problem less urgent for the global system, though individual countries remain seriously affected.

Increasingly, however, resource transfers of capital, as of technology and manpower, occur within multinational companies through foreign direct investment (FDI). Capital transfers may be needed to establish new industries or factories, or to expand existing operations, in the expectation for further benefit to the company through the transfer of profits to the parent company in its home country, usually in the developed world. The World Bank estimated that about one-third of all FDI by MNCs in 1993 was in developing countries, with 59 per cent of that proportion going to five countries – China, Mexico, Argentina, Malaysia, Thailand (World Bank, 1994a, 28).

Aid

Official development assistance (ODA), commonly known as Aid, is available to Third World countries either through multilateral institutions, such as the World Bank or the UN agencies, or through bilateral agencies, in direct government-to-government capital transfer, often for specific purposes (Plate 10.6). The United Nations has recommended that 0.7 per cent of the GNP in rich countries should be given as aid, but that proportion is only reached in a few countries, notably in Scandinavia and in Saudi Arabia and the United Arab Emirates (Table 10.5). Furthermore, the proportion has been consistently falling in most of these countries since the mid-1980s, as governments in

Plate 10.6 Road project, Nepal This is part of the 'Friendship Road' linking Nepal and China through the Himalayas, being maintained with the assistance of Swiss funds and the substantial knowledge of mountain roads of Swiss engineers. Note its use though by pedestrians, a woman carrying a headload, as well as an overcrowded lorry

Table 10.5 Official development assistance, 1990

	Total (US $mill.)	*As a % of donor's GNP*
OECD countries		
Norway	1,205	1.17
Netherlands	2,592	0.94
Denmark	1,171	0.93
Sweden	2,012	0.90
France	9,380	0.79
Finland	846	0.64
Belgium	889	0.45
Canada	2,470	0.44
Germany	6,320	0.42
Australia	995	0.34
Italy	3,395	0.32
Japan	9,069	0.31
Switzerland	750	0.31
United Kingdom	2,638	0.27
Austria	394	0.25
New Zealand	95	0.23
United States	11,394	0.21
Ireland	57	0.16
OPEC countries		
Saudi Arabia	3,692	3.9
United Arab Emirates	888	2.65
Nigeria	13	0.06
Algeria	7	0.03
Venezuela	15	0.03
Qatar	1	0.02
Libya	4	0.01

Note: ODA amounts include funds channelled through multilateral and regional channels, as well as in bilateral programmes.
Source of data: World Bank, 1992

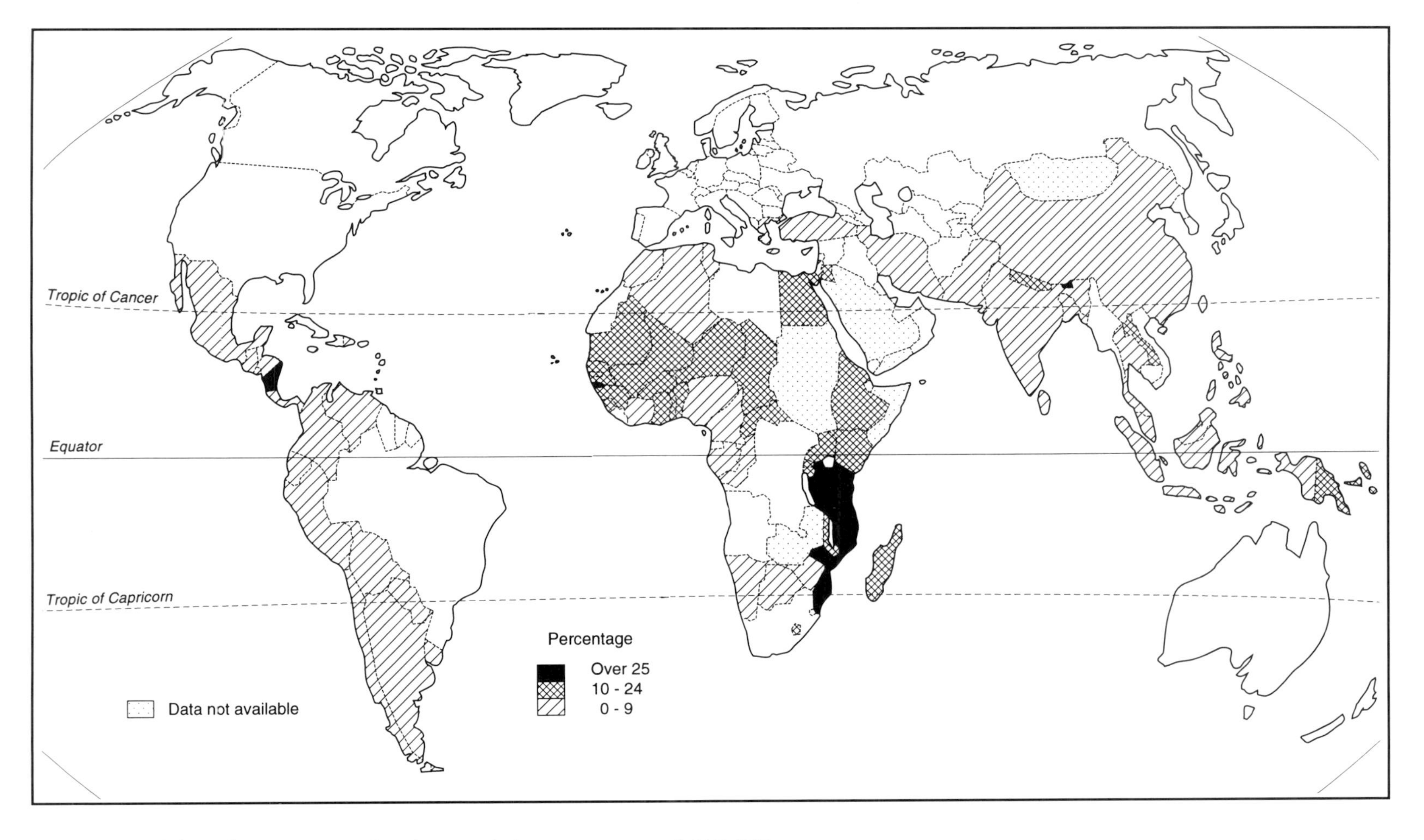

Figure 10.6 Aid dependency: Overseas Development Aid as a percentage of GNP, 1991
Source of data: World Bank, 1994a

Plate 10.7 Aid project, India Women and youths carrying rocks for the construction of a village dam, in a food-for-work project in Bihar

Europe and North America have sought to restructure their own spending priorities in the direction of reducing the activities of government.

Just as it is not necessarily the richest countries that give most aid (though there is a broad pattern of aid from rich to poor), so it is not the poorest countries that necessarily receive most aid (though clearly low-income countries taken as a whole are the main recipients). Figure 10.6, of official aid as a proportion of GNP of recipients, shows that a proportion of over 25 per cent is reached in a few extreme cases of severe and prolonged economic crisis, such as Mozambique, Tanzania and Guinea-Bissau. Overall Africa is clearly the most aid-dependent continent, as would be expected. However, within Africa Kenya, one of the richest countries, is among the largest recipients of ODA, which amounts to 11 per cent of its GNP, while Sierra Leone, one of the lowest-ranking countries on the UNDP Human Development Index, receives only 8 per cent of its GNP as ODA. Aid is given not only on the basis of absolute need, but on the ability of the recipient to use the capital transfer for development, and also on the political relationships between donor and recipient. Political disagreement and change of government can significantly affect the level and origins of aid received, as Sudan experienced, for example, during the Gulf War of 1991, when its much-needed food aid from the USA and the European Community was suddenly cut off as a penalty for its strong support for Saddam Hussein. In addition to regular aid programmes there is emergency aid, more specifically targeted, often through multilateral agencies such as the Red Cross and Red Crescent, to populations in areas of earthquakes and other natural disasters and to refugees in areas of political instability, usually through the UN High Commissioner for Refugees (UNHCR).

Aid is given in part for humanitarian and redistributive reasons, but also in part for

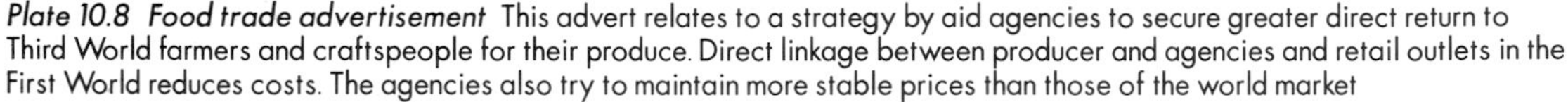

Plate 10.8 Food trade advertisement This advert relates to a strategy by aid agencies to secure greater direct return to Third World farmers and craftspeople for their produce. Direct linkage between producer and agencies and retail outlets in the First World reduces costs. The agencies also try to maintain more stable prices than those of the world market

strictly developmental reasons. The Brandt Commission in 1980 was particularly influential in making the mutuality argument about aid – that it brings mutual benefits to donor and recipient since it contributes further development to the recipient, its exports will expand and so will its imports, largely from donor countries since much developmental bilateral ODA is 'tied', in the sense that the donors specify purchases of materials and technical assistance must come from the donor country. There is regular criticism of large proportions of aid budgets being used to pay the high wages of specialists from the donor countries, so that the local multiplier effects are not necessarily large. More commonly, however, aid budgets are increasingly being used to provide additional support for established programmes of recipients rather than, as was more typical of the 1970s and 1980s, establishing separate and parallel development programmes and agencies directly and solely funded from external sources.

A still small but rapidly increasing proportion of aid does not pass through official government–government channels, but is provided privately through non-governmental organizations, such as Oxfam, Save the Children and the French medical charity, *Médecins sans frontières* (Plate 10.7). Their resources come directly from the public for the most part, though increasingly official bilateral agencies are directing funds through their national NGOs, and this has been particularly apparent in Britain in the 1990s when the programmes of the Overseas Development Agency, the government department responsible for aid, has directly supported the humanitarian and developmental work of a range of British charities to implement the objectives of national bilateral programmes. The NGOs are normally obliged to deal directly with the governments of the receiving countries, but increasingly here too there is a growing number and range of indigenous NGOs to which resources can be transferred (see Plate 10.8; Figure 5.4).

IDEAS

The media

The mass media provide one of the most powerful linkages between the developed and the Third World. Even in the period since the first edition of this book appeared, television has become a very potent force in bringing the condition of the Third World before audiences in the rich North. Reports of the impact of famine in the Sahel and the Horn of Africa, of natural catastrophes such as earthquakes and floods, of civil strife in Haiti or Rwanda, for example, or of environmental concerns such as rainforest destruction, provide familiar images on television sets in the First World, and prompt popular concern for the plight of the victims. There has been a 'globalization' of news, and television in particular has heightened awareness of the Third World and its problems, with agencies such as the US-based Cable News Network (CNN) bringing instantaneous and extensive reporting. It is appropriate to ask, however, whether such coverage is comprehensive: British coverage tends to focus on Africa. Is this because problems there are most extreme, or the images most graphic, or because of an historic interest in the region, not matched, say, for Latin America? Is such coverage 'news'; is it pro-active, intended to prompt concern and response; or is it merely some kind of intrusive and morbid curiosity?

Some critics within the Third World suggest that developed world media coverage is one-sided and negative, presenting only poverty and catastrophe. They claim that the images presented focus on disaster, natural or man-made, and rarely if ever relate to progress, to new developments, or even to offering a constructive analysis of development issues. More broadly, it has been argued that the media have created a 'cultural imperialism' in which the technology and hardware of communications, together with the 'software' of media products such as television programmes, films and advertising have created a new dependency within the Third World.

What is evident is that access to media sources has become more widespread. Access to radio from the 1930s and TV from the 1950s has meant greater awareness of a wider world within the Third World, and the advent of the transistor radio, video and the satellite dish have given freedom from power sources or transmitter catchment areas. In 1988 UNESCO recorded that there were only 39 countries without television services. Of these, 13 were in Africa, 8 in Latin America, 15 in Asia and Oceania, and 3 in Europe. Except in Africa, these were mainly small island territories, or small states such as San Marino and Bhutan. However, as Figure 10.7 shows, even provision of a television service does not guarantee access. In many countries, compared to the developed world, the number of TV sets per 1000 population is low – though in some cases communally owned sets may be installed in public places, or in bars and community institutions, and thus

Plate 10.9 Television advertisement, India The media message: the modern technology of satellite television is advertised by a hand-painted sign as people cycle to work, Varanasi

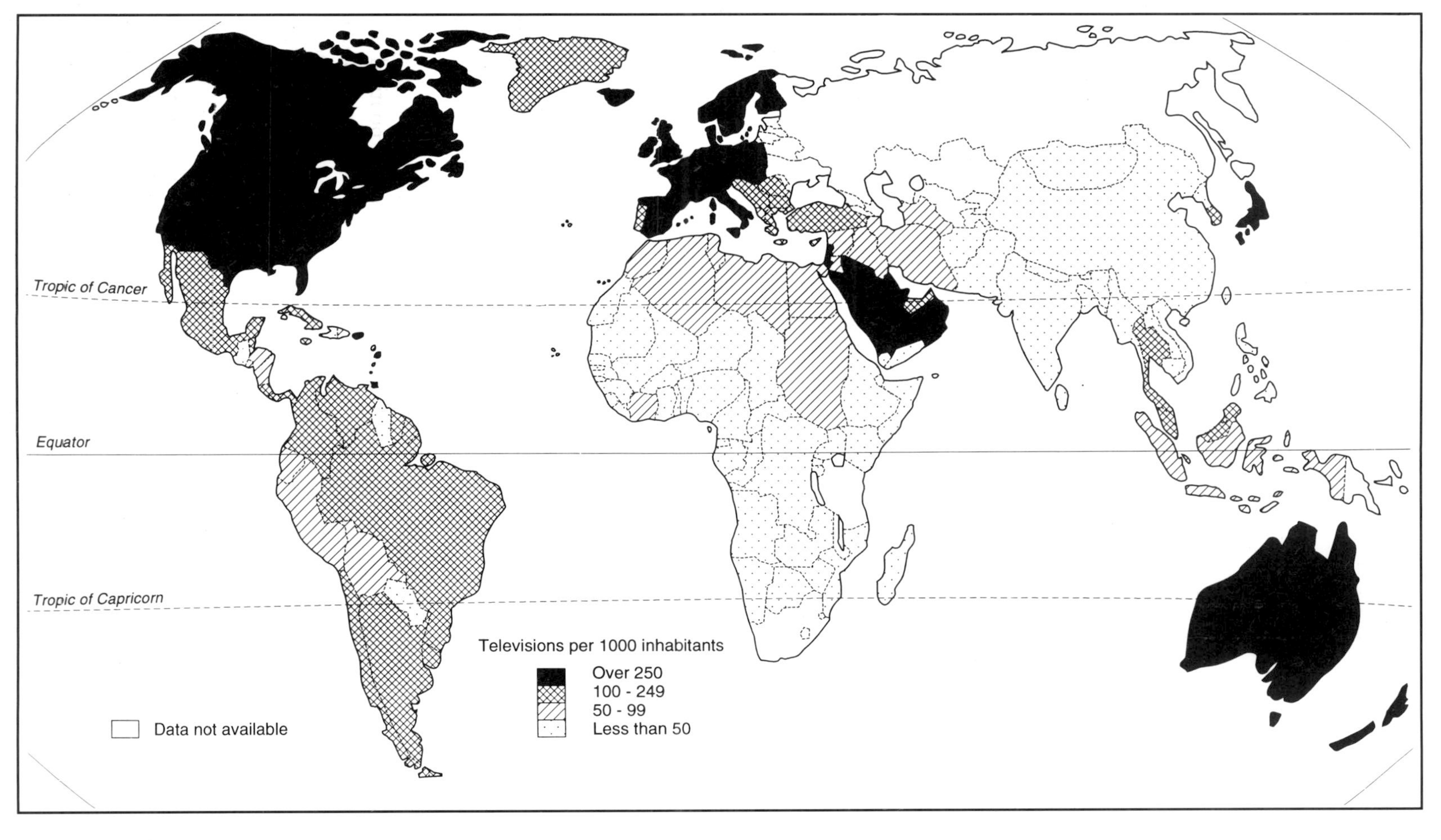

Figure 10.7 Televisions per 1000 inhabitants, 1989
Source of data: UN, 1992

give greater access to the population at large.

A similar imbalance can be seen with regard to the availability of newspapers. Leaving aside the question of levels of literacy, in many Third World countries there are few titles, and circulation levels are low. Several African states boast only a single daily newspaper, as opposed to more than 100 in Britain and Canada. Similarly, there are circulation estimates of over 400 papers per 1000 population in Britain, Sweden or Russia, as opposed to 10 or less in Bangladesh, Tanzania or Niger.

The media of television, radio and newspapers increasingly provide access for the world's population to news, information, entertainment, sport and advertising. As an extension of the notion of cultural imperialism, some observers see the intrusion of communications media as a negative influence, leading towards some kind of global homogenization (Plate 10.9). The commercial power of Western media means that it exercises a pervasive influence. People are exposed to developed world values and cultures; their aspirations are raised by the images from developed world programmes and advertising; conversely local cultures and traditions are not portrayed or appear inferior; *in extremis* the screening of *Twin Peaks*, *Neighbours* or *Baywatch*, dubbed into a foreign language and screened in a Latin American shanty town or an African village generates incongruity and contradiction.

In discussion of the role of the media as an intermediary between the haves and the have-nots, an important issue is that of control. Who determines what is printed, broadcast or screened? Who does so may also be variously interpreted. Are the BBC World Service or the Voice of America sources of neutral 'news' or agencies of capitalist propaganda? Many Third World critics argue that the global news agencies are all slanted towards the developed world and its interests, and somewhat biased in their coverage of the Third World – Associated Press and United Press International to the USA and Latin America, Agence-France-Presse to France and Francophone Africa, Reuters to Britain and Anglophone areas.

Increasingly the most powerful media, of the press and television, are controlled by MNCs; there are now global media corporations controlling television, newspapers and cinema. In 1989 UNESCO identified 78 major media corporations; of these, 48 were based in the USA or Japan, and the remainder in Europe, Canada and Australia. Not one was based in the Third World – but many had direct influence through subsidiaries or affiliated companies, or through satellite transmissions. Such companies determine what messages the media carry, about the Third World, and within it. As producers of news which is transmitted or programmes which are exported, they can powerfully influence the 'messages' exchanged between North and South. In addition, because they are commercial enterprises and depend in part on advertising for revenue, they convey developed world images and adverts to 'sell' the products of Western consumer society. They can be seen as disseminating a 'mass-produced culture'.

The power of the media has long been recognized in the Third World, as in the developed world. The early provision of radio and television was often by the national government, which provided subsidy and regulated output. In these circumstances they could be used by the state to convey news, create identity for newly independent states, sustain culture and provide a medium for education. An important debate, whether for national or global media systems, is the use to which it is put and how it is used. What constitutes neutral 'news' and what 'propaganda'? The media can be used to foster and encourage policies aimed at development – to cultivate new land or use new techniques, to practise birth control or increase population, or to prepare for war or subscribe to particular political or religious values. As such, control of the media is a potent weapon, as evidenced by the importance of securing the radio or TV station in coups and political disturbances.

In the very broadest terms, radio and television in Africa and Asia *tend* to be government owned and operated, while in Latin America they *tend*, except under military regimes, to be privately owned and operated. The developed world, with a commitment to democracy, sees a 'free press' as an essential ingredient in that process. In one-party states, dictatorships and military regimes such freedom may be deemed

BOX 10.2 THE MEDIA IN LATIN AMERICA: THREE EXAMPLES

After the Revolution of 1910–17 the Mexican government permitted a mixed structure of commercial and state-owned public radio. The state used broadcasting as a means of information, education and propaganda, but by 1941 it had left the medium entirely to the private sector. Mexico was the sixth country in the world to introduce television, and this was left entirely to the private sector. In consequence it became closely linked to US media networks, dependent on American technology, and heavily influenced in its programming and advertising by American values.

In Cuba, following the Revolution of 1959, the previously private media came under state control. Radio and TV stations, studios and theatres were nationalized. Television was consolidated from a previous six channels to two, one concerned with news and education, and the other with culture and entertainment.

In Brazil early radio stations were private but were not allowed to use advertising until 1932. From 1932 to 1945 they were subject to censorship. The first television station, set up in 1950, was a private company which was effectively a monopoly until 1962. In that year TV Globo, linked to an American media group, was established and is now the world's fourth largest television network. It owns or part owns more than a dozen TV stations and 30 radio stations, video and record companies, and publishes a major newspaper, books, magazines and comics. By 1988 it reached three-quarters of Brazil's *municipios* and served 95 per cent of the country's TV-owning households. During the military regime 1964–85 it was supportive of the government's nationalistic policies, and portrayed its achievements, but in the early 1980s became supportive of the return to democracy, and was influential in the election of a civilian president in 1985.

undesirable, such that there is close control of the press, radio and television. In the period of military governments in Latin America after 1960, such close control was exercised, in Uruguay from 1967 to 1980, in Peru from 1968 to 1980 and in Brazil from 1964 to 1984. The military used the press and television to sustain their views of nationalism and modernization, and to control the public availability of information. In one-party states such as China, Cuba, North Korea or Iraq, the media remain closely controlled, in terms of what is disseminated, and what, if any, access is given to outside journalists or the broadcasting of foreign material.

The media are clearly a potent and controversial force in relations between North and South. They provide a means of making the developed world aware of the problems and crises of the Third World; they expose the people of the Third World to the lifestyle and affluence of the First; they can portray traditional societies and cultures in their homelands – or undermine them through the portrayal of alternative values and opportunities. The 'globalization of the media' has become a powerful agent for the dissemination of verbal and visual images of the very different values and achievements of developed and developing countries. The growth and importance of the global media corporations would suggest that the dominant flow of traffic will be from the developed to the developing world, but there are some exceptions worth noting. Mexican and Brazilian television have been successful in exporting programmes abroad, partly, in the case of Mexico, to Hispanic populations in the USA, but Brazil's '*telenovelas*' (soap operas, but of distinctively different style to those of American, British and Australian television) have become popular not only in the former metropolitan power, Portugal, but in Eastern Europe and the Far East. Similarly, despite the global power of Hollywood, India has managed to sustain a major film industry, to meet a very large domestic demand and that of Indian communities overseas, but it has also achieved acceptance in international cinema.

CONCLUSION

International transfers of people, commodities, capital and ideas are the main forms of global interaction between the countries of the Third World, and the countries of the First and Second Worlds. Each of these interactions exhibits major structural effects that tend to distribute the benefits of the interactions unevenly, in favour of the richer and stronger partner. They are characterized by a weakness of redistributive mechanisms to ensure a more even distribution of benefit, such that the gap between rich and poor is sustained and in most cases is widened. The evidence presented in this chapter adds weight to the view, expressed in Chapters 1 and 2 of this book, that the countries of the Third World are enmeshed in dependency relationships of unequal exchange with the rich countries, and, despite much discussion and changing political conditions after a period of decolonization and national attempts to develop an indigenous development strategy, the principal features of the external relationships established in the first half of the twentieth century remain in place towards its end.

In the First World there is a strong presumption, derived in part from the Brandt Commission, that there is an inevitable mutual benefit to be derived from growing economies in the First World trading with countries of the Third World: in the medium term it is in the interests of the rich as well as of the poor that the interactions should not only continue, but should intensify as part of the global development process. Development is not to be seen as a 'zero-sum game' (that is, benefit to one group occurs as a result of an equally large 'disbenefit' in the other), but that rich and poor will benefit, and that the strength of the redistributive mechanisms will be critical in allocating the shares of that benefit. Such a view has been strongly criticized from the political Right – who argue that only with strong economic growth can redistribution be a practical reality – and from the political Left – who argue that the redistributive emphasis provides mere palliatives that will further the interest of corporate capitalism rather than the interests of the people of the Third World, and that only major structural changes in the nature of these exchanges will narrow the development gap. The optimism of Brandt in the early 1980s has faded in the face of the harsh realities of unequal global exchanges driven by monetarist rhetoric from the major international development agencies, such as the World Bank, and by increasingly assertive regional interests, as in the European Union and NAFTA.

The nature of global economic and social exchange and the processes operating to create and maintain international patterns of income and well-being have tended in the past to be less discussed by geographers than they might have been, as geographic work focused on national and regional patterns of growth and inequality. However, recent work has incorporated that global dimension much more easily than it did in the 1970s, most notably with a recognition that the world may be *A world in crisis?* (Johnston and Taylor, 1989), a book that develops Johnston's global view as presented at the beginning of this chapter. By incorporating the global perspective and global interactions into their approaches to development, geographers are more able fully to conceptualize and analyse the Third World.

SUMMARY

The increasing movements of people, commodities, capital and ideas between the Third World and the First and Second Worlds, in both directions, are the basic components of global interactions that operate to maintain, and indeed widen, the inequalities between the Third World as a whole and individual countries within it and the First World. Labour migration tends to be from poor to rich countries, providing skilled and unskilled labour at the destinations of the flow that bring to these areas more benefits than accrue to the source areas through remittances. The temporary flow of First World tourists to Third World destinations, by contrast, brings relatively little economic spin-off to tourist resorts relative to the profits from travel and hotels that MNCs can accumulate. Terms of international trade in the last decades have generally been in favour of industrial producers over agricultural exporters, so that traditional patterns of commodity exchange between Third World countries and other regions have favoured industrial countries, including the NICs as exporters of manufactures. Third World countries are increasingly indebted to First World commercial banks and to multilateral funding agencies, and the levels of official bilateral aid assistance, as a proportion of developed country incomes, going to Third World countries is less in the 1990s than it was in the 1970s. Globalization of the media has placed more control in the hands of developed world MNCs, and television in particular has raised the awareness of consumers in the Third World of the lifestyles of affluence in the rich countries, changing consumer tastes without bringing the means to achieve these levels of affluence any closer.

All these interactions seem to generate benefits disproportionately for the richer countries. However, since there are no sufficiently strong redistributive mechanisms to more nearly equalize the substantial benefits of these exchanges, they continue to widen the gap between the Third World and other parts of the world economic system.

11

Conclusion

The central argument of this book has been that the use of the term, 'The Third World', implies a degree of common characteristics amongst the countries that have been included within it. These countries have some similarities in their levels of income relative to the most affluent countries and as measured by various other criteria, but the major feature they share is the constraints imposed by the nature of their past and contemporary involvement in the world economy. They are brought together mainly by the relationship they have with the global economy in their dependence on the economic conditions and social and political processes dominated by countries of the developed world.

Yet the countries of the Third World also exhibit great diversity, in their physical environments, resource endowments, historical experience, cultural traditions, societal structures, economic organization and political practices. Diversity exists between these countries and within them. Moreover, there is not a great or self-evident gulf between the First World and the Third World. Annual per capita GNPs of US $60 in Mozambique and $36,080 in Switzerland represent the extremes, but between them is a continuum from poor to rich; the precise placing of subdivisions or 'break-points' along it is open to debate. The Third World does not consist of a precise number of uniform states; however defined, it possesses great internal diversity. There is no single measure or explanation of relative poverty or of the essential geographical and economic characteristics of each Third World country; nor is there a single key to progress which, if turned, will set them all on the same path to development and to levels of well-being equivalent to those of the USA, Sweden, Japan or West Germany, even if they wished to follow such a path.

Some Third World countries have already made considerable progress, and especially since the first edition of this book was written in 1981. At that time the special circumstances of the period since 1973 had created the distinctive group of oil-exporting countries which built up massive capital surpluses. Their spectacular growth slowed down in the 1980s, but they remain relatively rich countries. Their place as the most dynamic economies has now been occupied by another group of countries, which, if they have not leapt clearly into the First World (though Hong Kong and Singapore are listed by the World Bank as high-income countries), have at least progressed rapidly up the league table of per capita income. These are the NICs, found mostly in East and South-East Asia. These at least have achieved high rates of economic growth. Between 1980 and 1992 for example, Hong Kong, Singapore, Thailand, Malaysia and Indonesia all achieved annual rates of increase in GDP in excess of 5 per cent. As Table 11.1a shows, the poorest countries as a group achieved growth rates higher than those of the more advanced countries, but when that growth is related to population the pattern is less favourable, though it is still above the rate for high-income countries (Table 11.1b). The per capita figures provide further evidence of a growing gap between the poorest countries, predominantly in Africa, and middle-income countries, notably in East Asia and to a lesser extent in South Asia.

One can nevertheless point to some other signs of optimism and general progress. In much of the Third World birth and death rates have

Table 11.1 Economic growth rates, 1980–92

	(a) Average annual growth rate of GDP (%), 1980–92	*(b) Average annual growth rate of GNP per capita, 1980–92*
Low-income countries	6.1	3.9
Middle-income countries	2.6	0.8
High income countries	0.7	[illegible]
Sub-Saharan Africa	1.8	−0.8
East Asia and Pacific	7.7	6.1
South Asia	5.2	3.0
Middle East and North Africa	2.2	−2.3
Latin America and the Caribbean	1.8	−0.2

Source of data: World Bank, 1994a

fallen, infant mortality rates have declined and life expectancy has risen; access to health care and to education has improved; economic structures have diversified, reducing dependence on agriculture and a few primary product exports; some countries at least have become effective competitors in the world market for manufactured goods. Yet many countries and regions, and men, women and children, remain poor, in both relative and absolute terms.

Awareness of the problems of poverty has increased over the past three decades, on the part of both the less and the more developed countries and their peoples, and there is greater concern by both rich and poor for securing progress. The Third World has been persistent in its efforts in this direction; the developed world has been less consistent, as its contributions to aid and concessions to trade have been periodically sacrificed to domestic 'needs', fashions and pressures. Coverage by the media has made people more aware of the horrors of poverty, whether persistent or catastrophic, and has prompted widespread humanitarian response. The oil crisis brought home to many both the interdependence of the world economy in the 1970s, and in the 1990s the global environmental crisis is having a similar effect, identifying the potential influences which might be exercised in the interests of the Third World and in the mutual interests of humankind as a whole.

As part of this increasing interest in the need for development and in the processes by which it can be achieved there has been growing awareness of the potential available for advance in the Third World: greater knowledge of the resource endowment of soils, minerals and people, and increased utilization of innovations – new crops, new equipment, new concepts. One area of particular concern is the formulation and spread of techniques for development that are most appropriate to the needs and endowments of Third World countries. Underdevelopment of the poor South may have been partially due to its exploitation by the rich North, yet in its attempts to secure development the Third World has been heavily dependent on concepts, methods and technologies largely formulated in the developed countries to meet their needs, aspirations and endowments. Such borrowing may not necessarily be the most appropriate for Third World countries to secure development in the best interests of their populations as a whole. There is a great need to develop more appropriate, replicable and indigenous development techniques. The Third World has not been, and should not be, a passive recipient of First World ideas and technology, but rather an active participant in the processes of exchange, just as countries have actively sought to develop their own paths to economic growth, and not to be mere passive participants in a global economic system that seems to benefit the rich more than the poor.

This book has deliberately offered a geographical perspective on the Third World, in the belief that such a perspective has a positive contribution to make in identifying and analysing conditions affecting three-quarters of the world's population. Generalization about the Third World is inevitable in any brief book of this sort, yet we have sought to identify the main common structural elements of interest to geographers. Study of the geography of any one country of the Third World would, of course, require more detailed consideration of these and other themes, in the context of the specific physical, historical, social, economic and political circumstances of that country.

This second edition incorporates not only changes of detail since the early 1980s, but has also offered an opportunity to review the contributions of Geography and geographers to the study of the Third World. These can be made at a variety of scales: global, international, national, regional or local. Although we have defined the Third World primarily in terms of global relationships, and explored these at some depth in Chapters 2 and 10, it is evident that Geography has a larger contribution to make at the national and sub-national level, exploring structures, patterns and processes of spatial organization and the range of livelihoods of the national population. While it is true that many of the structural features of the Third World arise from and can be alleviated by international exchange and distribution mechanisms, a 'New International Order', however defined, could not in itself resolve many of the problems of resource availability, poverty and inequality that exist within Third World countries. It would need to be accompanied by changes and initiatives within each country and, given their internal diversity, would require many of the changes and initiatives to be specific to an individual country, or even to areas within it.

It is because of this very diversity that geographers may have special contributions to make to the study and process of development in four major but inter-related areas. The first is in the area of people–environment relations. Geographers have studied environment – soils, hydrology, climate, landforms – and people – distribution, society and economy – as geographical phenomena in their own right, but have also made important contributions, where people and environment interact, to such topics as land use, water management and resource utilization. It is at the local scale that human decisions about the use of environmental resources are most crucial. The new global awareness of 'environment' as a Third World issue, identifying the range of resource and pollution problems and the means by which they might be addressed, is an exciting area for the geographers' expertise to be applied in communities throughout the Third World.

The second major field is that of locational analysis. The conceptual revolution in geography in the 1960s and early 1970s pushed locational analysis to the forefront of geographical studies, particularly in the countries of the First World and, to a lesser extent, in the Second World. It also had an impact on the Third World. Studies of processes governing the location of plantations, factories, rural schools and towns, or the diffusion of disease or new technologies, are not merely matters for the geographical record but may contribute to improvements in patterns of provision, or checks to the spread of disease. Geographers are now less attracted to locational analysis than they were in the early 1980s, especially in their work in the Third World, for it can be seen to be in some sense symptomatic of the problem of the uncritical adaptation of possibly inappropriate models. However, geographers from the Third World – for example Akin Mabogunje of Nigeria, Milton Santos of Brazil – have themselves been leaders in the development of theories of locational analysis that can be and have been applied in Third World contexts.

Third, the geographer's traditional and continuing concern for regional analysis, both within and between regions, has brought geographical concepts and theories firmly into the centre of discussion of the nature and extent of spatial inequalities in Third World countries, and of means to diminish such imbalances. The broad thrust of concern within Third World studies derived from the existence of massive and growing international inequalities has inevitably been paralleled by a growing interest in the patterns and mechanisms of inequality within countries. Thus much work has been done and still remains to be done on the problems of regional development.

A fourth area for particular focus is one that has forcefully emerged in the decade between the first and second editions of this book. In their rediscovery of a sense of 'place', that individual localities in each country have their own unique identities, geographers have redirected their attention even more strongly than before to how people manage their livelihoods in these localities and develop their own sets of relationships with others in their household and community. Geographers' new concerns for social and gender relationships, and how these affect the nature of places, finds a very specific echo in the concern for poverty and 'lives of struggle' of the rural and urban poor of the Third World; how the poor – men and women, separately and together – manage the limited environmental and economic resources available to them.

In sum, the Third World offers geographers the opportunity to put their approaches, theories and techniques to practical use in the understanding and solution of at least some of the problems of a large proportion of the earth and its people. Part of that opportunity has been explored in this book.

Glossary

Agribusiness Large-scale agricultural enterprise which is part of the broad economic strategy of a multinational organization, usually with its headquarters in Europe or North America and involving processing, transport or marketing, as well as the production (or production under contract) of crops and livestock.

Autoconstruction Building of a house by its occupants, gradually upgrading its fabric – a major characteristic of many squatter settlements.

Backward/forward linkages Backward linkages represent the demands created by non-primary activities for inputs; forward linkages result from the generation of outputs which may be utilized by a consumer.

Backwash/spread effects Associated with the work of Gunnar Myrdal where, once growth has begun in a region, the flows of labour, capital and goods will focus on that region to the detriment of others (backwash); demands for resources such as minerals or agricultural produce from a growing region will generate 'spread' effects in the periphery. The terms 'polarization'/'trickling down' as used by Hirschman have similar implications.

Bazaar economy Small-scale commercial and manufacturing activity, synonymous with the informal sector.

Circulation Movements of people away from their place of residence for varying periods of time and over varying distances, followed by return; in contrast to the definitive changes in place of residence involved in migration.

Conditionality Major lending institutions such as the World Bank or the IMF, and bilateral donors, usually impose conditions on the loans they make. Common are 'market conditionality' (that the borrowing country seeks to reduce restrictions on the workings of a free market), 'green conditionality' (that the borrowing country seeks to introduce more effective environmental management controls), 'demographic conditionality' (that the borrowing country seeks to develop direct policies to reduce population growth).

Contraceptive prevalence rate (CPR) An index of the extent of usage of contraception, calculated as the proportion of all women age 15–49 currently using contraceptives, whether modern or traditional.

Core–periphery relationship Relationship of unequal exchange in which the interests of the periphery are subordinated to those of the core, of a country or of the world economy.

Dependency theory A set of inter-related propositions that identify the processes of unequal exchange in economic and social relations between the Third World and the industrialized countries, in which the economic performance and social structures of dependent countries are dominated by those of the richer countries.

Export processing zones (EPZs) Relatively small but well-defined areas established to attract factories, especially foreign-owned ones, where goods for export are manufactured or assembled, by offering a range of special incentives. The incentives may include exemption from tax or environmental and labour controls, or the provision of infrastructure such as roads, other services, and factory buildings.

Fordism/Post-Fordism 'Fordism' is a term applied to the networks and structures of industrial production, i.e. large-scale units of large firms producing specialist goods in bulk, that have dominated industrial production for most of the twentieth century, and named after the methods used in the automobile industry and associated with Henry Ford. In the late twentieth century, however, systems based on economies of scale are increasingly questioned, as firms and factories become small, producing a wider range of goods without a standard assembly line, but with more automation. Such flexible production systems are described as post-Fordist.

Formal/informal sector The formal sector consists of large-scale capitalist enterprises with fixed conditions of employment and job security, including government, manufacturing and services; the informal sector consists of small-scale labour-intensive activities with job insecurity, multiple occupations and low monetary returns.

G7 countries The seven largest economies (Japan, USA, Germany, France, UK, Italy, Canada) that together exercise a dominant role in the international economy.

Gross National Product (GNP)/Gross Domestic Product (GDP) GDP is the total value of all finished goods and services produced by an economy in a specific time period, usually one year; GNP is GDP *plus* income accruing to residents of a country from abroad, *less* incomes accruing to foreign residents investing in the country.

Intermediate industries Industries producing goods which are inputs for further processing.

Lumpenproletariat The urban marginal population of Marxist analysis, who are either unemployed or excluded from employment in the formal sector.

Mercantilism A term relating to seventeenth- and eighteenth-century policies of European countries seeking to maximize national wealth by control of trade, shipping and industrial activities.

Neo-colonialism The reassertion of economic control by advanced countries to replace colonial authority after independence.

Neo-Marxist The reinterpretation of Marxist philosophy in terms of modern manifestations of the struggle between the classes of a society.

New International Economic Order (NIEO) A loosely defined set of ideas aimed at redressing international inequalities and redistributing the benefits of international trade in favour of the Third World.

Penny capitalism Capitalist transactions of a small-scale type.

Primacy In many Third World countries the urban hierarchy is top heavy, with the largest city substantially larger in population than the other major cities.

Primary sector That part of an economy concerned with extractive activities – gathering, farming, lumbering, fishing and mining.

Rank–size rule An empirical rule whereby the population of any town can be estimated by dividing the population of the largest city in a country by the rank of the settlement under consideration.

Reserve army of labour A Marxist term used to describe a pool of unemployed people created by capitalism, the existence of which depresses wages through the excess demand for jobs.

Rising expectations The phenomenon by which demand for consumer goods is stimulated, so that people aspire to improved conditions in lifestyle and available goods.

Satisficer A person who may stop working as soon as immediate needs are met; in contrast a maximizer will seek the maximum return possible for his/her labour.

Secondary sector The part of an economy concerned with manufacturing – involving the transformation of raw materials and semi-finished goods into finished products.

Structural adjustment A term strongly associated with the lending of the World Bank and IMF which emphasizes the importance of Third

World countries making changes in their macroeconomic policies as part of managing the development process. Typically this has involved removing controls on exchange rates for the local currency, a rolling back of direct state involvement in the economy with privatization of state-run enterprises, reduction in the size of the civil service, and the passing on of some of the costs of services such as health and schooling directly to the consumer.

Sustainable development 'Development that meets the needs of the present without compromising the ability of future generations to meet their own needs' (Bruntland Commission, 1987, 43). This classic definition emphasizes inter-generational equity and also assumes the possibility of development that can be accompanied by environmental stability or even environmental improvement.

Terms of trade The ratio of the export earnings of a country to its import prices; worsening terms of trade reflects a fall in export prices and/or a rise in import prices – the common experience of much of the Third World in recent years.

Tertiary sector The sector of the economy concerned with neither extracting primary products nor processing them; instead tertiary activities involve a range of services such as retailing, administration, teaching, health care and finance.

Total Fertility Rate This is the most widely used comparative index of the level of fertility. It approximates to the total number of children a woman will be expected to have in her lifetime, given current fertility patterns, and is calculated by aggregating the recent age-specific fertility of women aged 15–49.

Underemployment Conditions where people are working less, per day, week or seasonally than they could and would wish to.

Questions and Topics for Discussion

CHAPTER 1 INTRODUCTION

1.1 Using the *World Development Report* or other source of data, examine the ways in which the extent of the 'Third World' might vary if you used different criteria of GNP per capita – for example $500; $1000; $3000 per head. Compare this with patterns using other criteria such as infant mortality, employment in agriculture, etc.

1.2 The 'Least Developed Countries' are defined by the UN as having one or more of the following characteristics:

- GNP per capita below $300
- being landlocked
- remote insularity
- desertification.

Consult Figure 1.19 and suggest the relative importance of these factors for specific countries.

1.3 Discuss the criteria you might use to include or exclude the following countries from the 'Third World': (a) Saudi Arabia, (b) China, (c) South Africa.

1.4 Outline the advantages and disadvantages of using GNP per head as a measure of 'development'. What alternative measures might be adopted?

CHAPTER 2 HISTORICAL

2.1 Examine the positive and negative legacies of colonial rule in Third World countries.

2.2 Discuss the influence of colonial rule, by countries such as Britain, France, Spain, Portugal and the Netherlands, on the way in which their former colonies have developed.

2.3 What are the main manifestations of neo-colonialism in Third World countries?

2.4 There has been much comment about the negative impact of economic development on the global environment. Is this necessarily the case? What are the environmental consequences of development at the more local level of particular countries and communities within the Third World?

CHAPTER 3 POPULATION

3.1 What factors have influenced the variations in the rates of decline in birth rates and death rates within the Third World?

3.2 What are the characteristics of the people migrating from rural areas to the cities? Consider the differences between these migrants in Africa, Asia and Latin America.

3.3 In what ways can education contribute to the development process? Where are levels of education provision most deficient and why should this be?

3.4 What indices could be used, and with what results, to examine the varying status of women in the Third World?

CHAPTERS 4/5 AGRICULTURE/RURAL

4.1 What have been the consequences of (a) population increase and (b) out-migration on the economy and society of rural areas?

4.2 Assess the contributions women make to agriculture and the rural economy. Are there variations within the Third World, and if so, why should this be?

4.3 It is claimed that traditional agricultural systems were often well adjusted to the physical environment. Illustrate this statement and consider the factors which have threatened the stability of this adjustment in the last 50 years.

4.4 Why must rural development involve more than just increase in food production?

4.5 Discuss the statement that 'For many Third World farmers, the choice of which crops to grow is limited by many factors other than price.'

4.6 What impacts can NGOs have on rural development at the local level? How and why might they be different from those of government agencies?

4.7 How far can it be claimed that agribusiness is to the contemporary Third World what the plantation was to the colonial World?

CHAPTER 6 MINING AND MANUFACTURING

6.1 What are the reasons for the importance of multinational corporations in the mining industry in Third World countries? What adverse effects might they have on the economies and societies of these countries?

6.2 Comment on the patterns of energy production and consumption in the Third World, and assess their significance for the development process.

6.3 Explain why only a few countries in Asia and Latin America have become 'Newly Industrializing Countries'.

6.4 Which resources for industrialization are most lacking in the Third World? Which countries appear to have the best potential for joining the 'Newly Industrializing Countries'?

CHAPTER 7 URBAN

7.1 What are the likely consequences of rapid urbanization for (a) urban dwellers, (b) the provision of jobs and services, (c) the natural environment?

7.2 Apply the characteristics of the informal sector indicated in Tables 7.5 and 7.6 to the urban experience of particular countries.

7.3 Why is the level of urbanization higher in East Asia and Latin America than in Africa and South Asia, and with what consequences for national development prospects?

7.4 To which sectors of the urban economy do women make their major contributions? Are there variations between different parts of the Third World? Why?

CHAPTER 8 INTERNAL INTERACTION

8.1 Use an atlas to produce maps of the road systems of two or three Third World countries. In what ways do they differ? What are the implications of the road patterns for the geography of development?

8.2 Why have Third World countries sought to improve interaction between urban and rural areas? Are the benefits of any exchanges equally distributed between the towns and the countryside?

CHAPTER 9 NATIONAL ECONOMIC MANAGEMENT

9.1 What might be the consequences of World Bank emphasis on 'market forces' for the poor regions and poorest people of Third World countries?

9.2 Development plans have tended to favour the relatively *more developed* parts of Third World countries. Is this an inevitable part of the development process? What might be done to assist the more 'backward' regions?

9.3 Examine, with examples, the relative advantages of 'top-down' and 'bottom-up' strategies for national development.

CHAPTER 10 EXTERNAL LINKAGES

10.1 Discuss the advantages and disadvantages of foreign aid. What are the relative benefits of aid from bilateral, multilateral and NGO sources?

10.2 What are the costs and benefits of international migration for (a) source countries in the Third World, (b) receiving countries in the Third World, (c) receiving countries in the developed world?

10.3 The development of tourism in the Third World has so far concentrated in a few regions and countries (Figure 10.2). How might other areas develop a tourist sector? What are their potential attractions for First World tourists?

10.4 Using information from an aid agency such as Oxfam or Christian Aid, outline a development strategy for a country of your choice.

Notes on the Authors

Colin Clarke is University Lecturer in Geography and Fellow of Jesus College, Oxford, and formerly Reader in Geography and Latin American Studies at the University of Liverpool. He is the author of books on *Kingston, Jamaica* (1975), *East Indians in a West Indian Town* (1986), and has edited *Geography and Ethnic Pluralism* (with D. Ley and C. Peach, 1984) and *South Asians Overseas* (with C. Peach and S. Vertovec, 1990). His main research interests are in the social geography of the West Indies and Mexico.

John Dickenson is Reader in Geography and Latin American Studies at the University of Liverpool. He has also held visiting professorships at the universities of Pittsburgh and Minas Gerais. He is the author of two books on *Brazil* (1978 and 1982), and has published numerous articles on the economic geography of Brazil and Latin America. His current research interests are in the cultural landscapes and the scientific exploration of Brazil.

Bill Gould is Professor of Geography at the University of Liverpool, and has been a consultant for the World Bank, UNESCO, and to several Third World governments. He is the author of *People and Education in the Third World* (1993) and editor of *Population Migration and the Changing World Order* (with A. Findlay, 1994) and *Planning for People* (with R. Lawton, 1986). His field research has been mainly in Africa, on education planning, migration and people–environment relationships, while his consultancy work has extended to Asia and the Caribbean.

Sandra Mather is Cartographer in the Graphics Unit of the Department of Geography, University of Liverpool. She is responsible for the cartographic and graphics work of the Department, and her work has appeared in a wide range of books and journals.

Mansell Prothero is Emeritus Professor of Geography at the University of Liverpool and Visiting Senior Fellow at the Liverpool School of Tropical Medicine, and has held academic appointments in Scotland and Nigeria, as well as visiting appointments in Australia, Israel, Mexico and the USA. He has served as a consultant to the Overseas Development Administration, World Bank and World Health Organization. He has wide interests in the Third World, particularly in Africa, relating to the fields of population, migration and health. He is the author of *Migrants and Malaria* (1965), *A Geography of Africa* (1969), and edited *Geographers and the Tropics* (with R.W. Steel, 1964) and *Circulation in Population Movement* (with M. Chapman, 1985), and numerous publications on population and health issues in the Third World.

David Siddle is Senior Lecturer in Geography at the University of Liverpool. He has also taught at the universities of Sierra Leone and Zambia, and has served as a consultant to the British government and aid agencies on rural development issues. His interests are primarily in rural and development issues in Africa. He is the author (with K. Swindell) of *Rural Change in Tropical Africa* (1990).

Clifford Smith is Emeritus Professor of Latin American Geography at the University of Liverpool, and formerly Fellow of St John's College, Cambridge. His interests are particularly in the geography of Peru, especially in historical geography, and agrarian, regional and urban development. He is co-editor (with H. Blakemore) of *Latin America: Geographical Perspectives* (2nd edn 1983), and of *An Historical Geography of Western Europe before 1800* (2nd edn 1978).

Elizabeth Thomas-Hope is Professor of Environmental Management at the University of the West Indies, and formerly Senior Lecturer in Geography at the University of Liverpool. She has been consultant to a number of United Nations agencies concerned with development and demographic issues, and is the author of *Explanation in Caribbean Migration* (1992) and *The Impact of Immigration on Receiving Countries: the United Kingdom* (1994).

List of Boxes

List of Plates

List of Figures

List of Tables

Suggestions for Further Reading

A valuable complement to this volume is the Routledge Introductions to Development Series (all London: Routledge):

Chandra, R. (1992) *Industrialization and Development in the Third World.*
Cole, J. (1987) *Development and Underdevelopment.*
Dixon, C. (1990) *Rural Development in the Third World.*
Drakakis-Smith, D. (1987) *The Third World City.*
Elliott, J. (1993) *An Introduction to Sustainable Development.*
Findlay, A.M. and Findlay, A. (1987) *Population and Development in the Third World.*
Gupta, A. (1988) *Ecology and Development in the Third World.*
Lea, J. (1988) *Tourism and Development in the Third World.*
Momsen, J. (1991) *Women and Development in the Third World.*
Parnwell, M. (1993) *Population Movements in the Third World.*
Soussan, J. (1988) *Primary Resources and Energy in the Third World.*

Binns, T. (1993) *Tropical Africa.*
Drakakis-Smith, D. (1991) *Pacific Asia.*
Findlay, A.M. (1994) *The Arab World.*
Gilbert, A. (1990) *Latin America.*

GENERAL

Brydon, L. and Chant, S. (1989) *Women in the Third World: Gender Issues in Rural and Urban Areas*, Aldershot and New Brunswick, New Jersey: Elgar and Rutgers University Press. A very good review of the place of women in Third World economies and societies. It explores the issue of gender in the household, in rural areas and in the city.
Corbridge, S. (ed.) (1995) *Development Studies: A Reader*, London: Arnold. A collection of the classic papers on Third World development, particularly from an economic perspective.
Kitching, G. (1989) *Development and Underdevelopment in Historical Perspective*, London and New York: Routledge. A useful review of theories of development from the 1940s.
Mabogunje, A. (1989) *The Development Process: A Spatial Perspective* (2nd edn), London and Boston: Hutchinson. An outstanding study of development by one of the Third World's leading geographers. It is especially valuable for its analyses of the nature of rural and urban development, and of processes of national integration.
Norwine, J. and Gonsalez, A. (eds) (1988) *The Third World: States of Mind and Being*, London and Washington: Unwin Hyman. A rather eclectic but very stimulating collection of essays by geographers, exploring the 'meaning' of the Third World, a range of systematic topics, and regional case studies.
Unwin, T. (ed.) (1994) *Atlas of World Development*, Chichester: Wiley. A very useful collection of over 100 maps and accompanying text on the environment, demography, economy and culture of the Third World. It includes both broad scale and case study material; a good source of material for classwork.

CHAPTER 1 INTRODUCTION

Berry, A. (1987) 'Poverty and inequality in Latin America', *Latin American Research Review*, 22, 202–14.
Hettne, B. (1995) *Development Theory and the Three Worlds* (2nd edn), London: Longman. A very up-to-date review of development theories, including explorations of the implications of the break-up of the 'Second World', and of the Rio Conference, on the environment.
Kay, C. (1989) *Latin American Theories of Development*, London: Routledge.
Prion, C. (1990) 'Changes in income distribution in poor agricultural nations: Malawi and Madagascar', *Economic Development and Cultural Change*, 39, 23–45.
Rampal, S. (1985) 'A world turned upside down',

Geography, 70, 193–205.
Schuurman, F.J. (ed.) (1993) *Beyond the Impasse: New Directions in Development Theory*, London: Zed Press.
Slater, D. (1993) 'The geopolitical imagination and the enframing of development theory', *Transactions of the Institute of British Geographers*, 18, 419–37.
Taylor, P. (1992) 'Understanding global inequalities: a world-systems approach', *Geography*, 77, 10–21.
Wolf-Philips, L. (1987) 'What "Third World"?: origin, definition and usage', *Third World Quarterly*, 8/4, 1311–20.

CHAPTER 2 HISTORICAL PERSPECTIVE

Blaut, J.M. (1993) *The Colonizer's Model of the World*, London and New York: Guilford Press. A critical perspective on the processes and consequences of European colonialism.
Cleary, M.C. (1992) 'Plantation agriculture and the formulation of native land rights in British North Borneo *c*. 1880–1930', *Geographical Journal*, 158, 170–81.
Corbridge, S. (1986) *Capitalist World Development*, Totowa, NJ.: Rowman and Littlefield.
Dixon, C. and Heffernan, M. (eds) (1991) *Colonialism and Development in the Contemporary World*, Rutherford: Mansell.
Fieldhouse, D.K. (1981) *Colonialism 1870–1945: An Introduction*, London: Weidenfeld and Nicolson. A useful review of the Victorian high point of colonialism and its consequences.
Said, E. (1991) *Orientalism*, London: Penguin. A dense but fascinating and critical exploration of Western attitudes towards the Orient.

CHAPTER 3 POPULATION

Barrett, H. and O'Hare, G. (1992) 'India counts its people', *Geography*, 77, 170–4.
Chant, S. (ed.) (1992) *Gender and Migration in Developing Countries*, London: Belhaven Press.
Gould, W.T.S. and Findlay, A. (eds) (1994) *Population Migration and the Changing World Order*, Chichester: Wiley. Case studies of migration between the First and Third Worlds, and within the Third.
Jowett, J. (1993) 'China's population: 1,133,709,738 and still counting', *Geography*, 78, 401–19.
Leinbach, T.R., Watkins, J.F. and Bowen, J. (1992) 'Employment behaviour and the family in Indonesian transmigration', *Annals of the Association of American Geographers*, 82, 23–47.
Skeldon, R. (1990) *Population Mobility in Developing Countries*, London: Belhaven Press. A broad history and explanation of migration, and case studies of Peru and Papua New Guinea.
Thomas-Hope, E. (1992) *Explanation in Caribbean Migration: Perception and Image – Jamaica, Barbados, St Vincent*, London: Macmillan.

CHAPTER 4 FARMING SYSTEMS AND AGRICULTURAL PRODUCTION

Adams, W. (1993) 'Indigenous use of wetlands and sustainable development in West Africa', *Geographical Journal*, 159, 209–18.
Adams, W.M. (1991) 'Large-scale irrigation in northern Nigeria: performance and ideology', *Transactions of the Institute of British Geographers*, 16, 287–300.
Bebbington, A. and Carney, J. (1990) 'Geography in the International Research Centers: theoretical and practical concerns', *Annals of the Association of American Geographers*, 80, 34–48.
Grigg, D. (1992) 'World agricultural production and productivity in the late 1980s', *Geography*, 77, 97–108.
Liverman, D.M. (1990) 'Drought impacts in Mexico: climate, agriculture, technology, and land tenure in Sonora and Puebla', *Annals of the Association of American Geographers*, 80, 49–72.
Morgan, W.B. and Solarz, J.A. (1994) 'Agricultural crisis in Sub-Saharan Africa: development constraints and policy problems', *Geographical Journal*, 160, 57–73.
Rigg, J. (1989) 'The Green Revolution and equity: who adopts the new rice varieties and why', *Geography*, 74, 144–50.
Young, L.J. (1991) 'Agricultural changes in Bhutan: some environmental questions', *Geographical Journal*, 157, 172–8.
Zimmerer, K.S. (1991) 'Wetland production and smallholder persistence: agricultural change in a highland Peruvian region', *Annals of the Association of American Geographers*, 81, 443–63.

CHAPTER 5 AGRARIAN STRUCTURES AND RURAL DEVELOPMENT

Bebbington, A. (1993) 'Modernization from below: an alternative indigenous development', *Economic Geography*, 69, 274–92.
Bebbington, A. (1993) 'Governments, NGOs and agricultural development: perspectives on changing inter-organizational relationships', *Journal of Development Studies*, 2, 199–219.
Becker, L.C (1990) 'The collapse of the family farm in West Africa? Evidence from Mali', *Geographical Journal*, 156, 313–22.
Kabeer, N. (1991) 'Gender dimensions of rural poverty: analysis from Bangladesh', *Journal of*

Peasant Studies*, 18, 241–62.

Kloos, H. (1991) 'Peasant irrigation development and food production in Ethiopia', *Geographical Journal*, 157, 295–306.

Lipton, M. with Longhurst, R. (1989) *New Seeds and Poor People*, London: Unwin Hyman.

Powell, S.G. (1992) *Agricultural Reform in China: From Communes to Commodity Economy 1978–90*, Manchester: Manchester University Press.

Preston, D. (1990) 'From hacienda to family farm: changes in environment and society in Pimampiro, Ecuador', *Geographical Journal*, 156, 31–8.

Rigg, J. (1994) 'Redefining the village and rural life: lessons from South-East Asia', *Geographical Journal*, 160, 123–35.

Rydzewski, J.R. (1990) 'Irrigation: a viable development strategy?' *Geographical Journal*, 156, 175–81.

Siddle, D. and Swindell, K. (1990) *Rural Change in Tropical Africa: from Colonies to Nation-States*, Oxford: Blackwell.

CHAPTER 6 MINING, ENERGY AND MANUFACTURING

Airriess, C. (1993) 'Export-oriented manufacturing and container transport in ASEAN', *Geography*, 78, 31–42.

Auty, R. (1992) 'Industrial policy reform in China: structural and regional imbalances', *Transactions of the Institute of British Geographers*, 17, 481–94.

Gwynne, R.N. (1985) *Industrialisation and Urbanisation in Latin America*, London and Sydney: Croom Helm. A detailed survey of the industrialization process in Latin America, including industrialization policies and spatial patterns of manufacturing. Despite its title, it says little about urban issues, but does provide a good case study of Chilean industrialization.

Hodder, R. (1990) 'China's industry – horizontal linkages in Shanghai', *Transactions of the Institute of British Geographers*, 15, 487–503.

Hussey, A. (1993) 'Rapid industrialization in Thailand 1986–1991', *Geographical Review*, 83, 14–28.

Sagawe, T. (1989) 'Mining as an agent for regional development: the case of the Dominican Republic', *Geography*, 74, 69–71.

South, R.B. (1990) 'Transnational "maquiladora" location', *Annals of the Association of American Geographers*, 80, 549–70.

Storper, M. (1991) *Industrialization, Economic Development and the Regional Question in the Third World*, London: Pion. A study of the polarization of development and the potential of regional development policy to counter this. Depends mainly on the Brazilian experience.

CHAPTER 7 URBANIZATION

Amis, P. and Lloyd, P. (eds) (1990) *Housing Africa's Poor*, Manchester: Manchester University Press.

Dutt, A.K., Monroe, C.B. and Vakamudi, R. (1986) 'Rural–urban correlates for Indian urbanization', *Geographical Review*, 76, 173–83.

Forbes, D. and Thrift, N. (eds) (1987) *The Socialist Third World: Urban Development and Territorial Planning*, Oxford: Blackwell. The notion of a socialist Third World is perhaps now a little dated, but this provides a useful range of case studies of countries in Asia, Africa and Latin America pursuing socialist development strategies in the 1980s.

Ford, L. (1993) 'A model of Indonesian city structure', *Geographical Review*, 83, 374–96.

Gilbert, A. (1994) *The Latin American City*, London: Latin American Bureau. An excellent portrait of the Latin American city, though focusing mainly on the urban poor, their housing, and their work.

Gilbert, A. and Gugler, J. (1992) *Cities, Poverty and Development: Urbanization in the Third World* (2nd edn), Oxford and New York: Oxford University Press.

Godfrey, B.J. (1991) 'Modernizing the Brazilian city', *Geographical Review*, 81, 18–34.

Harris, N. (ed.) (1992) *Cities in the 1990s: The Challenge for Developing Countries*, London: University College London Press.

Higgins, B.R. (1990) 'The place of housing programs and class relations in Latin American cities: the development of Managua before 1980', *Economic Geography*, 66, 378–88.

Kirby, R.J.R. (1985) *Urbanization in China: Town and Country in a Developing Economy 1949–2000 AD*, London: Croom Helm.

Lowder, S. (1986) *Inside Third World Cities*, London: Croom Helm. A good historical study and exploration of issues of housing and society in the Third World city.

Potter, R. (1993) 'Urbanization in the Caribbean and trends of global convergence/divergence', *Geographical Journal*, 159, 1–21.

Setchell, C.A. (1995) 'The growing environmental crisis in the world's mega-cities: the case of Bangkok', *Third World Planning Review*, 17, 1–18.

Simon, D. (1992) *Cities, Capital and Development: African Cities in the World Economy*, New York: Wiley.

Tan, K.C. (1986) 'Revitalized small towns of China', *Geographical Review*, 76, 138–48.

CHAPTER 8 INTERNAL INTERACTION

Leinbach, T.R. and Sien, C.L. (1989) *South-East Asian Transport: Issues in Development*, New York: Oxford University Press.

Scaraci, J.L. (1991) 'Primary-care decentralization in

the Southern Cone: shantytown health care as urban social movement', *Annals of the Association of American Geographers*, 81, 103–26.
Taylor, D.R.F. (1981) 'Conceptualizing development space in Africa', *Geografiska Annaler B*, 24, 87–93.

CHAPTER 9 NATIONAL ECONOMIC MANAGEMENT

Becker, B.K. and Egler, C.A.G. (1992) *Brazil: A New Regional Power in the World-Economy*, Cambridge: Cambridge University Press. A detailed study of the external relations and internal structures of the economic geography of Brazil, from a world-systems perspective.
Drake, C. (1989) *National Integration in Indonesia: Patterns and Policies*, Honolulu: University of Hawaii Press.
Drake, C. (1992) 'National integration in China and Indonesia', *Geographical Review*, 87, 295–312.
Linge, G. and Forbes, D.K. (eds) (1990) *China's Spatial Economy: Recent Developments and Reforms*, Hong Kong: Oxford University Press.
Lo, C. (1989) 'Recent spatial restructuring in Zhujiang Delta, South China: a study of socialist regional development strategy', *Annals of the Association of American Geographers*, 79, 293–308.
Logan, I.B. and Mengisteab, K. (1993) 'IMF–World Bank adjustment and structural transformation in Sub-Saharan Africa', *Economic Geography*, 69, 1–24.
Simon, D. (ed.) (1990) *Third World Regional Development: A Reappraisal*, London: Paul Chapman. A useful review of the regional dimension of development, with a particular emphasis on case studies of rural development strategies.
Wall, D. (1993) 'Spatial inequalities in Sandinista Nicaragua', *Geographical Review*, 83, 1–13.

CHAPTER 10 EXTERNAL RELATIONSHIPS

Auty, R. (1995) *Patterns of Development*, London: Arnold. A wide-ranging review of development issues, including urban primacy, the mining economy, and the state, with examples from Asia, Africa and Latin America.
Helleiner, G.K. (1992) 'The IMF, the World Bank and Africa's adjustment and external debt problems: an official view', *World Development*, 15, supplement.
Tarrant, J.R. (1985) 'A review of international food trade', *Progress in Human Geography*, 9, 235–54.

REGIONAL STUDIES

Beaumont, P., Blake, G.H. and Wagstaff, J.M. (1988) *The Middle East: A Geographical Study* (2nd edn), New York: Halsted.
Blouet, B. and Blouet, O. (eds) (1993) *Latin America and the Caribbean: A Systematic and Regional Survey* (2nd edn), Chichester and New York: Wiley.
Cannon, T. and Jenkins, A. (eds) (1990) *The Geography of Contemporary China*, London: Routledge.
Chapman, G.P. and Baker, K.M. (1992) (eds) *The Changing Geography of Asia*, London and New York: Routledge.
Gleave, M.B. (ed.) (1992) *Tropical African Development: Geographical Perspectives*, London: Longman.
King, R. (1989) '40 years of the People's Republic of China', *Geography*, 74, 339–68.
Preston, D. (ed.) (1996) *Latin American Development: Geographical Perspectives* (2nd edn), Harlow: Longman.
Rigg, J.D. (1991) *Southeast Asia: A Region in Transition*, London: Unwin Hyman.
Stock, R. (1995) *Africa South of the Sahara: A Geographical Interpretation*, Harlow: Longman.

'POPULAR' SOURCES OF USE TO PUPILS

NGOs such as Oxfam, Christian Aid and CAFOD frequently produce material relating to their work, with a strong emphasis on poverty, environmental crises, disaster relief and self-help projects.

Oxfam's publications include project development handbooks, studies of environmental issues, trade and aid, health care, gender and country studies. Such agencies may also be sources of film and video material.

Coote, B. (1992) *The Trade Trap*, London: Oxfam.
MacDonald, N. (1991) *Brazil: A Mask Called Progress*, London: Oxfam.
Mosse, J.C. (1992) *Half the World, Half a Chance. An Introduction to Gender and Development*, London: Oxfam.
Parker, B. (1995) *Ethiopia: Breaking New Ground*, London: Oxfam.
Sattaur, O. (1995) *Nepal: New Horizons?* London: Oxfam.
Watkins, K. (1995) *The Oxfam Poverty Report*, London: Oxfam.

Publishers such as Earthscan and Latin American Bureau produce popular studies on Third World issues, for example:

Dwyer, A. (1994) *On the Line: Life on the US–Mexican Border*, London: Latin American Bureau.

Lappé, F. and Schurman, R. (1989) *Taking Population Seriously*, London: Earthscan.

Sarré, P., Smith, P. and Morris, E. (1991) *One World for One Earth: Saving the Environment*, London: Earthscan.

The Open University also has books linked to its course on Third World development, of interest to college and university students, for example:

Allen, T. and Thomas, A. (eds) (1992) *Poverty and Development in the 1990s*, Oxford: Open University/Oxford University Press.

Bernstein, H., Crow, B. and Johnson, H. (eds) (1992) *Rural Livelihoods: Crisis and Response*, Oxford: Open University/Oxford University Press.

Hewitt, T., Johnson, H. and Wield, D. (eds) (1992) *Industrialization and Development*, Oxford: Open University/Oxford University Press.

Sarré, P. (ed.) (1994) *Environment, Population and Development*, London: Open University/Hodder and Stoughton.

Wuyts, M., MacKintosh, M. and Hewitt, T. (eds) (1992) *Development Policy and Public Action*, Oxford: Open University/Oxford University Press.

Penguin Books also have numerous Third World titles. These tend to take a populist, eco-crisis perspective, for example:

Harrison, P. (1993) *The Third Revolution*, Harmondsworth: Penguin.

Harrison, P. (1994) *Inside the Third World*, Harmondsworth: Penguin.

Jackson, B. (1994) *Poverty and the Planet*, Harmondsworth: Penguin.

JOURNALS

Bulletin of Latin American Studies A useful source of case study material, but with a strong emphasis on economic and political, rather than geographical material.

Cahiers d'Outre Mer An important French-language geographical journal on the Third World, with a strong focus on former French colonial territories.

Economic Development and Cultural Change is concerned with the impact of economic development on society and culture.

Geography Aimed mainly at the teaching profession, its 'Changing world' section is a useful source of up-to-date case studies.

Geographical Magazine A 'popular' journal of geography, with a useful focus on environmental and cultural issues. Well illustrated.

Journal of Developing Areas Concerned with issues of regional development. A good source of case study material.

Journal of Modern African Studies African material from a range of the social sciences.

Malay Journal of Tropical Geography Good case study material on environmental issues and the human geography of the Third World.

New Internationalist Topical material on aid and development issues.

Progress in Human Geography carries an annual review of work in the field of Third World studies.

Singapore Journal of Tropical Geography Problems and issues from the Third World, in both physical and human geography.

Third World Planning Review A scholarly journal with a strong emphasis on issues of urban and regional planning.

World Development A major multi-disciplinary journal, though with a strong emphasis on economics.

STATISTICAL SOURCES

Useful annual data sources are:

UNDP, *Human Development Report*, Oxford and New York: Oxford University Press.

United Nations, *United Nations Statistical Yearbook*, New York: UN. The 'standard' data reference. More comprehensive, but rather less up-to-date than the World Bank Report.

The World 1995/6 A Third World Guide, London: The Third World Institute/Oxfam (8th edn). Reference book with a wide range of country profiles, with data and thematic topics such as debt, environmental issues, trade and children.

World Bank, *World Development Report*, London and New York: World Bank/Oxford University Press. This is an essential source of recent social and economic data, to update material in this volume. It also contains substantive commentary, and often explores a particular theme.

References

Amin, S. (1973) *Neo-Colonialism in West Africa*, London: Penguin.

Auty, R. (1987) 'Large capital transfers to developing countries: the lessons of Indonesian oil windfall development', *Singapore Journal of Tropical Geography*, 8, 1–14.

Auty, R. (1995) *Patterns of Development*, London: Arnold.

Benton, J. (1987) 'A decade of change in the Lake Titicaca region', in J. Momsen and J. Townsend (eds) *Geography of Gender in the Third World*, London: Hutchinson, 212–22.

Birkbeck, C. (1979) 'Garbage, industry and the "vultures" of Cali, Colombia', in R. Bromley and C. Gerry (eds) *Casual Work and Poverty in Third World Cities*, Chichester: John Wiley, 161–83.

Boserup, E. (1970) *Women's Role in Economic Development*, London: Allen and Unwin.

Brandt, W. (1980) *North–South: A Programme for Survival*, London: Pan.

Bromley, R. and C. Gerry (1979) (eds) *Casual Work and Poverty in Third World Cities*, Chichester: John Wiley.

Browett, J.G. (1980) 'Development, the diffusionist paradigm and Geography', *Progress in Human Geography*, 4, 57–77.

Bruntland, G.H. (1987) *Our Common Future. The World Commission on Environment and Development*, Oxford: Oxford University Press.

Brydon, L. and Chant, S. (eds) (1989) *Women in the Third World; Gender Issues in Rural and Urban Areas*, London: Elgar.

Buang, A. (1993) 'Development and factory women: negative perceptions from a Malaysian source area', in J. Momsen and V. Kinnaird (eds) *Different Places, Different Voices: Gender and Development in Africa, Asia and Latin America*, London: Routledge, 197–210.

Buchanan, K.M. (1964) 'Profiles of the Third World', *Pacific Viewpoint*, 2, 97–126.

Buchanan, K.M. (1967) *The Southeast Asian World*, London: Bell.

CAPART (Council for Advancement of People's Action and Rural Technology) (n.d.) *Directory of Voluntary Agencies*, Delhi: CAPART.

CEMIG (Centrais Eletricas de Minas Gerais) (1952–93) *Relatorio Anual*, Belo Horizonte: CEMIG.

Chant, S. (1991) *Women and Survival in Mexican Cities: Perspectives on Gender, Labour Markets and Low-Income Households*, Manchester: Manchester University Press.

Chant, S. (1992) *Gender and Migration in Developing Countries*, Manchester: Manchester University Press.

Clarke, C.G. (1975) *Kingston, Jamaica: Urban Development and Social Change*, Berkeley: University of California Press.

Clarke, C.G. (1992) 'Migration and urban–rural differentiation in Oaxaca, Mexico', in M. Skoczek (ed.) *America Latina Local y Regional, Memorias del II Simposio Internacional de la Universidad de Varsovia sobre America Latina, Tomo 4*, Warsaw: CESLA, 163–78.

Clarke, C.G., Peach, C. and Vertovec, S. (1990) *South-East Asians Overseas*, Cambridge: Cambridge University Press.

Cohen, R. (1987) *The New Helots: Migrants in the International Division of Labour*, Aldershot: Gower.

Corbridge, S. (1986) *Capitalist World Development*, London: Macmillan.

Courtenay, P. (1965) *Plantation Agriculture*, London: Bell.

Courtenay, P. (1980) *Plantation Agriculture* (2nd edn), London: Bell.

de Soto, H. (1989) *The Other Path: The Invisible Revolution in the Third World*, London: I.B. Tauris.

Drakakis-Smith, D. (1993) 'Human development indicators', in T. Unwin (ed.) *Atlas of World Development*, Chichester: Wiley, 34–8.

Driant, J.-C. (1991) *Les Barriadas de Lima: Historia e Interpretación*, Lima: DESCO.

Dyson, T. (1994) 'Population growth and world food production: recent global and regional trends', *Population and Development Review*, 20, 2.

Economist (1993) 'Wealth in its grasp. A survey of Indonesia', 17 April.

FAO (Food and Agriculture Organization) (1988)

Production Yearbook, Washington: FAO.

Findlay, A. (1994) 'Return to Yemen; the end of the old order migration in the Arab world', in W.T.S. Gould and A. Findlay (eds) *International Migration and the Changing World Order*, Chichester: Wiley, 215–24.

Frank, A.G. (1969) *Capitalism and Underdevelopment in Latin America*, New York: Monthly Review Press.

Franklin, H. Montes de (1979) 'The housing problem in Caracas, Venezuela', unpublished PhD thesis, University of Liverpool.

FRB (Federative Republic of Brazil) (1974) *II National Development Plan*, Rio de Janeiro: IBGE.

Friedmann, J. (1966) *Regional Development Policy*, Cambridge, MA: MIT Press.

Friedmann, J. (1992) 'The end of the Third World', *Third World Planning Review*, 14, iv–viii.

Garcia, J.Z. (1991) 'Peru and Bolivia', in J.K. Black (ed.) *Latin America: Its Problems and its Promise*, Boulder: Westview Press, 446–67.

Gilbert, A. (1993) *In Search of a Home: Rental and Shared Housing in Latin America*, London: University College London Press.

Gilbert, A. and Varley, A. (1991) *Landlord and Tenant: Housing the Poor in Urban Mexico*, London: Routledge.

Gilbert, A. and Ward, P.M. (1985) *Housing, the State and the Poor: Policy and Practice in Three Latin American Cities*, Cambridge: Cambridge University Press.

Gould, W.T.S. (1985) 'Migration and development in Western Kenya, 1971–82: a retrospective analysis of school leavers', *Africa*, 55, 262–85.

Gould, W.T.S. (1990) 'Structural adjustment, decentralization and development planning in Ghana', in D. Simon (ed.) *Third World Regional Planning: A Reappraisal*, London: Longman, 21–5.

Gould, W.T.S. (1993) *People and Education in the Third World*, London: Longman.

Gould, W.T.S. (1994a) 'Population growth, migration and environmental stability in Western Kenya: from Malthus to Boserup', in J.I. Clarke and B. Zaba (eds) *Environment and Population Change*, Liège: Ordina Editions, 247–68.

Gould, W.T.S. (1994b) 'The brain drain', in T. Unwin (ed.) *The Atlas of World Development*, Chichester: Wiley, 222–3.

Gould, W.T.S. (1994c) 'Population movements and the changing world order: an introduction', in W.T.S. Gould and A. Findlay (eds) *International Migration and the Changing World Order*, Chichester: Wiley, 3–14.

Grigg, D. (1993) *The World Food Problem* (2nd edn), Oxford: Blackwell.

Habitat (United Nations Centre for Human Settlements) (1987) *Global Report on Human Settlements 1986*, Oxford and New York: Oxford University Press.

Hart, K. (1973) 'Informal income opportunities and urban employment in Ghana', *Journal of Modern African Studies*, 11, 61–81.

Hennessey, A. (1978) *The Frontier in Latin American History*, London: Edward Arnold.

Hirschman, A. (1958) *The Strategy of Economic Development* New Haven: Yale University Press.

Holston, J. (1989) *The Modernist City: An Anthropological Critique of Brasília*, Chicago and London: University of Chicago Press.

Johnston, R.J. (1984) 'The world is our oyster', *Transactions of the Institute of British Geographers*, 9, 443–59.

Johnston, R.J. and Taylor, P. (eds) (1989) *A World in Crisis?*, Oxford: Blackwell.

Keynes, J.M. (1936) *The General Theory of Employment, Interest and Money*, New York: Harcourt Brace.

King, R. and Knights, M. (1994) 'Bangladeshis in Rome: a case of migratory opportunism', in W.T.S. Gould and A. Findlay (eds) *International Migration and the Changing World Order*, Chichester: Wiley, 127–44.

Knox, P. and Agnew, J. (1994) *The Geography of the World Economy* (2nd edn), London: Arnold.

Lipton, M. (1977) *Why Poor People Stay Poor*, London: Temple Smith.

Lipton, M. and Longhurst, R.M. (1989) *New Seeds and Poor People*, London: Unwin Hyman.

LONRHO (1992) *Annual Report*, London: Lonrho.

Mabogunje, A. (1968) *Urbanisation in Nigeria*, London: University of London Press.

Makannah, T.J. (1990) 'Policy measures for stemming urban migration', in Union for African Population Studies, *The Role of Migration in African Development*, Dakar, 82–95.

Mangin, W. (1967) 'Latin American squatter settlements: a problem and a solution', *Latin American Research Review*, 2 (3), 65–98.

Manning, C. (1988) 'Rural employment creation in Java: lessons from the Green Revolution and oil boom', *Population and Development Review*, 14, 47–80.

Meadows, D.H., Meadows, D.L., Randers, J. and Behrens, W.W., III (1972) *The Limits to Growth*, London: Potomac Associates.

Momsen, J. and Kinnaird, V. (eds) (1993) *Different Places, Different Voices: Gender and Development in Africa, Asia and Latin America*, London: Routledge.

Momsen, J. and Townsend, J. (eds) (1987) *Geography of Gender in the Third World*, London: Hutchinson.

Mortimore, M. (1989) *Adapting to Drought: Farmers, Famines and Desertification in West Africa*, Cambridge: Cambridge University Press.

Murphy, A.D. and Stepick, A. (1991) *Social Inequality in Oaxaca: A History of Resistance and Change*, Philadelphia: Temple University Press.

Myrdal, G. (1963) *Economic Theory and Underdeveloped Regions*, London: Methuen.

Norwine, J. and Gonsalez, A. (eds) (1988) *The Third World: States of Mind and Being*, Boston: Unwin Hyman.

Parnwell, M. (1993) *Population Movements and the Third World*, London: Routledge.

Peil, M. (1991) *Lagos: The City is the People*, London: Belhaven Press.

Perroux, F. (1971) 'Note on the concept of growth poles', in I. Livingstone (ed.) *Economic Policy for Development*, Harmondsworth: Penguin, 278–99.

Riddell, B. (1992) 'Things fall apart again: structural adjustment programmes in Sub-Saharan Africa', *Journal of Modern African Studies*, 30, 53–68.

Rimmer, D. (1992) *Staying Poor: Ghana's Political Economy, 1950–1990*, Oxford: Pergamon.

Roberts, B. (1995) *The Making of Citizens: Cities of Peasants Revisited*, London: Edward Arnold.

Robey, B., Rutstein, S.E., Morris, L. and Blackburn, R. (1992) 'The reproductive revolution: new survey findings', *Population Reports, Series M*, 11. Baltimore: Johns Hopkins University Press.

Robinson, M.A. (1991) 'Evaluating the impact of NGOs in rural poverty alleviation: India country study', *ODI Working Paper 49*. London: Overseas Development Institute.

Rodney, W. (1972) *How Europe Underdeveloped Africa*, London: Bogle L'Ouverture Publications.

Rostow, W.W. (1960) *The Stages of Economic Growth*, Cambridge: Cambridge University Press.

Sage, C. (1991) 'The discourse on drugs in the Americas', *Bulletin of Latin American Research*, 10, 325–32.

Santos, M. (1979) *The Shared Space: Two Circuits of the Urban Economy in Underdeveloped Countries*, London and New York: Methuen.

Schumacher, E. (1974) *Small is Beautiful*, London: Abacus.

Seager, J. and Olson, A. (1986) *Women in the World: An International Atlas*, London: Pan.

Sjoberg, G. (1960) *The Pre-industrial City – Past and Present*, New York: Free Press.

Soussan, J. (1988) *Primary Resources and Energy in the Third World*, London: Routledge.

South Commission (1990) *The Challenge to the South*, Oxford: Oxford University Press.

Spate, O.H.K. (1957) *India and Pakistan* (2nd edn), London: Methuen.

Spencer, J.E. and Thomas, W.L. (1978) *Introducing Cultural Geography*, New York: John Wiley.

STT (Société Tunisienne de Topographie) (n.d.) *Environs de Sousse, Flle 4*, Tunis: STT.

Taaffe, E.J., Morrill, R. and Gould, P.R. (1963) 'Transport expansion in under-developed countries: a comparative analysis', *Geographical Review*, 53, 503–29.

Taylor, D.R.F. (1981) 'Conceptualizing development space in Africa', *Geografiska Annaler B*, 24, 87–93.

Thiesenhusen, W.C. (1989) *Searching for Agrarian Reform in Latin America*, Boston: Unwin Hyman.

Turner, J. and Fichter, R. (eds) (1972) *Freedom to Build*, New York: Macmillan.

United Nations (1986) *Urban and Rural Population Projections 1950–2025*, New York: United Nations.

United Nations (1992) *United Nations Statistical Yearbook*, United Nations: New York.

United Nations (1993) *United Nations Statistical Yearbook*, United Nations: New York.

United Nations Development Programme (UNDP) (1992) *Human Development Report*, New York: UNDP.

United States Department of Commerce (USDC) (1991) *World Population Profile*, Washington: USDC.

Varley, A. (1985) 'Urbanization and agrarian law: the case of Mexico City', *Bulletin of Latin American Research*, 4, 1–16.

Wallerstein, E. (1974) *The Modern World System: Capitalist Agriculture and the Origins of the European World Economy in the Sixteenth Century* (3rd edn 1989), London: Academic Press.

Ward, P.M. (ed.) (1982) *Self-Help Housing: A Critique*, London: Mansell.

Ward, P.M. (1986) *Welfare Politics in Mexico: Papering over the Cracks*, London: Allen and Unwin.

Ward, P.M. (1990) *Mexico City: The Production and Reproduction of an Urban Environment*, London: Belhaven.

World Bank (1980) *World Development Report*, New York: Oxford University Press.

World Bank (1983) *World Development Report*, New York: Oxford University Press.

World Bank (1992) *World Development Report*, New York: Oxford University Press.

World Bank (1993) *World Development Report*, New York: Oxford University Press.

World Bank (1994a) *World Development Report*, New York: Oxford University Press.

World Bank (1994b) *Annual Report*, Washington: World Bank.

Index

Figures, tables and plates are listed in *italic*.